Springer Series in Optical Sciences

Volume 228

Springer Series in Optical Sciences is led by Editor-in-Chief William T. Rhodes, Florida Atlantic University, USA, and provides an expanding selection of research monographs in all major areas of optics:

- lasers and quantum optics
- ultrafast phenomena
- optical spectroscopy techniques
- optoelectronics
- information optics
- applied laser technology
- industrial applications and
- other topics of contemporary interest.

With this broad coverage of topics the series is useful to research scientists and engineers who need up-to-date reference books.

More information about this series at http://www.springer.com/series/624

Nahid Talebi

Near-Field-Mediated Photon–Electron Interactions

 Springer

Nahid Talebi
Stuttgart Center for Electron
Microscopy (StEM)
Max Planck Institute for Solid
State Research
Stuttgart, Baden-Württemberg, Germany

Institute of Experimental
and Applied Physics
Christian-Albrechts University in Kiel
Kiel, Germany

ISSN 0342-4111 ISSN 1556-1534 (electronic)
Springer Series in Optical Sciences
ISBN 978-3-030-33818-3 ISBN 978-3-030-33816-9 (eBook)
https://doi.org/10.1007/978-3-030-33816-9

This Springer imprint is published by the registered company Springer Nature Switzerland AG
The registered company address is: Gewerbestrasse 11, 6330 Cham, Switzerland

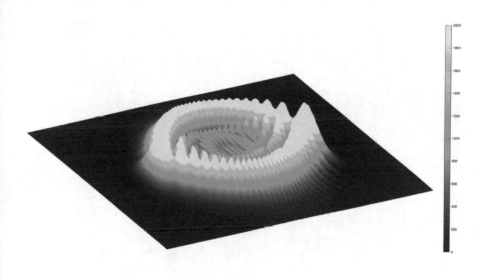

Preface

Nanooptics is the physics of light–matter interactions at the nanoscale. It concerns the control of the dynamics of the oscillating charges and vibrational motions of ions and molecules. The applications of such control mechanisms are magnificent, transcending various fields such as energy conversion, the control of chemical reactions, optically induced phase transitions, quantum cryptography, and data processing.

Many approaches might be utilized to enhance the light–matter interaction, in other words, increasing the coupling efficiency of the optical sources to the confined volume around and within the nanostructure. The region around nanostructures within which all the momentum and energy conversion might take place is called the near-field zone and extends only a few nanometres around the nanoobjects. This is the zone where quantum emitters such as quantum dots should be placed as well to provide the possibility of exciting the optical modes of nanostructures. Ubiquitously, the diffraction of light by nanoscale objects can be an additional driver of the energy–momentum conversion. However, the background-free high-resolution characterization of the nanoscale excitations to allow mapping of the spatio-spectral and spatio-temporal evolutions of optical excitations at the nanoscale demands more exotic techniques. Among them, methods utilizing electron probes have attracted increasing attention. Thanks to the de Broglie wavelength of matter waves, electron beams theoretically possess sufficiently high resolution to be able to probe the dynamics of electronic orbitals and motions when an appropriate time-resolved spectroscopy technique is incorporated, e.g. pumping by laser beams and probing by electron waves. However, this demands tremendous control over the motion of the electron probe itself and the ability to magically phase-lock the laser and electron pulses.

With this motivation, this book provides the first attempts of the author and his co-workers to

(i) understand the physics behind electron–light interactions by using electron microscopes,

(ii)   provide methodologies to advance electron microscopy towards controlling the electron dynamics with better time resolution, and

(iii) realize analytical and numerical techniques to explore the theoretical basis of both classically and quantum mechanically.

Many different aspects of electron–light interactions are covered, spanning elastic and inelastic interactions, electron energy loss spectroscopy, and mechanisms of the radiation from moving electrons interacting with thin films and nanostructures. We also elucidate the physics of polaritons from the point of view of classical electromagnetism, though by considering more generalized matter, including 2D electron gases (surface conductivity) and magnetoelectric effects. We will show that, in some particular cases, anisotropic materials, particularly hyperbolic materials, might be used to enhance the electron-induced radiation. This effect will also be used to realize an electron-driven photon source with an enhanced mechanism of radiation in comparison with surface-plasmon-based metamaterial sources. In the final chapter, the attempts of the author to develop a numerical toolbox by combining the Schrödinger and Maxwell equations will be outlined. We apply this toolbox to understand various concepts, such as the quantum walk of a single-electron wave packet in an optical lattice that is generated by two propagating optical paths at a certain angle with respect to the direction of the propagation of the electron.

When writing this book in 2019, I was fortunate to write it at a time when the study of the coherent electron–light interactions in electron microscopy has already started flourishing with the execution of wonderful experiments and reports of interesting results worldwide. These experiments in particular will provide an impetus to the development of a theoretical basis to further understand the theoretical models involved, to benchmark the approximations, and to develop even more characterization tools based on coherent electron–light interactions. These theoretical developments are a keen interest of the author—some of the results focused on this aspect will be highlighted in more detail.

Stuttgart/Kiel, Germany   Nahid Talebi

Acknowledgements

Many people have directly and indirectly influenced my path in general, culminating in the exposition of my knowledge in the form of the present habilitation. The most influential person, with whom I recently had the opportunity of collaborating, is my habilitation mentor Prof. Harald Giessen. His sustained faith in promoting young scientists, support, and advice guided me to continue my path into science in general and to complete my habilitation as a result. I am honoured to recognize his hard-working and spiritual habits, which have been a source of inspiration for me during the last two years.

My research career as a postdoc started with the generous support of the Alexander von Humboldt (AvH) Foundation that lasted for a duration of three years. AvH has been quite successful in supporting individual keen scientists in developing their career paths. I am also particularly thankful to my host institutes during this period, the Max Planck Institute for Intelligent Systems and later the Max Planck Institute for Solid State Research, for providing me with both the platform and infrastructure to pursue the realization of my ideas and the development of my knowledge in a quite independent way as a principal investigator. In this sense, my direct group, the Stuttgart Center for Electron Microscopy, has had a great impact by their continuous support. Head of the group, Prof. Peter van Aken, has been supportive of my ideas and critical to my scientific output. I also warmly acknowledge my student Dr. Surong Guo for her persistent hard-working attitude in turning each provided idea into its best possible realized performance and my colleagues Kersten Hahn, Caroline Heer, Julia Deuschle, and Marion Kelsch for their generous assistance and consultations in daily life. I also particularly and gratefully acknowledge the European Research Council for granting me an ERC starting grant that has enabled me with great ease to further realize my ideas. Some of the results reported here were made possible only by the ERC starting grant NanoBeam.

I am specifically grateful to my mentor and long-term collaborator, Prof. Christoph Lienau. His passionate way of performing research, his continuous guidance, his excellent patience and flexibility in addressing my scientific questions, and, in particular cases, his help in confronting frustrations in my scientific path let me develop my independent identity as a scientist and as a physicist.

Some of the results presented in this thesis are the outcomes of collaborative efforts with the groups of Albert Polman (AMOLF, Amsterdam), Christoph Lienau (Carl von Ossietzky Universität Oldenburg), Harald Giessen (University of Stuttgart), and Mathieu Kociak (Université Paris-Sud, Orsay). I am particularly thankful to Mario Hentschel for his excellent help in these collaborative attempts and to Sophie Meuret for her cathodoluminescence studies.

Last and most important, my family has never let me feel alone on this path and has always provided me with their continuous encouragement and warm support. In the current era, when mobility within the scientific community has become imperative to achieving success, having such a wonderful impetus for my career and long-lasting and limitless source of confidence and support is not a thing to be appreciated by just a few words here. Thank you all!

Contents

Acronyms

CL	Cathodoluminescence
CR	Cherenkov radiation
EEGS	Electron energy-gain spectroscopy
EEL	Electron energy loss
EELS	Electron energy loss spectroscopy
EFTEM	Energy-filtered transmission electron microscopy
FDTD	Finite-difference time-domain method
h-BN	Hexagonal boron nitride
KDE	Kapitza–Dirac effect
LHM	Left-handed material
ME	Magnetoelectric
MREEL	Momentum-resolved electron energy loss
MREELS	Momentum-resolved EELS
PDOS	Photonic density of states
PEEM	Photoemission electron microscopy
PINEM	Photon-induced near-field electron microscopy
PLDOS	Photonic local density of states
PPM	Point projection electron microscopy
SNOM	Scanning near-field optical microscopy
SPP	Surface plasmon polariton
SPR	Smith–Purcell radiation
TR	Transition radiation
Tr2PPEEM	Time-resolved two-photon photoemission electron microscopy

Chapter 1
Introduction

Abstract Here, a short survey on the physics and applications of electron beams interacting with light and polaritons is provided. The focus is on the approaches that will be used throughout the current dissertation to understand and control electron-light interactions. This chapter is organized to systematically cover the sub-systems and fields that are subjects of the investigations and developments in this book: polaritons, electron energy-loss spectroscopy (EELS), cathodoluminescence (CL), and numerical methods for simulating the interactions of electrons with light and nanostructures.

1.1 Plasmon Polaritons

Polaritons are bosonic quasi-particles resulting from the strong interaction of electromagnetic waves with dipolar excitations in matter. Perhaps the most well-known member of the class of polaritons is the plasmon polariton. In metals, conduction electrons form quasi-free electron waves that oscillate coherently in response to an applied electric field. The frequency of such oscillations, known as the plasma frequency (ω_p), is directly related to the density of free electrons in metals (n_e) as $\omega_p = \sqrt{n_e e^2 / m_e \varepsilon_0}$, where m_e and e are the elementary electron charge and mass, respectively, and ε_0 is the permittivity of free space. The optical response of several metals, including Al and alkali elements, can be understood by the free-electron-gas model. In this way, the permittivity of these materials is also described by the Drude function as

$$\varepsilon_r(\omega) = 1 - \frac{\omega_p^2}{\omega(\omega + i\Gamma)} \tag{1.1}$$

where $\Gamma = \tau^{-1}$ is a phenomenological damping rate and τ, the relaxation time, is related to the DC conductivity of the metal as $\sigma_0 = ne^2\tau / m_0$ [1]. The real part of the permittivity approaches zero at $\omega = \sqrt{\omega_p^2 - \Gamma^2}$ and becomes negative at frequencies

© Springer Nature Switzerland AG 2019
N. Talebi, *Near-Field-Mediated Photon–Electron Interactions*,
Springer Series in Optical Sciences 228,
https://doi.org/10.1007/978-3-030-33816-9_1

of $\omega < \sqrt{\omega_p^2 - \Gamma^2}$. In this frequency range, plasmon polariton waves are supported and can propagate at the surface of a metal, with the evanescent tails entering the metal and the surrounding medium. The dispersion of these polaritons can be determined by using Maxwell's equations. Throughout this dissertation, analytical treatments of electromagnetic systems are carried out using the vector-potential approach, which helps us to deal with anisotropic systems with magnetoelectric effects as well [2]. We briefly outline this approach in Chap. 2. Here, we mention only the main concepts and a few outcomes of this approach.

The wave function associated with surface plasmon polaritons (SPPs) bound to a metal/air interface located at $z = 0$ and propagating along the x-direction is given by $\psi_n = \psi_{0n} \exp(-\kappa_n |z|) \exp(-i\gamma x)$, with a propagation constant $\gamma = k_0 \sqrt{\varepsilon_r / (1 + \varepsilon_r)}$ and evanescent ratios $\kappa_1 = k_0 \sqrt{(\varepsilon_r - 1) / \varepsilon_r + 1}$ and $\kappa_2 = k_0 \sqrt{-\varepsilon_r^2 / \varepsilon_r + 1}$ for domains $z > 0$ (air region) and $z < 0$ (metallic region), respectively. The subscript n takes values of 1 for $z > 0$ and 2 for $z < 0$. To support bound modes, we should have $\mathrm{Re}\{\kappa_\alpha\} > 0$, which restricts the frequency range to those frequencies at which $\varepsilon_r < -1$, i.e., $\omega < \sqrt{0.5\omega_p^2 - \tau^{-2}} \approx \omega_p / \sqrt{2}$.

The propagation constant γ is a complex-valued function of the frequency; the real part of it is the phase constant $\beta = \mathrm{Re}\{\gamma\}$, and its imaginary part is the attenuation constant α. The dispersion diagram for SPPs propagating at the Al/air interface is shown in Fig. 1.1a. SPP waves are supported, as expected, at frequencies of $\omega < \omega_p / \sqrt{2}$, where the plasma energy for Al is

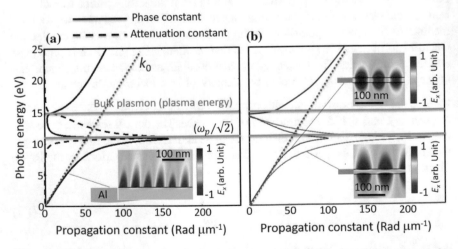

Fig. 1.1 Plasmon polaritons at the interface and in thin films. **a** Dispersion of SPPs supported at the interface between Al and air. The spatial distribution of the x-component of the electric field at an energy of $E = 6$ eV is shown in the inset. **b** Dispersion of plasmon polaritons supported by an Al film with a thickness of 20 nm. Two symmetric and antisymmetric modes are formed as a result of the hybridization between SPPs supported by the two surfaces. The spatial distribution of the x-component of the electric field at an energy of $E = 7$ eV is shown in the inset for each mode

$E_p = \hbar\omega_p = 14.3\,\mathrm{eV}$. The frequency $\omega_s = \omega_p\big/\sqrt{2}$ is called the static plasmon frequency [3]. The attenuation constant also peaks at this frequency. Another branch of plasmon polaritons, known as transverse polaritons [4], happens at frequencies of $\omega > \omega_p$, and within the frequency range $\omega_s < \omega < \omega_p$, the dissipation losses cause the domination of the attenuation constant over the phase constant; hence, a bandgap for SPPs is formed.

The formation of SPPs with demonstrated dissipation is understood to be a result of the strong interaction between two harmonic oscillators, namely, collective longitudinal plasmon oscillations and plane wave photons in free space. This interaction happens at the interface between air and metals. Moreover, the coupling between two metallic surfaces, such as those supported by thin films, causes the SPPs to hybridize into symmetric and antisymmetric (even and odd symmetry) distributions (Fig. 1.1b). This behaviour is very similar to the hybridization schemes in diatomic molecular orbitals [5] and has been investigated in various coupled metallic nanoparticles as well [6–9].

SPPs can be localized in nanoparticles as small as only a few nanometres with applications in strong light-matter interactions and the enhancement of electric field amplitudes. Both the resonant energies and enhancement factors can be nicely tuned by a careful choice of the material and by engineering the shape of the host nanoparticles [10]. In particular, linear nanorods known as nanoantennas deliver remarkable case of design and tunability [11, 12]. Various forms of high-quality nanoantennas have been produced using different metals, such as gold and silver [13, 14].

1.2 Other Forms of Interface Electromagnetic Waves

In addition to plasmon polaritons, there exist several other forms of polaritonic waves. The criterion mentioned above for supporting bound waves at the interface between air and metals can be generalized to include anisotropic materials with a magnetoelectric effect, layered van der Waals materials, and two-dimensional systems [15–20]. Indeed, the group of two-dimensional materials has recently become an individual topic with promise in the realization of polaritonic systems with lower dissipation losses, an extreme level of localization, and the high-speed data transformation of electromagnetic waves far beyond the reach of bulk metals [17, 21, 22]. The most widely investigated class in this family is the Dirac plasmons in graphene [23, 24]. Graphene plasmons offer a large degree of parametric tunability via controlled chemical and electrical doping, thus emerging as a promising tool for electro-optical modulation [25], sensors [26] and transformation optics [27].

In addition, hyperbolic materials such as hexagonal boron nitride (h-BN) possess intriguing characteristics [28, 29]. A hyperbolic material has a uniaxial crystalline structure with two distinguished permittivity components, namely, in-plane ($\varepsilon_{xx} = \varepsilon_{yy} = \varepsilon_{\parallel}$) and normal ($\varepsilon_{zz} = \varepsilon_{\perp}$) components, where the coordinate system is positioned along the principal axes of the crystal. In addition, in some frequency ranges, the signs of the two permittivity components are not the same. In other words, for some certain polarizations of the incident light, the material behaves like a metal, whereas for other polarizations, the optical response exhibits a dielectric-like behaviour. The dispersion relation for a plane wave propagating in the bulk of a hyperbolic material in an arbitrary direction with the wave vector $\vec{k} = \left(k_x, k_y, k_z\right)$ is decomposed into two groups: ordinary and extraordinary rays. The isofrequency surface of the extraordinary rays (transverse magnetic waves) in particular is given by [30]

$$\frac{k_x^2 + k_y^2}{\varepsilon_{r\perp}(\omega)} + \frac{k_z^2}{\varepsilon_{r\parallel}(\omega)} = \frac{\omega^2}{c^2} \tag{1.2}$$

which, for the case of $\varepsilon_{r\perp}\varepsilon_{r\parallel} < 0$, forms a hyperboloid (Fig. 1.2). Additionally, two distinguished types of hyperboloids are expected: for a material with $\varepsilon_{r\parallel} < 0$ and $\varepsilon_{r\perp} > 0$, the isofrequency surface exhibits a gap, and this material is referred to as hyperbolic type I, whereas for a material with $\varepsilon_{r\parallel} > 0$ and $\varepsilon_{r\perp} < 0$, there is no gap in the isofrequency surface, and such a material is called hyperbolic type II.

Why are hyperbolic materials interesting? In addition to these materials being a simpler case of a metamaterial [31], there exist several applications for hyperbolic materials. They can be used for the enhancement of the Purcell factor and spontaneous emission [32] and enhancement of the photonic density of states (PDOS) [33], and they have an extreme confinement factor for the optical energy at the nanoscale because of the large effective refractive indices of the optical modes that they sustain [34]. Moreover, there are natural materials with hyperbolic dispersion covering distinct regions of the electromagnetic spectrum, from the terahertz range to the ultraviolet range [35] (Fig. 1.3).

Additionally, many materials, including topological insulators, support topologically protected surface states with optical signatures that are distinguished from the

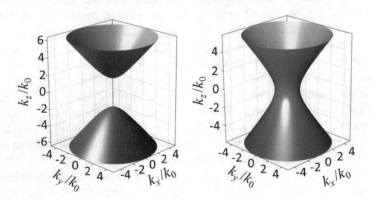

Fig. 1.2 Isofrequency surfaces of plane waves propagating inside a hyperbolic medium. (left) $\varepsilon_{r\parallel} < 0$ and $\varepsilon_{r\perp} > 0$ and (right) $\varepsilon_{r\parallel} > 0$ and $\varepsilon_{r\perp} < 0$

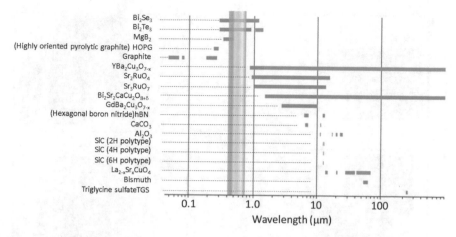

Fig. 1.3 Natural hyperbolic materials and the frequency ranges they cover. Adapted from [35]. Blue and orange correspond to the bands with one and two negative components in the diagonal dielectric permittivity tensor, respectively

bulk responses. Moreover, they can be a platform for the realization of the topological magnetoelectric (ME) effect and axion electrodynamics [36]. The generalization of the constitutive relations, Lorentz gauge theory, and the dispersion of interface polaritons to a form suitable for investigating the electromagnetic response of these materials have been studied elsewhere [2].

1.3   Detection of Plasmon Polaritons

The most commonly used mechanisms for characterization of the propagating and localized polaritons are based on the interaction of electron beams and light waves with nanostructures. On the excitation level, either electron beams or light can be used to induce coherent collective polarizations inside materials. On the detection level, again, either scattered electrons or photons can be used (Fig. 1.4). On this basis, the characterization techniques are categorized into different types, as described in the following:

1st—Purely optics-based methods such as scanning near-field electron microscopy (SNOM), which offers resolutions at a level not reachable with conventional far-field optical detection schemes (Fig. 1.4a) [37, 40–43]. Belonging to the category of scanning probe techniques [44], this technique makes use of atomically sharp tips and tapers, giving access to a highly focused and broadband field distribution localized at the apex of such tips. In this way, a large distribution of wave vectors at the apex, far beyond what is possible with free-space plane waves, provides spatial resolutions at the level of only a few nanometres. A variant of this technique, which benefits

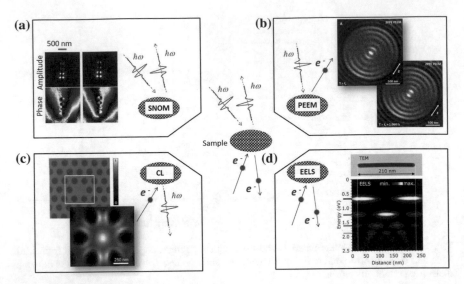

Fig. 1.4 Characterization systems for localized polaritons in nanostructures. **a** Purely optical techniques, such as SNOM. The inset shows the amplitude and phase of an optically excited gold Yagi-Uda nanoantenna [37]. Reproduced by permission from Nano Letters. **b** In PEEM, the sample is excited by light, and the emitted photoelectrons are probed to resolve the polaritons. The inset shows the PEEM image at two selected delay times between the pump and probe laser excitations [38]. Reproduced by permission from Sci. Adv. **c** CL is the luminescence of a material under electron irradiation. CL can be used to probe the radiative photonic density of states with high spatial and energy resolution. Shown in the inset is the CL map of a photonic crystal cavity etched in a 200-nm-thick Si_3N_4 membrane at a wavelength of 650 nm [39]. © 2015, Materials Research Society. **d** The EELS signal is related to the PLDOSs projected along the electron trajectory. Presented in the inset is the EELS energy-distance map of a linear gold nanoantenna [12], where multipolar excitations and their resonant energies are overserved. © 2013, Springer Nature

from adiabatic nanofocusing [45–47], delivers better control of the excitation and detection of polaritonic waves at the nanoscale [48, 49].

2nd—Combined electron-optical characterization techniques in which lasers are used to excite probe electrons from the sample (Fig. 1.4b). Photoemission electron microscopy (PEEM) has been applied recently as a tool for the characterization of plasmon polaritons [38, 50–54] in addition to investigating the electronic states. In this technique, coherent optical excitation results in the emission of electrons from the valence and conduction bands into the continuum. The emitted electrons are controlled and detected using electron optics. By exploiting high-intensity and few-cycle XUV pulses, tremendous control over the emitted wavefunction of single-electron wave packets from solids is attainable via quantum path interferometry [55, 56]. Recently, PEEM was broadly used for exploring the mechanisms of coherent control in above-threshold atomic ionization and high harmonic generation [57, 58]. This field is by itself a growing independent topic beyond the scope of this book.

3rd—Combined electron-optical characterization techniques in which electrons are used to generate probe photons from the sample (Fig. 1.4c). CL is the luminescence exhibited by a sample upon electron-beam irradiation [59–61]. Depending on the material under investigation and the parameters of the electron beam itself (duration, kinetic energy, and pulse shape), CL emission can be coherent or incoherent [62]. Coherent CL emission has a fixed phase relation with the near-field zone of the electron probe itself and happens when the electron probe generates coherent collective excitations and polarization currents [63]. Such coherent CL emission may happen at the surface of metals due to the sudden annihilation of the dipole created by the electron probe and its image charge [64]. This phenomenon is called transition radiation. In addition, electrons traversing thin metallic films can induce coherent SPPs at the surface of metals. Although SPPs in general do not radiate, their interaction with engineered or natural defect centres and gratings can lead to their coupling to the radiation continuum. Another source of coherent radiation is Cherenkov radiation, which happens when the velocity of an electron moving inside a bulk dielectric becomes larger than the light velocity in that medium.

In contrast, incoherent CL does not have any specific phase correlation with the electron beam and cannot interfere with coherent sources of radiation. The incoherent CL generated by electron-hole recombination in semiconductors, however, is much stronger than the coherent CL and is used spectroscopically in materials science, mineralogy, and semiconductor physics for material characterization [65–67].

CL detectors are becoming routinely installed in scanning and transmission electron microscopes. Due to the available space in scanning electron microscopes (SEMs), the CL detectors might be equipped with imaging, spectroscopy, angle-resolved, and polarimetry apparatuses as well [68], facilitating the precise characterization of the polarization states, frequency, and directionality of the emission.

4th—Purely electron-based excitation and detection methods such as EELS, where the inelastic scattering of electrons with a sample is used to probe the electronic and photonic densities of states of the sample (Fig. 1.4d). Thus, in EELS, the loss of energy from the electron beam is acquired using a spectrometer. Compared to CL, EELS can be considered a primary detection algorithm that probes the full photonic local density of states (PLDOSs) projected along the trajectory of the incident electron beam [63, 69], albeit considering a momentum-conservation criterion [70]. EELS and CL detectors can both be installed in a transmission electron microscope (TEM), providing vital tools as analytical electron microscopy techniques for probing both the electronic and photonic local densities of states (Fig. 1.5a). Considering EELS in particular, the electron energy-loss (EEL) spectrum provides a tremendous amount of information over an ultra-broad bandwidth (Fig. 1.5b). Higher energy-loss ranges, at energies above 100 eV, carry information about the electronic band structure of the core electrons in individual elements of the material. In contrast, in lower energy ranges—the so-called low-loss regime—valence and conduction electrons contribute the most to the EELS signal. Thus, in this regime, collective excitations such as plasmons, magnons, and phonons (in materials that can support them) are the dominant response of the sample [4, 71–73].

The relaxation channels for the electron-induced transitions inside the material include both radiative and non-radiative decay. In the former case, the relaxation of

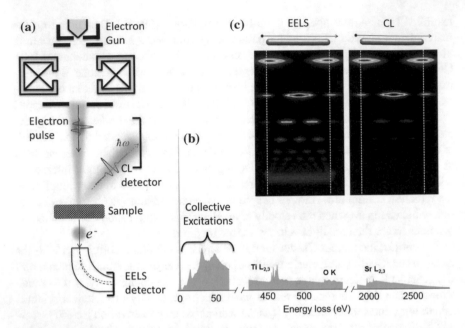

Fig. 1.5 EELS and CL as electron microscopy techniques for the characterization of PLDOS. **a** In a TEM, an electron can undergo elastic and/or inelastic interactions with the sample. During the inelastic interactions, the electron will lose energy, and the energy loss can be measured using an electron spectrometer. **b** The electron energy-loss spectrum covers an ultra-broadband energy range and has contributions from the dynamics of conduction/valence electrons as well as the transition rates between the various electronic band diagrams of the material. **c** CL and EELS can be complementary to each other, unravelling the coupling efficiency of the different photonic modes of the sample to the radiation continuum. Reprinted by permission from [74]; Copyright © 2015 American Chemical Society

the excited electrons results in the emission of photons to the far field. Depending on the wavelength of the emitted photons, either CL detectors or X-ray detectors might be used to acquire the spectrum of the processes involved. The nonradiative yield includes the emissions of backscattered/secondary electrons and Auger electrons and might include the nonradiative pathways of the excited photonic density of states (optical losses) through heat or photochemical reactions. Thus, CL probes only those photonic modes that contribute to the radiative channels, whereas EELS is a probe of both radiating and non-radiating optical excitations (Fig. 1.5c) [74].

Imaging the PLDOS with EELS detectors can be facilitated by both scanning transmission electron microscopy (STEM) and energy-filtered transmission electron microscopy (EFTEM) techniques. Using STEM, electron beams are focused onto a spot with a radius of a few angstroms and used to raster scan the sample. In contrast, for EFTEM, the electron beams can be approximated as plane waves (with a spot size larger than the dimensions of the sample). The differences between these techniques are in the acquisition time and transverse coherence of the imaging principles. Nevertheless, for both techniques, a 3D data cube is obtained, including a transverse

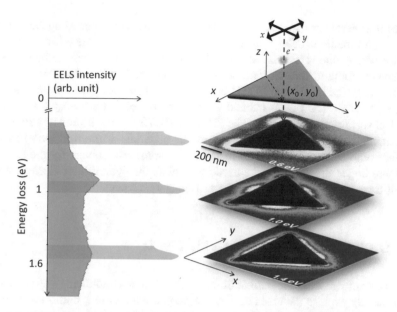

Fig. 1.6 3-Dimensional data cube obtained using STEM-EELS or EFTEM techniques. Either a parallel electron beam (plane wave electron) or focused beam is used to obtain the electron energy-loss spectrum at each impact parameter (x_0, y_0). This data cube will be then used to obtain the spatial distribution of the resonances in polaritonic or photonic systems. The figure shows the spectrum and the EFTEM series of a mesoscopic gold plate at selected energies, in which the excitation of wedge plasmon polaritons is observed [75]

spatial coordination and longitudinal energy-loss axis (Fig. 1.6). This data cube will then be used to extract both the resonances and the spatial distribution of the optical modes in the selected energy ranges. In Chaps. 5 and 6, we use these techniques to investigate the photonic modes of various systems, including toroidal moments and 3-dimensional gold tapers.

1.4   Numerical Methods

The numerical methods included in this book are divided into three different sets of techniques: 1st—classical numerical techniques for simulating optical systems based on Maxwell's equations [76], 2nd—self-consistent numerical techniques obtained by combining the Maxwell and the Lorentz equations [77], and 3rd—self-consistent numerical techniques obtained by combining the Maxwell and the Schrödinger equation [78, 79].

The classical treatment of the electron energy-loss spectroscopy of plasmon polaritons can nicely reproduce the experimental EFTEM and EEL spectra, with multitudes of reports presented in the literature. Many techniques have been used to simulate the electron energy-loss spectra of complicated nanostructures, including the discrete dipole approximation (DDA), the generalized multipole technique (GMT) [80, 81], finite element methods (FEMs) [29, 82], the discontinuous Galerkin method (DGM) [83], and the finite-difference time-domain (FDTD) method [76, 84]. Among these methods, only the DGM and FDTD are within the time domain. Throughout this book, we will use the FDTD method to calculate the electron energy-loss spectra and CL of various nanostructures. The FDTD method is a versatile tool that allows an efficient modelling of nonlinear and anisotropic materials. Moreover, we will show how the combination of the FDTD method with numerical solvers for the Lorentz and Schrödinger equations can be used to simulate the evolution of electron wave packets traversing nanostructures and electromagnetic waves.

1.5 Outline and Structure of This Thesis

Here, we briefly discuss the outline of this habilitation and summarize our attempts to extensively explore the areas of electron-photon interactions, both in free space and in the near-field zone of nanostructures (Fig. 1.7). For this purpose, our research concerns 1st—the characterization of plasmon and hyperbolic polaritons in nanostructures using state-of-the-art analytical electron microscopy, 2nd—the development of new microscopy techniques to better understand the correlations in coupled nanophotonic systems and the use of these correlations for developing spectral interferometry techniques, and 3rd—the advancement of numerical techniques for exploring novel classical and quantum mechanical phenomena happening during the interaction of slow and fast electrons with nanostructures and metamaterials and the use of the nanostructures for electron-beam shaping. The feasibility of using correlation functions as an indirect tool for performing spectral interferometry using electron microscopes becomes more apparent with the introduction of *electron-induced*, *photon-induced* (EDPHS), and *photon-assisted* domains. This lays the groundwork for the structure of the current habilitation book as well.

In Chap. 2, we briefly mention the theory of electron-plasmon interactions. We also discuss the formalisms for calculating the EELS and CL spectra and outline the mechanisms by which the radiation of electron beams interacts with photonic systems, but only those mechanisms that will be used in this habilitation are described. In other words, we will not describe the Larmor radiation, cyclotron radiation, Bremsstrahlung, and free-electron lasers, as they are extensive individual topics beyond the scope of the present work. Instead, we focus on the Cherenkov radiation (CR), the transition radiation (TR), and the Smith-Purcell radiation (SPR).

In Chap. 3, we investigate the electron-induced domain and provide various examples of nanophotonic systems investigated using EELS, EFTEM, and CL techniques. Additionally, the optical signatures of more complicated systems, including anisotropic materials and hyperbolic phonon polaritons, are studied. A topological insulator such as Bi_2Se_3 at the reststrahlen band includes some of these exotic behaviours [19]. With the advent of high-energy-resolution electron microscopes

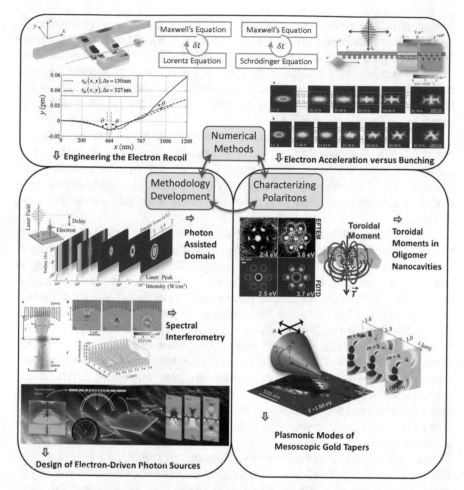

Fig. 1.7 Systems and concepts covered by this book. Characterizing Polaritons: EELS, EFTEM, SNOM, and CL techniques are used to characterize toroidal moments in oligomer nanocavities [85–89] and the plasmonic modes of mesoscopic gold tapers [70, 90]. Methodology Development: The photon-assisted domain will be explored to understand the correlations between photonic systems [76]. Based on this domain, the possibility of performing spectral interferometry with electron microscopes will be discussed [91]. It will be shown that spectral interferometry can be advanced by introducing a precisely engineered EDPHS to facilitate the control of the directionality and intensity of the emissions [92]. Numerical Methods: Self-consistent numerical toolboxes are introduced and used for the purpose of exploring classical and quantum electron recoil and the shape of the electron wave packets interacting with light and nanostructures [76, 77, 93]

with angstrom spatial resolution and meV energy resolution [94], it will soon become possible to experimentally investigate such interesting features.

In Chap. 4, we develop a self-consistent numerical tool, by combining Maxwell and Lorentz equations. We use this toolbox to explore the plasmon-induced electron recoil. We also explore mechanisms to be able to engineer this recoil using the concept of self-intereference. Furthermore, we discuss the relation between the CL and EELS using the Poynting theorem of the Maxwell's equations. In Chap. 5, optical modes of oligomer nanocavities are investigated [85–89]. It will be shown that these structures, realized by considering the Babinet's principles, can support a loop of oscillating magnetic moments; therefore support the excitation of toroidal moments. Coupling of toroidal moments to each other as well as the radiation of toroidal moments are extensively studied.

In Chap. 6, we discuss the interaction of electrons with three dimensional gold tapers [70, 90]. We show that two different scenarios, originating from the optical modes with higher order angular momentums and reflection from the apex contribute to the observed resonances in electron energy-loss spectra. We elucidate more on the analytical formalism describing the interaction of electrons with the optical modes of fibers.

In Chap. 7, we first briefly review photon-induced near-field electron microscopy (PINEM) [95]. Then, we introduce the photon-assisted domain and discuss the possibility of observing interferences in time-energy EELS maps [76]. We also propose methods for observing the interference maps in an indirect way, i.e., by exploring correlations between the desired sample and an electron-driven photon source (EDPHS) [91, 96]. We further propose design principles for engineering an EDPHS that facilitates the production of enhanced, focused, and directional radiation, which is desirable for the future of spectral interferometry with electron microscopes [92]. Finally, in Chap. 8, we discuss our attempts for the development of numerical toolboxes to help us to better understand the evolution of electron wave packets traversing electromagnetic radiation and nanostructures [79, 93]. We use our developed toolboxes to understand the classical recoil that an electron undergoes upon passing the near-field zone of a nanostructure [77]. Next, we discuss the quantum effects in linear particle accelerators (LINACs)—the so-called inverse Smith-Purcell effect [97]. We further study the free-space electron-light interactions within the framework of the Kapitza-Dirac effect (KDE) in more detail. We will show that the KDE can be generalized by the inclusion of two different light pathways to the case where a quantum coherent interference effect between the absorptive and ponderomotive channels is observed. Finally, we will show that point-projection electron microscopy can be an ideal tool for investigating such interference effects [98].

References

1. N.W. Ashcroft, N.D. Mermin, *Solid State Physics* (Thomson Learning Inc., USA, 1976)
2. N. Talebi, Optical modes in slab waveguides with magnetoelectric effect (in English). J. Opt. UK **18**(5), 055607 (2016). https://doi.org/10.1088/2040-8978/18/5/055607
3. J.M. Pitarke, V.M. Silkin, E.V. Chulkov, P.M. Echenique, Theory of surface plasmons and surface-plasmon polaritons (in English). Rep. Prog. Phys. **70**(1), 1–87 (2007). https://doi.org/10.1088/0034-4885/70/1/R01
4. P. Schattschneider, *Fundamentals of Inelastic Electron Scattering* (Springer, Wien, Austria, 1986)

5. R. Eisberg, R. Resnick, *Quantum Mechanics of Atoms, Molecules, Solids, Nuclei, and Particles*, Lecture Notes in Chemistry (Wiley, New York, USA, 1974)
6. N.J. Halas, S. Lal, W.S. Chang, S. Link, P. Nordlander, Plasmons in strongly coupled metallic nanostructures (in English). Chem. Rev. **111**(6), 3913–3961 (2011). https://doi.org/10.1021/cr200061k
7. J. Christensen, A. Manjavacas, S. Thongrattanasiri, F.H.L. Koppens, F.J.G. de Abajo, Graphene plasmon waveguiding and hybridization in individual and paired nanoribbons. ACS Nano **6**(1), 431–440 (2012). https://doi.org/10.1021/nn2037626
8. P. Nordlander, E. Prodan, Plasmon hybridization in nanoparticles near metallic surfaces. Nano Lett. **4**(11), 2209–2213 (2004). https://doi.org/10.1021/nl0486160
9. N. Liu, H. Guo, L. Fu, S. Kaiser, H. Schweizer, H. Giessen, Plasmon hybridization in stacked cut-wire metamaterials. Adv. Mater. **19**(21), 3628–3632 (2007). https://doi.org/10.1002/adma.200700123
10. X. Lu, M. Rycenga, S.E. Skrabalak, B. Wiley, Y. Xia, Chemical synthesis of novel plasmonic nanoparticles. Annu. Rev. Phys. Chem. **60**(1), 167–192 (2009). https://doi.org/10.1146/annurev.physchem.040808.090434
11. J. Dorfmüller et al., Fabry-Pérot resonances in one-dimensional plasmonic nanostructures. Nano Lett. **9**(6), 2372–2377 (2009). https://doi.org/10.1021/nl900900r
12. M. Bosman et al., Surface plasmon damping quantified with an electron nanoprobe. Sci. Rep. **3**, 1312 (2013) (online). https://doi.org/10.1038/srep01312. https://www.nature.com/articles/srep01312#supplementary-information
13. B.J. Wiley, S.H. Im, Z.Y. Li, J. McLellan, A. Siekkinen, Y.N. Xia, Maneuvering the surface plasmon resonance of silver nanostructures through shape-controlled synthesis (in English). J. Phys. Chem. B **110**(32), 15666–15675 (2006). https://doi.org/10.1021/jp0608628
14. V. Giannini, A.I. Fernandez-Dominguez, S.C. Heck, S.A. Maier, Plasmonic nanoantennas: fundamentals and their use in controlling the radiative properties of nanoemitters (in English). Chem. Rev. **111**(6), 3888–3912 (2011). https://doi.org/10.1021/cr1002672
15. V. Dziom et al., Observation of the universal magnetoelectric effect in a 3D topological insulator. Nat. Commun. **8**, 15197 (2017) (online). https://doi.org/10.1038/ncomms15197. https://www.nature.com/articles/ncomms15197#supplementary-information
16. T. Morimoto, A. Furusaki, N. Nagaosa, Topological magnetoelectric effects in thin films of topological insulators (in English). Phys. Rev. B **92**(8), 085113 (2015). https://doi.org/10.1103/physrevb.92.085113
17. T. Low et al., Polaritons in layered two-dimensional materials (in English). Nat. Mater. **16**(2), 182–194 (2017). https://doi.org/10.1038/NMAT4792
18. D.N. Basov, M.M. Fogler, F.J.G. de Abajo, Polaritons in van der Waals materials (in English). Science **354**(6309), aag1992 (2016). https://doi.org/10.1126/science.aag1992
19. J.S. Wu, D.N. Basov, M.M. Fogler, Topological insulators are tunable waveguides for hyperbolic polaritons (in English). Phys. Rev. B **92**(20), 205430 (2015). https://doi.org/10.1103/physrevb.92.205430
20. T. Stauber, Plasmonics in Dirac systems: from graphene to topological insulators (in English). J Phys Condens Mat **26**(12), 123201 (2014). https://doi.org/10.1088/0953-8984/26/12/123201
21. A. Manjavacas, F.J.G. de Abajo, Tunable plasmons in atomically thin gold nanodisks (in English). Nat. Commun. **5**, 3548 (2014). https://doi.org/10.1038/ncomms4548
22. F.J.G. de Abajo, A. Manjavacas, Plasmonics in atomically thin materials. Faraday Discuss. **178**, 87–107. https://doi.org/10.1039/c4fd00216d
23. A. Woessner et al., Highly confined low-loss plasmons in graphene-boron nitride heterostructures (in English). Nat. Mater. **14**(4), 421–425 (2015). https://doi.org/10.1038/NMAT4169
24. F.J.G. de Abajo, Graphene plasmonics: challenges and opportunities (in English). ACS Photonics **1**(3), 135–152 (2014). https://doi.org/10.1021/ph400147y
25. Y. Ding et al., Effective electro-optical modulation with high extinction ratio by a graphene–silicon microring resonator. Nano Lett. **15**(7), 4393–4400 (2015). https://doi.org/10.1021/acs.nanolett.5b00630

26. E.W. Hill, A. Vijayaragahvan, K. Novoselov, Graphene sensors (in English). IEEE Sens. J. **11**(12), 3161–3170 (2011). https://doi.org/10.1109/Jsen.2011.2167608

27. A. Vakil, N. Engheta, Transformation optics using graphene. Science **332**(6035), 1291–1294 (2011). https://doi.org/10.1126/science.1202691

28. F.J. Alfaro-Mozaz et al., Nanoimaging of resonating hyperbolic polaritons in linear boron nitride antennas. Nat. Commun. **8**, 15624 (2017) (online), https://doi.org/10.1038/ncomms15624. https://www.nature.com/articles/ncomms15624#supplementary-information

29. A.A. Govyadinov et al., Probing low-energy hyperbolic polaritons in van der Waals crystals with an electron microscope (in English). Nat. Commun. **8**, 95 (2017). https://doi.org/10.1038/s41467-017-00056-y

30. A. Poddubny, I. Iorsh, P. Belov, Y. Kivshar, Hyperbolic metamaterials (in English). Nat. Photonics **7**(12), 948–957 (2013). https://doi.org/10.1038/Nphoton.2013.243

31. R.S. Kshetrimayum, A brief intro to metamaterials. IEEE Potentials **23**(5), 44–46 (2005). https://doi.org/10.1109/MP.2005.1368916

32. Z. Jacob, I.I. Smolyaninov, E.E. Narimanov, Broadband Purcell effect: radiative decay engineering with metamaterials (in English). Appl. Phys. Lett. **100**(18), 181105 (2012). https://doi.org/10.1063/1.4710548

33. N. Talebi, C. Ozsoy-Keskinbora, H.M. Benia, K. Kern, C.T. Koch, P.A. van Aken, Wedge Dyakonov waves and Dyakonov plasmons in topological insulator Bi_2Se_3 probed by electron beams (in English). ACS Nano **10**(7), 6988–6994 (2016). https://doi.org/10.1021/acsnano.6b02968

34. Y.R. He, S.L. He, X.D. Yang, Optical field enhancement in nanoscale slot waveguides of hyperbolic metamaterials (in English). Opt. Lett. **37**(14), 2907–2909 (2012) (Online). Available: <Go to ISI>://WOS:000306709900046

35. E.E. Narimanov, A.V. Kildishev, Metamaterials naturally hyperbolic (in English). Nat. Photonics **9**(4), 214–216 (2015). https://doi.org/10.1038/nphoton.2015.56

36. F. Wilczek, Two applications of axion electrodynamics. Phys. Rev. Lett. **58**(18), 1799–1802 (1987). https://doi.org/10.1103/physrevlett.58.1799

37. J. Dorfmuller et al., Near-field dynamics of optical Yagi-Uda nanoantennas (in English). Nano Lett. **11**(7), 2819–2824 (2011). https://doi.org/10.1021/nl201184n

38. D. Podbiel et al., Imaging the nonlinear plasmoemission dynamics of electrons from strong plasmonic fields (in English). Nano Lett. **17**(11), 6569–6574 (2017). https://doi.org/10.1021/acs.nanolett.7b02235

39. T. Coenen, B.J.M. Brenny, E.J. Vesseur, A. Polman, Cathodoluminescence microscopy: optical imaging and spectroscopy with deep-subwavelength resolution. MRS Bull. **40**(4), 359–365 (2015). https://doi.org/10.1557/mrs.2015.64

40. L. Novotny, The history of near-field optics (in English). Prog. Opt. **50**, 137–184 (2007). https://doi.org/10.1016/S0079-6638(07)50005-3

41. F. Keilmann, R. Hillenbrand, Near-field microscopy by elastic light scattering from a tip. Philos. Trans. R. Soc. A Math. Phys. Eng. Sci. **362**(1817), 787–805 (2004). https://doi.org/10.1098/rsta.2003.1347

42. B. Knoll, F. Keilmann, Near-field probing of vibrational absorption for chemical microscopy. Nature **399**(6732), 134–137 (1999). https://doi.org/10.1038/20154

43. B. Knoll, F. Keilmann, Enhanced dielectric contrast in scattering-type scanning near-field optical microscopy. Opt. Commun. **182**(4–6), 321–328 (2000). https://doi.org/10.1016/s0030-4018(00)00826-9

44. R.C. Dunn, Near-field scanning optical microscopy. Chem. Rev. **99**(10), 2891–2928 (1999). https://doi.org/10.1021/cr980130e

45. M.I. Stockman, Nanoplasmonics: past, present, and glimpse into future (in English). Opt. Express **19**(22), 22029–22106 (2011). https://doi.org/10.1364/Oe.19.022029

46. M.I. Stockman, Nanofocusing of optical energy in tapered plasmonic waveguides (vol 93, 137404, 2004) (in English). Phys. Rev. Lett. **106**(1), 019901 (2011). https://doi.org/10.1103/physrevlett.106.019901

47. M.I. Stockman, Nanofocusing of optical energy in tapered plasmonic waveguides (in English). Phys. Rev. Lett. **93**(13), 137404 (2004). https://doi.org/10.1103/physrevlett.93.137404

48. D. Sadiq, J. Shirdel, J.S. Lee, E. Selishcheva, N. Park, C. Lienau, Adiabatic nanofocusing scattering-type optical nanoscopy of individual gold nanoparticles. Nano Lett. **11**(4), 1609–1613 (2011). https://doi.org/10.1021/nl1045457

49. P. Groß, M. Esmann, S.F. Becker, J. Vogelsang, N. Talebi, C. Lienau, Plasmonic nanofocusing—grey holes for light. Adv. Phys. **X**, 1–34 (2016). https://doi.org/10.1080/23746149.2016.1177469

50. A. Kubo, N. Pontius, H. Petek, Femtosecond microscopy of surface plasmon polariton wave packet evolution at the silver/vacuum interface. Nano Lett. **7**(2), 470–475 (2007). https://doi.org/10.1021/nl0627846

51. A. Kubo, K. Onda, H. Petek, Z. Sun, Y.S. Jung, H.K. Kim, Femtosecond imaging of surface plasmon dynamics in a nanostructured silver film. Nano Lett. **5**(6), 1123–1127 (2005). https://doi.org/10.1021/nl0506655

52. L. Douillard et al., Short range plasmon resonators probed by photoemission electron microscopy (in English). Nano Lett. **8**(3), 935–940 (2008). https://doi.org/10.1021/nl080053v

53. B. Frank et al., Short-range surface plasmonics: localized electron emission dynamics from a 60-nm spot on an atomically flat single-crystalline gold surface. Sci. Adv. **3**(7), e1700721 (2017). https://doi.org/10.1126/sciadv.1700721

54. M. Großmann et al., Light-triggered control of plasmonic refraction and group delay by photochromic molecular switches. ACS Photonics **2**(9), 1327–1332 (2015). https://doi.org/10.1021/acsphotonics.5b00315

55. W. Quan et al., Quantum interference in laser-induced nonsequential double ionization (in English). Phys. Rev. A **96**(3), 032511 (2017). https://doi.org/10.1103/physreva.96.032511

56. C.F.D. Faria, T. Shaaran, X. Liu, W. Yang, Quantum interference in laser-induced nonsequential double ionization in diatomic molecules: role of alignment and orbital symmetry (in English). Phys. Rev. A **78**(4), 043407 (2008). https://doi.org/10.1103/physreva.78.043407

57. D. Pengel, S. Kerbstadt, L. Englert, T. Bayer, M. Wollenhaupt, Control of three-dimensional electron vortices from femtosecond multiphoton ionization. Phys. Rev. A **96**(4), 043426 (2017). https://doi.org/10.1103/physreva.96.043426

58. L. Seiffert, T. Paschen, P. Hommelhoff, T. Fennel, High-order above-threshold photoemission from nanotips controlled with two-color laser fields (in English). J. Phys. B At. Mol. Opt. **51**(13), 134001 (2018). https://doi.org/10.1088/1361-6455/aac34f

59. T. Coenen, N.M. Haegel, Cathodoluminescence for the 21st century: learning more from light (in English). Appl. Phys. Rev. **4**(3), 031103 (2017). https://doi.org/10.1063/1.4985767

60. M. Kociak, L.F. Zagonel, Cathodoluminescence in the scanning transmission electron microscope (in English). Ultramicroscopy **176**, 112–131 (2017). https://doi.org/10.1016/j.ultramic.2017.03.014

61. R. Gómez-Medina, N. Yamamoto, M. Nakano, F.J.G. de Abajo, Mapping plasmons in nanoantennas via cathodoluminescence. New J. Phys. **10**(10), 105009 (2008). https://doi.org/10.1088/1367-2630/10/10/105009

62. B.J.M. Brenny, T. Coenen, A. Polman, Quantifying coherent and incoherent cathodoluminescence in semiconductors and metals. J. Appl. Phys. **115**(24), 244307 (2014). https://doi.org/10.1063/1.4885426

63. F.J.G. de Abajo, Optical excitations in electron microscopy. Rev. Mod. Phys. **82**(1), 209–275 (2010). https://doi.org/10.1103/revmodphys.82.209

64. V.L. Ginzburg, I.M. Frank, Radiation of a uniformly moving electron due to its transition from one medium into another. J. Phys. (USSR) **9**, 353–362 (1945)

65. A. Petersson, A. Gustafsson, L. Samuelson, S. Tanaka, Y. Aoyagi, Cathodoluminescence spectroscopy and imaging of individual GaN dots (in English). Appl. Phys. Lett. **74**(23), 3513–3515 (1999). https://doi.org/10.1063/1.124147

66. P.R. Edwards, R.W. Martin, Cathodoluminescence nano-characterization of semiconductors. Semicond. Sci. Technol. **26**(6), 064005 (2011). https://doi.org/10.1088/0268-1242/26/6/064005

67. L.F. Zagonel et al., Nanometer-scale monitoring of quantum-confined Stark effect and emission efficiency droop in multiple GaN/AlN quantum disks in nanowires. Phys. Rev. B **93**(20), 205410 (2016). https://doi.org/10.1103/physrevb.93.205410
68. T. Coenen, S.V. den Hoedt, A. Polman, A new cathodoluminescence system for nanoscale optics, materials science, and geology. Microsc. Today **24**(3), 12–19 (2016). https://doi.org/10.1017/S1551929516000377
69. F.J.G. de Abajo, M. Kociak, Probing the photonic local density of states with electron energy loss spectroscopy. Phys. Rev. Lett. **100**(10), 106804 (2008). https://doi.org/10.1103/physrevlett.100.106804
70. N. Talebi et al., Excitation of mesoscopic plasmonic tapers by relativistic electrons: phase matching versus eigenmode resonances. ACS Nano **9**(7), 7641–7648 (2015). https://doi.org/10.1021/acsnano.5b03024
71. R.F. Egerton, Electron energy-loss spectroscopy in the TEM (in English). Rep. Prog. Phys. **72**(1), 016502 (2009). https://doi.org/10.1088/0034-4885/72/1/016502
72. O.L. Krivanek et al., Vibrational spectroscopy in the electron microscope (in English). Nature **514**(7521), 209–+ (2014). https://doi.org/10.1038/nature13870
73. A. Konecna, et al., Vibrational electron energy loss spectroscopy in truncated dielectric slabs. Phys. Rev. B **98**, 205409 (2018)
74. A. Losquin, M. Kociak, Link between cathodoluminescence and electron energy loss spectroscopy and the radiative and full electromagnetic local density of states. ACS Photonics **2**(11):1619–1627 (2015)
75. L. Gu et al., Resonant wedge-plasmon modes in single-crystalline gold nanoplatelets (in English). Phys. Rev. B **83**(19), 195433 (2011). https://doi.org/10.1103/physrevb.83.195433
76. N. Talebi, W. Sigle, R. Vogelgesang, P. van Aken, Numerical simulations of interference effects in photon-assisted electron energy-loss spectroscopy. New J. Phys. **15**(5), 053013 (2013). https://doi.org/10.1088/1367-2630/15/5/053013
77. N. Talebi, A directional, ultrafast and integrated few-photon source utilizing the interaction of electron beams and plasmonic nanoantennas. New J. Phys. **16**(5), 053021 (2014). https://doi.org/10.1088/1367-2630/16/5/053021
78. N. Talebi, Schrödinger electrons interacting with optical gratings: quantum mechanical study of the inverse Smith-Purcell effect. New J. Phys. **18**(12), 123006 (2016). https://doi.org/10.1088/1367-2630/18/12/123006
79. N. Talebi, Electron-light interactions beyond the adiabatic approximation: recoil engineering and spectral interferometry. Adv. Phys. X **3**(1), 1499438 (2018). https://doi.org/10.1080/23746149.2018.1499438
80. L. Kiewidt, M. Karamehmedović, C. Matyssek, W. Hergert, L. Mädler, T. Wriedt, Numerical simulation of electron energy loss spectroscopy using a generalized multipole technique. Ultramicroscopy **133**, 101–108 (2013). https://doi.org/10.1016/j.ultramic.2013.07.001
81. T. Wriedt, Y. Eremin (eds.), *The Generalized Multipole Technique for Light Scattering: Recent Developments* (Springer International Publishing, Switzerland, 2018)
82. Q. Liang et al., Investigating hybridization schemes of coupled split-ring resonators by electron impacts. Opt. Express **23**(16), 20721–20731 (2015). https://doi.org/10.1364/oe.23.020721
83. C. Matyssek, J. Niegemann, W. Hergert, K. Busch, Computing electron energy loss spectra with the Discontinuous Galerkin Time-Domain method. Photonics Nanostruct. Fundam. Appl. **9**(4), 367–373 (2011). https://doi.org/10.1016/j.photonics.2011.04.003
84. Y. Cao, A. Manjavacas, N. Large, P. Nordlander, Electron energy-loss spectroscopy calculation in finite-difference time-domain package. ACS Photonics **2**(3), 369–375 (2015). https://doi.org/10.1021/ph500408e
85. S. Guo, N. Talebi, A. Campos, M. Kociak, P.A. van Aken, Radiation of dynamic toroidal moments. ACS Photonics **6**, 467–474 (2019). https://doi.org/10.1021/acsphotonics.8b01422
86. B. Öğüt, N. Talebi, R. Vogelgesang, W. Sigle, P.A. van Aken, Toroidal plasmonic eigenmodes in oligomer nanocavities for the visible. Nano Lett. **12**(10), 5239–5244 (2012). https://doi.org/10.1021/nl302418n

87. N. Talebi, S. Guo, A. van AkenPeter, Theory and applications of toroidal moments in electro-dynamics: their emergence, characteristics, and technological relevance. Nanophotonics **7**, 93 (2018)
88. S. Guo, N. Talebi, P.A. van Aken, Long-range coupling of toroidal moments for the visible. ACS Photonics **5**(4), 1326–1333 (2018). https://doi.org/10.1021/acsphotonics.7b01313
89. N. Talebi, B. Ögüt, W. Sigle, R. Vogelgesang, P.A. van Aken, On the symmetry and topology of plasmonic eigenmodes in heptamer and hexamer nanocavities. Appl. Phys. A **116**(3), 947–954 (2014). https://doi.org/10.1007/s00339-014-8532-y
90. S. Guo et al., Reflection and phase matching in plasmonic gold tapers. Nano Lett. **16**(10), 6137–6144 (2016). https://doi.org/10.1021/acs.nanolett.6b02353
91. N. Talebi, Spectral interferometry with electron microscopes. Sci. Rep. **6**, 33874 (2016) (online). https://doi.org/10.1038/srep33874. https://www.nature.com/articles/srep33874# supplementary-information
92. N. Talebi et al., Merging transformation optics with electron-driven photon sources. Nat. Commun. (2019) (accepted)
93. N. Talebi, Interaction of electron beams with optical nanostructures and metamaterials: from coherent photon sources towards shaping the wave function. J. Opt. UK **19**(10), 103001 (2017). https://doi.org/10.1088/2040-8986/aa8041
94. O.L. Krivanek et al., Vibrational spectroscopy in the electron microscope. Nature **514**, 209 (2014). https://doi.org/10.1038/nature13870. (online)
95. B. Barwick, D.J. Flannigan, A.H. Zewail, Photon-induced near-field electron microscopy. Nature **462**, 902 (2009) (online). https://doi.org/10.1038/nature08662. https://www.nature. com/articles/nature08662#supplementary-information
96. N. Talebi Sarvari, Method and devices for time-resolved pump-probe electron microscopy. USA Patent Appl. US20170271123A1, 2018
97. K. Mizuno, J. Pae, T. Nozokido, K. Furuya, Experimental evidence of the inverse Smith-Purcell effect. Nature **328**(6125), 45–47 (1987). https://doi.org/10.1038/328045a0
98. J. Vogelsang et al., Plasmonic-nanofocusing-based electron holography. ACS Photonics **5**(9), 3584–3593 (2018). https://doi.org/10.1021/acsphotonics.8b00418

Chapter 2
Characterization Techniques for Nanooptical Excitations

Abstract Investigation of nanooptical systems by means of far-field optics suffers from the diffraction limit of light; hence, spatial resolutions beyond the optical wavelengths are hardly achievable. Many different approaches are employed to improve the spatial resolution. These methods have been shortly described in Chap. 1. Here, we briefly discuss in more details SNOM and PEEM, with the emphasis on the applicability of these methods to different optical systems and their advantages and limitations. EELS and CL will be more intensively described in the subsequent chapters.

2.1 Scanning Near-Field Optical Microscopy

SNOM is a powerful method for investigating the near-field zone of nanostructures. In SNOM, optical fields are focused at the apex of a sharp three-dimensional taper or tip, which can be either metallic or dielectric. The optical excitations at the apex are confined to volumes as small as a few nanometers, depending on the sharpness of the tip, or the size of the aperture, in the case aperture SNOM is used (Fig. 2.1). This effect thus provides a nanoscale optical source to probe optical excitations with spatial resolutions of about 10 nm.

Various forms of SNOM tips can be utilized. These tips are usually categorized as (i) having an aperture or being apertureless (Fig. 2.2a, b), (ii) being either dielectric or metallic, as well as being dielectric with a metallic cladding (Fig. 2.2a–c), (iii) being operated with either cantilevers or tuning forks. Moreover, two different approaches might be utilized to excite the tip. Normally, tips are excited using the far-field light with an appropriate polarization and an incidence angle to minimize the parasitic signal background [1–3]. In contrast, and in particular cases when appropriate metallic tapers are introduced as the operational tips, light can be focused by means of adiabatic nanofocusing [4, 5]. In the latter case, plasmons are excited at the shaft of a metallic taper, by using for example gratings, at the distance of several micrometers from the apex (Fig. 2.2d). The excited plasmons then couple to the long-range propagating modes of the tapers all the way towards the apex, and are partially localized at the apex, and partially reflected. Thanks to the broadband nature of the plasmon

© Springer Nature Switzerland AG 2019
N. Talebi, *Near-Field-Mediated Photon–Electron Interactions*,
Springer Series in Optical Sciences 228,
https://doi.org/10.1007/978-3-030-33816-9_2

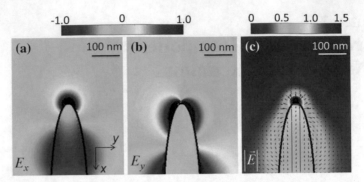

Fig. 2.1 Optical excitation in a hyperbolic gold taper. Demonstrated is the **a** x-component and **b** y-component of the electric field, as well as **c** the electric-field magnitude, when the tip is excited with a y-polarized far-field pulsed light at the carrier wavelength of 800 nm [6]. Re-printed by permission from [6] under the CC BY license

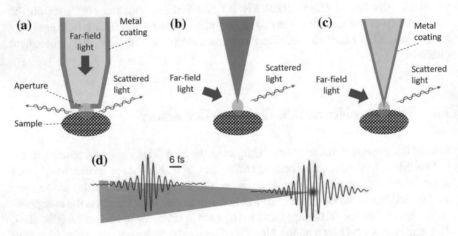

Fig. 2.2 Categorization of SNOM setups based on the excitation mechanism and the utilized tip. **a** SNOM setup with an aperture tip, which is composed of a dielectric taper coated with a thin metallic layer. The apex is not sharp, but an aperture is inserted, to couple the optical modes of the taper to the optical modes of the sample (panel 1d is reprinted from [6]). Re-printed under the CC BY license

polaritons, optical pulses as short as 10 femtoseconds can be delivered to the apex (incident pulse has the temporal duration of 6 fs) [6].

SNOM has been widely used to measure the near-field spatial distribution of single nanostructures. In particular, metallic nanoantennas have been intensively studied. For example, Fabry–Perot resonances in linear gold nanoresonators and in hexagonal boron nitride nanorods have been demonstrated (Fig. 2.3) [7]. Thanks to the broadband nature of localized apex excitations, optical excitations from far-infrared to the visible range have been investigated using SNOM. Particularly, local-ized plasmon excitations in gold nanorods within the visible range (Fig. 2.3a) [7]

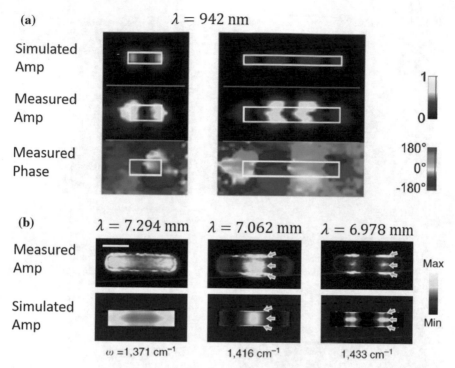

Fig. 2.3 Optical excitations in linear nanoantennas probed by scattering-type apertureless SNOM. Simulated and measured near-Field distributions (electric-field component normal to the surface) in a linear **a** (top) 140 nm-length and (bottom) 520 nm-length gold [7]—Copyright © 2009 American Chemical Society—and **b** hexagonal boron nitride, at depicted wavelengths [8]—re-printed under the CC BY license. Scale bar is 400 nm

and localized hyperbolic phonon polaritons in hexagonal boron nitride in the far-infrared range [8] were both intensively studied. Moreover, combination of SNOM and heterodyne interferometry [9] or phase-shifting algorithm [10] allows for both amplitude and phase resolved measurements, which have not yet been demonstrated for EELS. In addition, aperture-SNOM allows for measuring simultaneously various field components [11].

SNOM has been also used to measure ultrafast responses of polaritons using the pump-probe spectroscopy technique, in the visible range [12] (Fig. 2.4a) and in the infrared range [13] (Fig. 2.3b). The former setup was used to probe the propagation mechanisms and the relaxation of exciton polaritons in WSe2 thin films, as an example, whereas the latter was used to investigate Dirac plasmons in single and few-layer graphene.

Depending on the introduced tips, two different interaction types for SNOM are normally recognized; namely, week-interaction and strong-interaction regimes. Dielectric-type SNOM tips are only weakly interacting with the nanolocalized optical modes of the sample. Therefore, the tip is often approximately modeled as a dipole

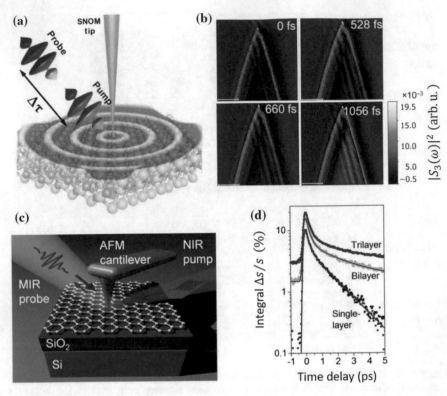

Fig. 2.4 Pump-probe SNOM. **a** Visible-range setup with both pump and probe signals are within the optics window, which is used to investigate the exciton-polariton propagation dynamics in the WSe2 thin films [12]. Demonstrated in **b** is the SNOM signal demodulated at the third harmonic of the tip-resonance frequency, at selected delay times; **a** and **b** panels reprinted by slight modifications under the CC BY license. **c** Infrared-range SNOM setup with pump signal located at the near-infrared and probe at the mid-infrared ranges, used to investigate the dynamics of Dirac polaritons in the exfoliated graphene [13]. **d** Spectrally-integrated and normalized signal changes versus time delay between pump and probe, measured upon single-layer, bilayer, and trilayer graphene. Panels c and d are reprinted by permission from [13]; Copyright © 2014 American Chemical Society

positioned at the apex and polarized along the symmetry axis of the tip (Fig. 2.5) [7, 14–16]. In this way, the scattered light to the far-field is also often assumed to be the electric-field component parallel to the symmetry axis of the tip. Thus, dielectric tips combined with interferometry techniques are considered as perfect probes for the amplitude and phase of the electric field component normal to the scanning surface. The same approximation is also used for SNOMs with metallic tips, when they are operated at the cross—polarization scheme; i.e., exciting the combined system of tip/sample with s-polarized light and detecting the light at a polarization perpendicular to that of the excitation (p-polarized far-field light).

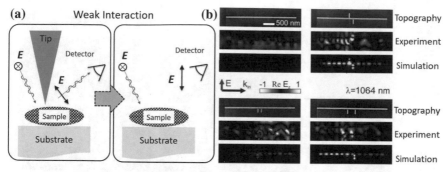

Fig. 2.5 Cross-polarized SNOM operating within the week-interaction regime. **a** Considering the week interaction between tip and sample, the detected light is approximated as the electric-field component aligned with the tip axis. **b** Within this approximation, SNOM can probe the near-field distribution around nanostructures, and the results can be reproduce numerically by the z-component of the near-field distribution) z is normal to the plane). Demonstrated is the z-component of the electric field at certain times, probed by the SNOM experimentally and calculated numerically using FDTD, for a system of a plasmonic rib waveguide coupled to two nanoantennas positioned at specific locations to be able to trigger directional propagations in the waveguide. Reprinted by permission from [16]; Copyright © 2013 WILEY-VCH Verlag GmbH & Co. KGaA, Weinheim

The situation is however different for p-polarized light exciting metallic tips coupled to metallic samples. Due to the strong interaction between tip and sample for this case, simplified approximations of the sort introduced above are no longer valid [4]. Hence, a multiple-scattering approach based on the dyadic Green's functions are often used to model the detected light. In this case, the detected light is a complicated mixture of various components of the near-field polarization. Spectral features acquired using such systems demonstrate a typical Fano-type resonance [17]. It is to be noticed that this multiple scattering algorithm is based on the fact that approximating the tip with a point-like dipolar scatterer with the polarization perpendicular to the surface (parallel to the tip axis) is an over-simplification, particularly for the cases when a rather blunt tip with more than 5 nm radius is used. This fact is already noticed in Fig. 2.1, when both the z- and y-components of the field at the apex have similar strengths. In this way, the dipolar excitation in the sample, induces another dipole in the apex (see Fig. 2.6). These strong interferences between different dipole excitations cause a Fano-type resonance and asymmetric spatial distribution of the SNOM signal in the detected spectral images [17, 18] (Fig. 2.6b).

2.2 Photoemission Electron Microscopy of Surface Plasmons

Two-photon photoemission electron microscopy (2PPEEM), particularly time-resolved 2PPEEM (tr2PPEM), has been proven to be a highly sensitive probe of plasmonic excitations at surfaces and interfaces [19–30]. In 2PPEEM the sample is

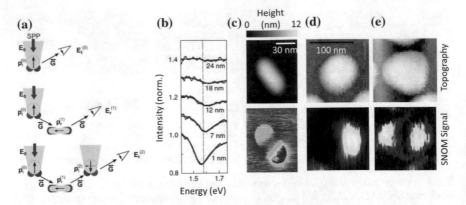

Fig. 2.6 Strong interactions between tip and sample in SNOM. **a** Multiple scattering between tip and sample can be treated using a Green's function formalism. Here, up to second-order scattering is schematically represented [17]. **b** These multiple scatterings between sample and tip causes a Fano-type resonance. Demonstrated is the acquired spectra form an elliptical gold sample, at certain sample-tip distances. Strong coupling can also lead to asymmetric spatial distributions, represented for elliptical [17] **c** and **d** circular gold nanostructures [18]. **e** The asymmetrical pattern (Panel d) reported for metallic tips has to be compared with the symmetrical SNOM signals when a carbon nanotube tip is used [18]. Reproduced by permission from [17, 18]; Copyright © 2019 Springer Nature; Copyright © 2009 American Physical Society

excited by laser light and the lateral distribution of photoemitted electrons is detected [31]. However, mainly due achromatic aberration, PEEM has less spatial resolution compared to EELS and SNOM. Nevertheless, the unprecedented power of PEEM is its ability of attosecond temporal resolution in mapping collective excitations. This ability, particularly, has rendered itself in new activities in manifesting the propagation dynamics of interface plasmons and near-field distributions in Bi_2Te_3 flakes [32], plasmons in nanostructures [21, 33, 34], optical modes with certain angular momentum [26], and in controlling the dispersion of surface plasmons by means of photochromic molecular layers [35] (Fig. 2.7).

The ability of PEEM to resolve collective electronic excitations was first demonstrated in 2002 [20]. This technique is based on a nonlinear two-photon process, hence, in an interferometric approach two laser beams are used as pump and probe, to excite the conduction/valence band electrons and probe it. Time-resolved two-photon PEEM (tr2PPEEM) signal is thus related to the third order nonlinear response of the sample. Typical setup for tr2PPEEM are demonstrated in Fig. 2.8. The differences in tr2PPEEM are based on the orientations of the pump and probe excitations and the detection scheme. To the best of our knowledge, two different schemes might be used: (i) normal-incidence mode, where pump excitation is normal to the interface [36], and (ii) oblique excitation mode [31]. Moreover, two different collection schemes might be used: (i) the excitation and detection taking place at the same surface (Fig. 2.8a), (ii) the sample being excited from one side of a planar geometry, while the photoelectrons are collected from the other side (Fig. 2.8b) [23].

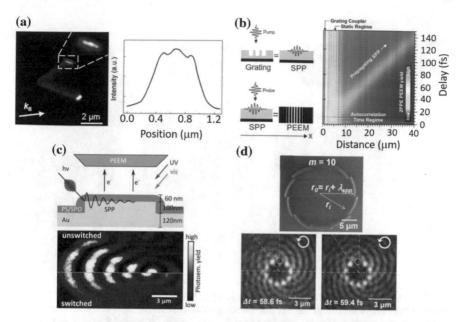

Fig. 2.7 Characterization of optical modes at surfaces and interfaces using PEEM. **a** Optical wedge modes supported along Bi$_2$Te$_3$ flakes. (Left) PEEM image under p-polarized laser illumination; (Right) PEEM intensity along the line profile shown by red, where a Fabry-Perot like excitation is obvious. **b** Time resolved PEEM used to demonstrate the excitation and propagation of surface plasmons along a gold surface. (Right) PEEM intensity versus time and distance from the grating coupler. **c** Switching behavior of plasmon polaritons propagating at the interface between gold and spirophenanthrooxazine molecule. The latter is panchromatic material which changes its optical index under UV illumination (365 nm LED light) and is deactivated with visible light (532 nm CW laser wavelength). **d** Excitation of plasmonic vortices by means of a plasmonic vortex generator with geometrical order $m = 10$ milled into a gold layer (Top). The dynamics of the generated vortex beams via propogation at the surface and its handedness has been demonstrated by time-resolved PEEM Images at different time delays between pump and probe (Down)

In general, the detection scheme for surface plasmons using Tr2PPEEM is performed within an autocorrelative approach, where both pump and probe beams have similar strength and wavelengths. In [37] however, another approach is theoretically investigated where a weak probe and strong pump is utilized. Similar to holography experiments, both the reconstruction scheme for acquiring the field amplitudes and also temporal profile of the plasmon excitation are nevertheless much easier when a strong-pump and week-probe scheme is considered. This is due to the fact that the two-photon PEEM signal is directly related to the two-photon absorption processes, which the latter depends on the third-order nonlinearity of the electric susceptibility of the material (i.e., $\chi^{(3)*}$) [37]. The coherent two photon absorption process hence is modelled as $\mathrm{Im}\left(\vec{P}^{*}.\vec{E}\right) = \varepsilon_0 \chi^{(3)*}\left|\vec{E}\right|^{4}$, where $\vec{P}$ is the nonlinear polarizability, ε_0 is the free-space permittivity, and $\vec{E} = \vec{E}_{\mathrm{Pu}} + \vec{E}_{\mathrm{Pr}}e^{i\omega\tau}$ is the total electric field, given by the summation of pump (Pu) and probe (Pr) electric fields, and τ is the temporal

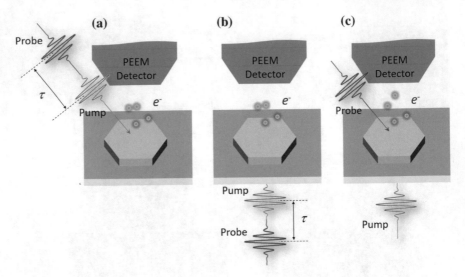

Fig. 2.8 Different setups for PEEM. Collinear pump-probe excitation and detection **a** both taking place on one side of the planar sample, and **b** taking place at different sides of the sample. **c** a laser pump excites the near-field distribution at one side, whereas the probe and detection scheme takes place at the other side

delay between them. Considering a weak probe, the time-dependent PEEM signal is then expanded as [37]

$$
I^{\text{PEEM}} \propto \left| \vec{E}_{\text{Pu}} + \vec{E}_{\text{Pr}} e^{i\omega\tau} \right|^4
$$

$$
= \left(\left| \vec{E}_{\text{Pu}} \right|^2 + \left| \vec{E}_{\text{Pr}} \right|^2 + 2\text{Re}\left(\vec{E}_{\text{Pu}}^* \cdot \vec{E}_{\text{Pr}} e^{i\omega\tau} \right) \right)^2
$$

$$
\approx \left| \vec{E}_{\text{Pu}} \right|^2 \left(\left| \vec{E}_{\text{Pu}} \right|^2 + 4\text{Re}\left(\vec{E}_{\text{Pu}}^* \cdot \vec{E}_{\text{Pr}} e^{i\omega\tau} \right) \right) \qquad (2.1)
$$

The detected signal is then the time-averaged response, which can be obtained theoretically by integrating (2.1) over the entire detection time, which for simplicity is taken to be infinite. Tr2PhPEEM signal is then composed of a static background and an interference term that depends on the temporal beam profiles and the displacement in time of probe and pump signals.

2.3 Summary

In this chapter, we briefly studied two characterization techniques for investigating the optical near fields in the vicinity of single nanoparticles. SNOM is a widely-used high-resolution technique for characterization of optical near-fields based on linear optics and scattering mechanisms, therefore it is flexible for studying various forms of materials. Also, SNOM has been developed to study nonequilibrium dynamics of exciton polaritons and surface waves. PEEM is in contrast depends on the nonlinear responses of materials, therefore is restricted to materials with sufficient third-order electric susceptibilities, like metals such as gold and silver—nevertheless, it was also exploited for probing the optical modes in materials like Bi_2Te_3. Thanks to the nonlinear process involved in tr2PhPEEM, this method is highly suitable for studying correlations and nonequilibrium dynamics, by developing a highly efficient pump-probe scheme with attosecond temporal resolution.

References

1. F. Keilmann, R. Hillenbrand, Near-field microscopy by elastic light scattering from a tip. Philos. Trans. R. Soc. Math. Phys. Eng. Sci. **362**(1817), 787–805 (2004). https://doi.org/10.1098/rsta. 2003.1347
2. A. Bek, R. Vogelgesang, K. Kern, Apertureless scanning near field optical microscope with sub-10 nm resolution. Rev. Sci. Instrum. **77**(4), 043703 (2006). https://doi.org/10.1063/1.2190211
3. T. Neuman, P. Alonso-González, A. Garcia-Etxarri, M. Schnell, R. Hillenbrand, J. Aizpurua, Mapping the near fields of plasmonic nanoantennas by scattering-type scanning near-field optical microscopy. Laser Photonics Rev. **9**(6), 637–649 (2015). https://doi.org/10.1002/lpor. 201500031
4. D. Sadiq, J. Shirdel, J.S. Lee, E. Selishcheva, N. Park, C. Lienau, Adiabatic nanofocusing scattering-type optical nanoscopy of individual gold nanoparticles. Nano Lett. **11**(4), 1609–1613 (2011). https://doi.org/10.1021/nl1045457
5. S. Schmidt et al., Adiabatic nanofocusing on ultrasmooth single-crystalline gold tapers creates a 10-nm-sized light source with few-cycle time resolution. ACS Nano **6**(7), 6040–6048 (2012). https://doi.org/10.1021/nn301121h
6. P. Groß, M. Esmann, S.F. Becker, J. Vogelsang, N. Talebi, C. Lienau, Plasmonic nanofocusing—grey holes for light. Adv. Phys. **X**(1–34), 2016 (2016). https://doi.org/10.1080/23746149.2016. 1177469
7. J. Dorfmüller et al., Fabry-Pérot resonances in one-dimensional plasmonic nanostructures. Nano Lett. **9**(6), 2372–2377 (2009). https://doi.org/10.1021/nl900900r
8. F.J. Alfaro-Mozaz et al., Nanoimaging of resonating hyperbolic polaritons in linear boron nitride antennas. Nat. Commun. **8**, 15624. https://doi.org/10.1038/ncomms15624, https://www. nature.com/articles/ncomms15624#supplementary-information
9. T. Taubner, R. Hillenbrand, F. Keilmann, Performance of visible and mid-infrared scattering-type near-field optical microscopes. J. Microsc. **210**(3), 311–314 (2003). https://doi.org/10. 1046/j.1365-2818.2003.01164.x
10. B. Deutsch, R. Hillenbrand, L. Novotny, Near-field amplitude and phase recovery using phase-shifting interferometry. Opt. Exp. **16**(2), 494–501 (2008). https://doi.org/10.1364/oe.16.000494
11. B. le Feber, N. Rotenberg, D.M. Beggs, L. Kuipers, Simultaneous measurement of nanoscale electric and magnetic optical fields. Nat Photonics **8**, 43. https://doi.org/10.1038/nphoton. 2013.323, https://www.nature.com/articles/nphoton.2013.323#supplementary-information
12. M. Mrejen, L. Yadgarov, A. Levanon, H. Suchowski, Transient exciton-polariton dynamics in WSe_2 by ultrafast near-field imaging. Sci. Adv. **5**(2), eaat9618 (2019). https://doi.org/10.1126/sciadv.aat9618
13. M. Wagner et al., Ultrafast and nanoscale plasmonic phenomena in exfoliated graphene revealed by infrared pump-probe nanoscopy. Nano Lett. **14**(2), 894–900 (2014). https://doi.org/10.1021/nl4042577

14. R. Hillenbrand, F. Keilmann, P. Hanarp, D.S. Sutherland, J. Aizpurua, Coherent imaging of nanoscale plasmon patterns with a carbon nanotube optical probe. Appl. Phys. Lett. **83**(2), 368–370 (2003). https://doi.org/10.1063/1.1592629
15. T. Zentgraf et al., Amplitude- and phase-resolved optical near fields of split-ring-resonator-based metamaterials. Opt. Lett. **33**(8), 848–850 (2008). https://doi.org/10.1364/ol.33.000848
16. M. Esslinger, W. Khunsin, N. Talebi, T. Wei, J. Dorfmüller, R. Vogelgesang, K. Kern, Phase engineering of subwavelength unidirectional plasmon launchers. Adv. Opt. Mater. **1**(6):434–437 (2013)
17. M. Esmann et al., Vectorial near-field coupling. Nat. Nanotechnol. **14**(7), 698–704 (2019). https://doi.org/10.1038/s41565-019-0441-y
18. A. García-Etxarri, I. Romero, F.J. García de Abajo, R. Hillenbrand, J. Aizpurua, Influence of the tip in near-field imaging of nanoparticle plasmonic modes: weak and strong coupling regimes. Phys. Rev. B **79**(12), 125439 (2009). https://doi.org/10.1103/physrevb.79.125439
19. A. Kubo, N. Pontius, H. Petek, Femtosecond microscopy of surface plasmon polariton wave packet evolution at the silver/vacuum interface. Nano Lett. **7**(2), 470–475 (2007). https://doi.org/10.1021/nl0627846
20. A. Kubo, K. Onda, H. Petek, Z. Sun, Y.S. Jung, H.K. Kim, Femtosecond imaging of surface plasmon dynamics in a nanostructured silver film. Nano Lett. **5**(6), 1123–1127 (2005). https://doi.org/10.1021/nl0506655
21. B. Frank et al., Short-range surface plasmonics: Localized electron emission dynamics from a 60-nm spot on an atomically flat single-crystalline gold surface. Sci. Adv. **3**(7), e1700721 (2017). https://doi.org/10.1126/sciadv.1700721
22. C. Lemke et al., Mapping surface plasmon polariton propagation via counter-propagating light pulses. Opt. Exp. **20**(12), 12877–12884 (2012). https://doi.org/10.1364/oe.20.012877
23. A. Klick et al., Femtosecond time-resolved photoemission electron microscopy operated at sample illumination from the rear side. Rev. Sci. Instrum. **90**(5), 053704 (2019). https://doi.org/10.1063/1.5088031
24. M. Aeschlimann et al., Adaptive subwavelength control of nano-optical fields. Nature **446**(7133), 301–304 (2007). https://doi.org/10.1038/nature05595
25. P. Klaer et al., Polarization dependence of plasmonic near-field enhanced photoemission from cross antennas. Appl. Phys. B **122**(5), 136 (2016). https://doi.org/10.1007/s00340-016-6410-3
26. G. Spektor et al., Revealing the subfemtosecond dynamics of orbital angular momentum in nanoplasmonic vortices. Science **355**(6330), 1187 (2017). https://doi.org/10.1126/science.aaj1699
27. Y. Gong, A.G. Joly, D. Hu, P.Z. El-Khoury, W.P. Hess, Ultrafast imaging of surface plasmons propagating on a gold surface. Nano Lett. **15**(5), 3472–3478 (2015). https://doi.org/10.1021/acs.nanolett.5b00803
28. M. Shibuta, K. Yamagiwa, T. Eguchi, A. Nakajima, Imaging and spectromicroscopy of photocarrier electron dynamics in C60 fullerene thin films. Appl. Phys. Lett. **109**(20), 203111 (2016). https://doi.org/10.1063/1.4967380
29. T. Leißner, O. Kostiučenko, J.R. Brewer, H.-G. Rubahn, J. Fiutowski, Nanostructure induced changes in lifetime and enhanced second-harmonic response of organic-plasmonic hybrids. Appl. Phys. Lett. **107**(25), 251102 (2015). https://doi.org/10.1063/1.4938007
30. M. Dąbrowski, Y. Dai, A. Argondizzo, Q. Zou, X. Cui, H. Petek, Multiphoton photoemission microscopy of high-order plasmonic resonances at the Ag/vacuum and Ag/Si interfaces of epitaxial silver nanowires. Acs Photonics **3**(9), 1704–1713 (2016). https://doi.org/10.1021/acsphotonics.6b00353
31. O. Schmidt et al., Time-resolved two photon photoemission electron microscopy. Appl. Phys. B **74**(3), 223–227 (2002). https://doi.org/10.1007/s003400200803
32. X. Lu et al., Observation and manipulation of visible edge plasmons in Bi_2Te_3 nanoplates. Nano Lett. **18**(5), 2879–2884 (2018). https://doi.org/10.1021/acs.nanolett.8b00023
33. P. Kahl et al., Direct observation of surface plasmon polariton propagation and interference by time-resolved imaging in normal-incidence two photon photoemission microscopy. Plasmonics **13**(1), 239–246 (2018). https://doi.org/10.1007/s11468-017-0504-6

34. D. Podbiel et al., Imaging the nonlinear plasmoemission dynamics of electrons from strong plasmonic fields. Nano Lett. **17**(11), 6569–6574 (2017). https://doi.org/10.1021/acs.nanolett.7b02235
35. M. Großmann et al., Light-triggered control of plasmonic refraction and group delay by photochromic molecular switches. Acs Photonics **2**(9), 1327–1332 (2015). https://doi.org/10.1021/acsphotonics.5b00315
36. P. Kahl et al., Normal-incidence photoemission electron microscopy (NI-PEEM) for imaging surface plasmon polaritons. Plasmonics **9**(6), 1401–1407 (2014). https://doi.org/10.1007/s11468-014-9756-6
37. T.J. Davis, B. Frank, D. Podbiel, P. Kahl, F.-J. Meyer zu Heringdorf, H. Giessen, Subfemtosecond and nanometer plasmon dynamics with photoelectron microscopy: theory and efficient simulations. Acs Photonics **4**(10), 2461–2469 (2017). https://doi.org/10.1021/acsphotonics.7b00676

Chapter 3
Electron-Light Interactions

Abstract Swift electrons can undergo inelastic interactions with single electrons as well as collective electron excitations within the sample, such as plasmon and phonon polaritons, as a result of which they lose energy (Garcia de Abajo in Rev. Mod. Phys. 82:209–275, 2010 [1]). Within the classical formalism, EEL spectra are theoretically rationalized by a simple but intuitive interpretation that has a direct correspondence with first principles, demanding that all inelastic signals are collected (Ritchie and Howie, in Philos. Mag. A 58:753–767, 1988 [2].

In the non-recoil approximation, assuming that the electron is travelling parallel to the z-axis, the EEL spectrum is provided as $\Gamma^{\mathrm{EELS}}(\omega) = (e/\pi\hbar\omega)\mathrm{Re}\int \mathrm{d}z\, E_z^{\mathrm{sca}}(R_{\mathrm{el}}, z; \omega) \times \exp(-i\, z\omega/v_{\mathrm{el}}) = (e/\pi\hbar\omega)\mathrm{Re}\tilde{E}_z^{\mathrm{sca}}(R_{\mathrm{el}}; k_z = \omega/v_{\mathrm{el}}; \omega)$, where e is the elementary charge, $\hbar$ is the reduced Planck constant, ω is the angular frequency of the light, $R_{\mathrm{el}} = (x_{\mathrm{el}}, y_{\mathrm{el}})$ is an electron impact parameter, $\vec{v}_{\mathrm{el}} = v_{\mathrm{el}}\hat{z}$ is the velocity of the electron, k_z is the z-component of the wave vector, and $\tilde{E}_z^{\mathrm{sca}}$ is the Fourier-transformed z-component of the electric field. Moreover, by measuring the amount of energy loss from the electrons using an electron energy-loss (EEL) detector, we have direct access to the photonic local density of states projected along the electron trajectory [4]. Considering this formalism, many aspects of the optical near-field distribution of nanostructures, such as dark versus bright modes [5–8], localized surface plasmons [9–12], void plasmons [13–16], and wedge plasmons [17–19], can be understood by EELS. The specific selection of the wave vector ($k_z = \omega/v_{\mathrm{el}}$) by swift electrons takes into account the energy–momentum conservation criterion, which is crucial for mapping the whole local density of mesoscopic structures such as gratings [20] and conical tapers [21, 22].

Portions of the text of this chapter have been re-published with permission from [3], Copyright 2017 IOP Publishing Ltd.

© Springer Nature Switzerland AG 2019
N. Talebi, *Near-Field-Mediated Photon–Electron Interactions*,
Springer Series in Optical Sciences 228,
https://doi.org/10.1007/978-3-030-33816-9_3

As a result of the interaction of an electron with the photonic modes of the sample, the electromagnetic far-field radiation is important, as this radiation can be analysed using CL spectroscopy [23–25]. CL is complementary to EELS, in the sense that only the radiative modes are detected by CL. However, optically dark modes such as bulk plasmons still contribute to the electron-induced radiation [26] because the emitter (here, the electron) traverses the material. Moreover, whenever an electron beam crosses an interface between a metal and a dielectric, TR will contribute to the far-field radiation, which is mutually coherent with the evanescent near-field zone of the electron [1].

In addition to spectroscopy and the static imaging of PDOSs, the time-resolved responses of samples can be probed as well using PINEM [27–29]. In PINEM, samples are excited using focused high-intensity pulsed-laser beams of a given polarization, and the excited states are probed using pulsed electron excitations. Hence, PINEM is a pump-probe time-resolved spectroscopic tool, with lasers serving as the pump and pulsed electrons acting as probes. PINEM will be discussed in more detail in Chap. 8.

Indeed, many aspects of the interaction of the electrons with nanostructures are well understood and have been investigated in the literature, covering a broad range of material science, plasmonics, and acceleration physics. This chapter provides a rather general overview of the whole field, with an emphasis on provoking more research on the interaction of electron beams with carefully engineered materials, such as gratings and metamaterials [30, 31], either to create novel few-photon sources or to shape the electron beam but also to provide new characterization methodologies by combining these two apparently distinct aspects of the field, as will be discussed in Chap. 8.

3.1 First Principles

The interactions between electrons and optical plane waves in infinite free space are mostly restricted to elastic scattering, a prominent example of which is the Kapitza--Dirac effect (KDE), as pointed out by Kapitza and Dirac in 1933 [32]. In the KDE, an electron beam interacts with a standing wave light pattern in a vacuum. As a result of this interaction, the electron beam is elastically diffracted, mainly because of a two-photon absorption and emission process [33]. Similar elastic scattering effects for ionic beams have been utilized in matter wave experiments, using standing-wave optical patterns as gratings to efficiently split matter waves into the required beam paths [34, 35]. Travelling-wave-assisted beam splitters have also been proposed and theoretically investigated, demonstrating that the effect is not necessarily restricted to standing wave patterns in free space [36]. However, because of the small values of the Thomson scattering cross section, very high laser beam intensities are required to achieve intense diffracted matter waves. For this reason, the experimental verification of the KDE effect had to wait for the development of the laser [37]. Figure 3.1

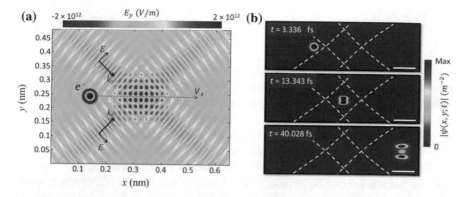

Fig. 3.1 A Gaussian electron wave packet interacting with two propagating Gaussian optical waves, demonstrating the KDE. **a** Spatial profile of the *y*-component of the electric field at a given time. The wavelength of the light wave is $\lambda_0 = 30$ nm, and the polarization state is depicted in the frame. The electron wave packet traverses the optical waves at their intersection at a velocity of 0.04*c*. **b** The magnitude of the electron wave function at the times depicted at each frame. The scale bar is 100 nm. Reprinted by permission from [3]; © 2017 IOP Publishing Ltd

demonstrates the diffraction of a Gaussian electron wave function from two propagating Gaussian optical waves at a wavelength of 30 nm. The processes were simulated with the first-principles numerical toolbox explained in Chap. 8. The electron has a velocity of 0.04*c* and longitudinal and transverse widths of 5 nm and is travelling through the standing wave pattern of two counter-propagating Gaussian beams, as shown in Fig. 3.1a. The amplitude of the electric field is $E_0 = 1.5 \times 10^{12}$ V/m. The evolution of the electron wave function is shown in a spatial representation in Fig. 3.1b, which clearly demonstrates the diffraction of the wave function into three peaks by light.

However, for an *inelastic* electron–photon interaction to occur, one needs a mediator to satisfy both the energy and momentum conservation criteria. For the case of the emission of a single photon by an electron, these criteria are given by

$$\begin{cases} E_f = E_i + \hbar\omega \\ \vec{p}_f = \vec{p}_i + \hbar\vec{k} \end{cases} \tag{3.1}$$

assuming that other selection rules, as for angular momentum, are also satisfied. In (3.1), E_α is the energy of the electron, $\alpha \in (i, f)$ denotes the initial and final states of the electron before and after the interaction, ω is the angular frequency of the emitted photon, $\vec{k}$ is the photon wave vector, and $\vec{p}_i$ and $\vec{p}_f$ are the linear momenta of the electron before and after the interaction, respectively. Considering the relativistic energies $E_\alpha = \sqrt{(m_0^2 c^4 + p_\alpha^2 c^2)}$ and momenta $\vec{p}_\alpha = \gamma_\alpha m_0 \vec{v}_\alpha$, where γ_α is the Lorentz factor and $\vec{v}_\alpha$ is the electron velocity, we obtain the critical angle

$$\cos(\theta) = \frac{(\hbar k)^2 - (\hbar k_0)^2}{2 p_i \, \hbar k} \pm \frac{c \, k_0}{v_i k} \qquad (3.2)$$

where $\cos(\theta) = \vec{p}_i \cdot \vec{k} / p_i k$ and k_0 is the free-space wavenumber of the photon. It is obvious that in free space ($k = k_0$), (3.2) will simplify to $\cos(\theta) = \pm c/v_i$, which cannot lead to real values for θ. However, there exist several ways to satisfy (3.2) [38, 39], by manipulating either the momentum or the energies of the photon or the electron in either the longitudinal or transverse direction. Among these possibilities, slow-wave mediators incorporating optical devices for manipulating the momentum of the photons allow the most concise designs and are discussed here. To review devices that use Larmor-based mechanisms of radiation and that incorporate quasi-free electrons, the readers are referred to [40–43] for free electron lasers and [44, 45] for cyclotron radiation.

In the classical limit where $\hbar \to 0$ and at low photon energies where $\hbar \omega \ll p_i c$, the second term in (3.2) is dominant. To have $c/v_i \leq k/k_0$, one can use matter with a refractive index $n(\omega) \geq c/v_i$. This is the condition for CR [46]. This mechanism will be discussed in more detail in Sect. 3.2. Another method is to make use of the nearfield of materials that provides evanescent modes with wavenumbers large enough to satisfy (3.2). An example of such a topology is an electron travelling adjacent and parallel to a single interface [47]. Moreover, when an electron crosses an interface, this leads to TR [1, 48]. This radiation is understood by the coupling of the moving electron with its image charge inside the material, which together form a transient dipole. Examples of TR from metamaterials will be discussed in Sect. 3.3. The Smith–Purcell effect happens when an electron travels near a grating, as will be discussed in Sect. 3.4.

3.2   Cherenkov Radiation

CR takes place as a result of the interaction of a moving charge with a bulk material with a refractive index higher than c/v_{el}, where v_{el} is the electron velocity. Despite being similar to the acoustic waves created by a uniformly moving source at a velocity greater than the sound velocity, equivalent electromagnetic waves were not experimentally detected until the 1930s [49–51]. Evidently, this is because accelerators were needed to create sufficiently fast-moving charges.

Several publications have been devoted to the classical [46, 52] and quantum-mechanical [53–56] theoretical bases of CR and the applications of CR [57] in isotropic and homogeneous metals and dielectrics. Although the classical basis of CR is well understood, possible quantum effects were only recently discussed. Interestingly, the quantum mechanical aspects of CR directly include fundamental aspects as follows: 1st—Is it the matter or the electron that emits the photon [56]? 2nd—Which form of the photon momentum is to be exploited, the Minkowski or the Abraham representation [58]? Both of these debates are at the heart of quantum optics. In most practical situations, the classical treatment that excludes the quantization of the radiation agrees with the experimental results and explains the physical facts [55]. It is only at the extreme regimes of ultraslow electrons (below 5 eV kinetic energy) and ultrahigh photon energies (e.g., x-ray photons) that quantum-optical corrections are required [46]. However, when an electron wave packet with a certain spin and angular momentum is considered, novel aspects such as the splitting of the CR cone into two cones and radiation in the backward direction have been conjectured [54]. Hereafter, only the classical treatments are considered.

CR emission can be observed through a transmission electron microscope as a result of the interaction of the electron beam with the sample. CR leads to a resonance in the electron energy-loss spectra from various materials [59, 60], and the dispersion diagram associated with CR can be retrieved using momentum-resolved EELS (MREELS) [61]. The analytical expression for the electron energy-loss spectra associated with CR in a bulk material has been provided by Kröger [48]. For the purpose of a direct application to metamaterials, we have calculated the EEL spectra using a vector potential approach in which the solutions are constructed at the potential level. We assume here that $\vec{B} = \vec{\nabla} \times \vec{A}$, $\vec{E} = -\dot{\vec{A}} - \vec{\nabla}\varphi$, and we use the Lorentz gauge. Briefly, the current density of a moving electron with speed v_{el} along the z-axis is given by $J_z(\vec{r}, t) = -ev_{el}\delta(x)\delta(y)\delta(z - v_{el}t)$. The vector potential component A_z—n frequency domain—is derived by solving the inhomogeneous Helmholtz equation $\vec{\nabla}^2 A_z(\vec{r}, \omega) + \varepsilon_{rzz}(\omega)\mu_r k_0^2 A_z(\vec{r}, \omega) = -\mu_0\mu_r \tilde{J}_z(\vec{r}, \omega)$, where $\tilde{J}_z$ is the current density in the frequency domain, as

$$A_z(\vec{r}, \omega) = \frac{1}{4\pi^2} \int_{-\infty}^{+\infty}\int_{-\infty}^{+\infty} \tilde{A}_z(k_x, k_y, z; \omega)e^{-ik_y y}e^{-ik_x x}dk_x dk_y, \qquad (3.3)$$

with $\tilde{A}_z$ given by

$$\tilde{A}_z = e\mu_0\mu_r \frac{1}{\varepsilon_{rzz}(\omega)\mu_r k_0^2 - k_x^2 - k_y^2 - (\omega/v_{el})^2} \exp\left(i\frac{\omega}{v_{el}}z\right). \qquad (3.4)$$

Having the solutions to the vector potential, the field components are calculated as [62]

$$\vec{E}(\vec{r}, \omega) = -i\omega\vec{A}(\vec{r}, \omega) + \frac{[\hat{\varepsilon}_r(\omega)]^{-1}}{i\omega\varepsilon_0\mu_0\mu_r}\vec{\nabla}\left(\vec{\nabla} \cdot \vec{A}(\vec{r}, \omega)\right) \qquad (3.5)$$

for the electric field and

$$\vec{H}(\vec{r}, \omega) = \frac{1}{\mu_0\mu_r}\nabla \times \vec{A}(\vec{r}, \omega). \qquad (3.6)$$

Using the Poynting theorem, the power radiated to the far-field zone along the z-axis is calculated as $P_z = \frac{1}{2}\mathrm{Re}\int dx \int dy \left(E_x H_y^* - E_y H_x^*\right) = \int dk_x \int dk_y \tilde{p}_z(\omega; k_x, k_y)$. Here, $\tilde{p}_z(\omega; k_x, k_y) = \frac{1}{2}\mathrm{Re}\left(\tilde{e}_x \tilde{h}_y^* - \tilde{e}_y \tilde{h}_x^*\right)$ is the radiated power calculated in the reciprocal space. As a result of the interaction of the moving electron within a bulk material, P_z is given by

$$\tilde{p}_z(\omega; k_x, k_y) = \frac{e^2}{8\pi^2 \varepsilon_0 v_{\mathrm{el}}} \mathrm{Re}\left(\frac{k_x^2}{\varepsilon_{rxx}} + \frac{k_y^2}{\varepsilon_{ryy}}\right) \frac{1}{\left|\varepsilon_{rzz}\mu_r k_0^2 - k_x^2 - k_y^2 - \left(\omega/v_{\mathrm{el}}\right)^2\right|^2}$$

(3.7)

The EELS integral will be given by the action integral [63]:

$$\Gamma^{\mathrm{EELS}}(\omega) = \frac{-1}{\hbar\omega}\mathrm{Re}\iiint\limits_v \vec{E}(\vec{r}, \omega) \cdot \vec{J}^*(\vec{r}, \omega)\, dv$$

$$= \frac{e}{2\pi\hbar\omega}\mathrm{Re}\iint\limits_{k_x, k_y} dk_x dk_y \int\limits_{z=-\infty}^{+\infty} dz\, \tilde{E}_z(k_x, k_y; z; \omega)e^{-i\frac{\omega}{v_{\mathrm{el}}}z},$$

(3.8)

and the MREEL spectrum in diffraction is defined as

$$\Gamma_k^{EELS}(k_x, k_y; \omega) = \frac{e}{2\pi\hbar\omega}\int\limits_{z=-\infty}^{+\infty} dz\, \tilde{E}_z(k_x, k_y; z; \omega)e^{-i\frac{\omega}{v_{\mathrm{el}}}z},$$

(3.9)

By combining this with the electric field calculated using (3.5), the MREEL spectrum is obtained as

$$\frac{d\Gamma^{\mathrm{EELS}}(\omega; k_x, k_y)}{dz} = \frac{e^2}{4\pi^2\hbar\omega^2\varepsilon_0}\mathrm{Im}\left\{\frac{1}{\varepsilon_{rzz}}\frac{\left(\varepsilon_{rzz}\mu_r k_0^2 - (\omega/v_{\mathrm{el}})^2\right)}{\varepsilon_{rzz}\mu_r k_0^2 - k_x^2 - k_y^2 - (\omega/v_{\mathrm{el}})^2}\right\}.$$

(3.10)

The only nonzero components of the permittivity tensor are assumed to be ε_{rxx}, ε_{ryy}, and ε_{rzz} for the sake of simplicity. Moreover, $\vec{k} = (k_x, k_y, k_z = \omega/c)$ is the wave vector of the emitted light in the reciprocal space, e is the electron charge, and $k_0 = \omega/c$ and ε_0 are the free-space wavenumber and permittivity, respectively. Interestingly, the EEL spectrum of a bulk material is related only to the relative permittivity component ε_{rzz}, although for the emitted power in the z-direction, all permittivity components are relevant. Moreover, the EEL spectrum has two contributions from the longitudinal and transverse terms as

$$\frac{d\Gamma^{\mathrm{EELS}}(\omega; k_x, k_y)}{dz} = \frac{e^2}{4\pi^2\hbar\omega^2\varepsilon_0}\mathrm{Im}\frac{1}{\varepsilon_{rzz}}$$

$$+ \frac{e^2}{4\pi^2 \hbar \omega^2 \varepsilon_0} \mathrm{Im} \frac{1}{\varepsilon_{rzz}} \frac{k_x^2 + k_y^2}{\varepsilon_{rzz}\mu_r k_0^2 - k_x^2 - k_y^2 - (\omega/v_{el})^2} \tag{3.11}$$

where the first term in (3.9) is known as the longitudinal term and the second term is the transverse term. This nomenclature is because of the longitudinal and transverse current densities in a bulk material, both of which contribute to the energy loss experienced by the electron [63]. An examples of a longitudinal excitation is a bulk plasmon, whereas an example of a transverse excitation is CR. As EEL spectra are routinely employed to perform Kramers-Kronig analysis, care should be taken regarding the transverse part [61], which is also theoretically confirmed by using (3.7) (Fig. 3.2).

The Cherenkov radiation in specific materials can strongly interact with phonons, as a result of which level repulsion might occur. In particular, h-BN is a well-known uniaxial material belonging to the class of natural hyperbolic materials. Within the far-infrared energy range, this material supports optical phonons at both the lower and upper reststrahlen bands, i.e., the energy range between transverse and longitudinal optical phonons. The reason for this highly birefringent behaviour is the anisotropic

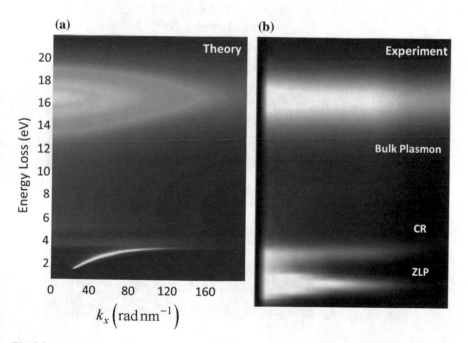

Fig. 3.2 Momentum-resolved EELS for bulk silicon, for which the CR and bulk plasmon contributions are apparent. **a** Theoretical EEL spectrum calculated using (3.5). **b** Experimental momentum-resolved EELS measured by Stöger-Pollach et al. [61]. Panel a is reprinted by permission from [3]; © 2017 IOP Publishing Ltd. Panel b is reprinted by permission from [61]; Copyright © 2006 Elsevier Ltd

structure of h-BN, where optical phonons are supported both along the c-axis and within ab-planes (see Fig. 3.3a) [64]. Within the short energy bands of 0.093 eV < E < 0.1 eV and 0.17 eV < E < 0.2 eV, the permittivity components along the ab- and c-axes change their signs, with the lower and upper bands distinguished as hyperbolic type II and type I, respectively. Within the entire ranges of energies outside these bands, the material is dielectric and supports the Cherenkov radiation explained above. Exactly within the given energy ranges, however, swift electrons with a kinetic energy of 200 keV can excite bulk longitudinal phonon excitations at energies where the permittivity crosses zero. Depending on whether the swift electron travels parallel

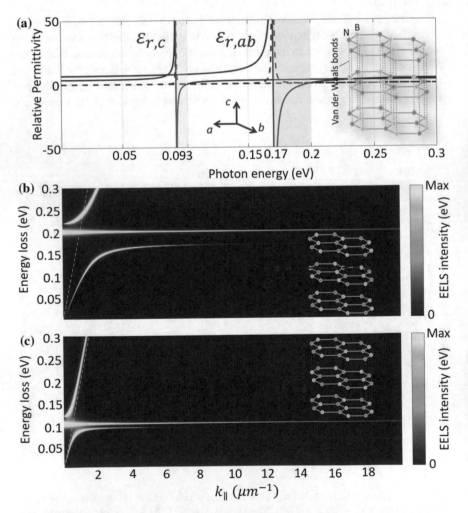

Fig. 3.3 Cherenkov radiation in h-BN. **a** Permittivity of h_BN, MREEL spectrum of an electron at the kinetic energy of 200 keV propagating, **b** within the ab-plane and **c** along the c-axis. The interaction of the Cherenkov radiation with longitudinal phonons causes a significant level repulsion

to the c-axis or within the ab-plane, either the loweror upper band phonons might be excited (Fig. 3.3b, c). As a result of the interaction of the Cherenkov radiation with phonons, a strong level repulsion is observed in the MREEL spectra.

3.3 Transition Radiation

A moving electron crossing an interface between two different materials generates radiation called TR, which is mutually coherent with the self-field of the electron. It is widely accepted that TR occurs because of the interaction of the moving electron with its image charge. In this regard, the following events occur: 1st—the creation of an image charge in the second medium when the electron is still moving in the first medium, 2nd—a gradual annihilation of the image when the electron approaches the interface, and 3rd—the screening of the electron charge when the electron has crossed the interface. This effect was first predicted by Ginzburg and Frank [65] and thereafter intensively studied both theoretically and experimentally by many groups. Ginzburg considered several ways to treat the problem theoretically either as a Larmor radiation created due to a *sharp* change in the velocity of the electron at the interface [66] or as a sudden change in the refractive index and hence the phase velocity of the light waves due to the inhomogeneity of the medium [67]. Interestingly, TR can also be considered within the category of radiation by a uniformly moving charge, and hence, a theoretical treatment based on the non-recoil approximation is suitable to describe this phenomenon. Similar to CR, TR also contributes to the EEL spectra. A detailed analytical treatment of the EELS probability was already performed by Kröger in 1968 [48] and more recently by Garcia de Abajo et al. [68], who provided more detail considering inclined excitations towards the surface. Taking into account the retardation effect and boundary conditions, such a treatment has been shown to be in good agreement with experimental results [68].

To describe the theory of TR step by step, we first consider an electron traversing an interface between vacuum and an anisotropic material with a linear permittivity tensor $\varepsilon_{r\alpha\beta} = \varepsilon_{r\alpha\beta}\delta_{\alpha\beta}$ and permeability μ_r ($\delta_{\alpha\beta}$ is the Kronecker delta function, and $\alpha \in (x, y, z)$). The surface lies at $z = 0$, and the electron impact is considered to be $\vec{R}_{el} = (x_{el}, y_{el}) = (0, 0)$. The solution for the vector potential will be constructed as $\vec{A} = (A_x, 0, A_z)$, where the A_α components are given by

$$
A_\alpha(\vec{r}; \omega) = \int\limits_{-\infty}^{+\infty} dk_x \int\limits_{-\infty}^{+\infty} dk_y
$$

$$
\left\{ \tilde{A}_1^\alpha e^{-ik_{z,1}z} e^{-ik_x x - ik_y y} + \delta_{\alpha z} e^{i\frac{\omega}{v_{el}}z} \frac{-e\mu_0}{2\pi} \frac{1}{k_0^2 - k_x^2 - k_y^2 - (\omega/v_{el})^2} \right\}
$$

$$
\tag{3.12a}
$$

For $z > 0$ and,

$$A_\alpha(\vec{r};\omega) = \int\limits_{-\infty}^{+\infty} dk_x \int\limits_{-\infty}^{+\infty} dk_y$$

$$\left\{ \tilde{A}_2^\alpha e^{-ik_{z,2}^\alpha z} e^{-ik_x x - ik_y y} + \delta_{\alpha z} e^{i\frac{\omega}{v_{el}} z} \frac{-e\mu_0\mu_r}{2\pi} \frac{1}{\varepsilon_{rzz}\mu_r k_0^2 - k_x^2 - k_y^2 - (\omega/v_{el})^2} \right\} \qquad (3.12b)$$

For $z < 0$. Here, $\left(k_{z,1}\right)^2 = k_0^2 - k_x^2 - k_y^2$, and $\left(k_{z,2}^{(\alpha)}\right)^2 = \varepsilon_{r\alpha\alpha}\mu_r k_0^2 - k_x^2 - k_y^2$. The unknown components $\tilde{A}_{1,2}^\alpha$ are solved by satisfying the boundary conditions. After some straightforward algebra, we find that the unknowns $\tilde{A}_2^\alpha$ should satisfy $\sum_{\alpha \in \{x,z\}} C_\alpha^\beta \tilde{A}_2^\alpha = I^\beta$ for $\beta = 1, 2$. Here, $C_x^1 = -\left[(k_0^2 - k_x^2) k_{z,2}^{(x)}/k_{z,1} + (\varepsilon_{rxx}\mu_r k_0^2 - k_x^2)/\varepsilon_{rxx} \right]$, $C_z^1 = -k_x\left(k_{z,1} + k_{z,2}^{(z)}/\varepsilon_{rxx}\right)$, $C_x^2 = k_x\left(k_{z,2}^{(x)}/k_{z,1} + \varepsilon_{ryy}^{-1}\right)$, and $C_z^2 = -\left(k_{z,1} + k_{z,2}^{(z)}/\varepsilon_{ryy}\right)$. In addition, the excitation term I^β takes the values

$$I^1 = \frac{-e\mu_0\mu_r}{2\pi} k_x \frac{\omega}{v_{el}} \left[\frac{1}{\varepsilon_{rxx}} \frac{1}{\varepsilon_{rzz}\mu_r k_0^2 - k_x^2 - k_y^2 - \left(\omega v_{el}^{-1}\right)^2} - \frac{1}{k_0^2 - k_x^2 - k_y^2 - \left(\omega v_{el}^{-1}\right)^2} \right]$$

$$+ \frac{-e\mu_0\mu_r}{2\pi} k_x k_{z,1} \left(\frac{1}{\varepsilon_{rzz}\mu_r k_0^2 - k_x^2 - k_y^2 - \left(\omega v_{el}^{-1}\right)^2} - \frac{1}{k_0^2 - k_x^2 - k_y^2 - \left(\omega v_{el}^{-1}\right)^2} \right) \qquad (3.13)$$

and

$$I^2 = \frac{-e\mu_0\mu_r}{2\pi} \frac{\omega}{v_{el}} \left[\frac{1}{\varepsilon_{ryy}} \frac{1}{\varepsilon_{rzz}\mu_r k_0^2 - k_x^2 - k_y^2 - \left(\omega v_{el}^{-1}\right)^2} - \frac{1}{k_0^2 - k_x^2 - k_y^2 - \left(\omega v_{el}^{-1}\right)^2} \right]$$

$$+ \frac{-e\mu_0\mu_r}{2\pi} k_{z,1} \left(\frac{1}{\varepsilon_{rzz}\mu_r k_0^2 - k_x^2 - k_y^2 - \left(\omega v_{el}^{-1}\right)^2} - \frac{1}{k_0^2 - k_x^2 - k_y^2 - \left(\omega v_{el}^{-1}\right)^2} \right) \qquad (3.14)$$

The remaining unknown amplitudes $\tilde{A}_1^\alpha$ are related to $\tilde{A}_2^\alpha$ by $\tilde{A}_1^x = -\tilde{A}_2^x k_{z,2}^{(x)}/\mu_r k_{z,1}$ and $\tilde{A}_1^z = \tilde{A}_2^z \mu_r^{-1} + (-e\mu_0/2\pi)\left(\left(\varepsilon_{rzz}\mu_r k_0^2 - k_x^2 - k_y^2 - (\omega/v_{el})^2\right)^{-1} - \left(k_0^2 - k_x^2 - k_y^2 - (\omega/v_{el})^2\right)^{-1} \right)$. Using the Poynting theorem, the power radiated to the far-field zone along the z-axis is calculated as $P_z = \frac{1}{2}\mathrm{Re} \int dx \int dy \left(E_x H_y^* - E_y H_x^*\right) = \frac{1}{2}\mathrm{Re} \int dk_x \int dk_y \left(\tilde{e}_x \tilde{h}_y^* - \tilde{e}_y \tilde{h}_x^*\right)$. For the vacuum side, we obtain $P_{+z,1} = \omega\left(4\pi k_0^2 \mu_0\right)^{-1} \int_{-\infty}^{+\infty} dk_x \int_{-\infty}^{+\infty} dk_y \, \tilde{p}(k_x, k_y)$, where $\tilde{p}(k_x, k_y)$ is given by

$$\tilde{p}(k_x, k_y) = \left|\tilde{A}_1^x\right|^2 (k_0^2 - k_x^2)k_{z,1} + \left|\tilde{A}_1^z\right|^2 (k_x^2 + k_y^2)k_{z,1}$$

$$- \left\{ \tilde{A}_1^x \left(\tilde{A}_1^z\right)^* + \tilde{A}_1^z \left(\tilde{A}_1^x\right)^* \right\} k_x k_{z,1}^2 \qquad (3.15)$$

The first two terms on the RHS of (3.15) might be observed as the intensity of the transverse magnetic TM_x and TM_z radiation, whereas the last term demonstrates an interesting interference effect between these waves. This interference is a prominent cause of some of the fundamental differences between the CL and EEL spectra [69]. In particular, the EEL spectrum for the same system is obtained as

$$
\begin{aligned}
\Gamma^{\text{EELS}}(k_x, k_z; \omega) = {} & \frac{e}{2\pi\hbar k_0^2} \text{Re}\left\{ \tilde{A}_1^x \frac{k_x k_{z,1}}{k_{z,1} + \omega/v_{\text{el}}} - \tilde{A}_1^z \frac{k_x^2 + k_y^2}{k_{z,1} + \omega/v_{\text{el}}} \right\} \\
& + \frac{-e}{2\pi\hbar k_0^2} \text{Re}\left\{ \tilde{A}_2^x \frac{k_x k_{z,2}^{(x)}}{\left(k_{z,2}^{(x)} - \omega/v_{\text{el}}\right)\varepsilon_{rzz}\mu_r} + \tilde{A}_2^z \frac{k_x^2 + k_y^2}{\left(k_{z,2}^{(z)} - \omega/v_{\text{el}}\right)\varepsilon_{rzz}\mu_r} \right\} \\
& + \frac{e^2\mu_0 L}{4\pi^2\hbar k_0^2} \text{Im}\left(\frac{1}{\varepsilon_{rzz}} \frac{\varepsilon_{rzz}\mu_r k_0^2 - (\omega/v_{\text{el}})^2}{\varepsilon_{rzz}\mu_r k_0^2 - k_x^2 - k_y^2 - (\omega/v_{\text{el}})^2} \right)
\end{aligned}
\tag{3.16}
$$

The first and second terms are the contributions of the scattered fields in the $z > 0$ and $z < 0$ domains to the EEL spectra, respectively. The last term is the bulk contribution, which has been derived in (3.6) as well. Here, L is the propagation length of an electron in the material. Apparently, the types of interference effects obtained using CL spectroscopy are not directly covered by EELS.

Electrons traversing the surface of a metal such as Au not only cause TR but excite SPPs as well, though the latter do not contribute to the radiation (Fig. 3.4). The SPPs at the surface of isotropic metals are in the form of degenerate TM_x and TM_z waves. Radiation pattern of the TR for metals is very similar to the radiation pattern of a dipole positioned at the interface with, however, a slightly different orientation of the maximum radiation intensity [1].

The scattering theory described above is based on the inhomogeneous system of equations $\sum_{\alpha\in\{x,z\}} C_\alpha^\beta \tilde{A}_2^\alpha = I^\beta$. Considering the homogenous system $\sum_{\alpha\in\{x,z\}} C_\alpha^\beta \tilde{A}_2^\alpha = 0$, one can obtain the optical modes of the system. Obviously, we obtain a solution only if $\left| C^\beta \right| = 0$.

Thus, the characteristic equation for the optical modes supported by a single interface will be obtained as

$$
\left(\varepsilon_{ryy} + \frac{k_{z,2}^{(z)}}{k_{z,1}} \right)\left(\mu_r + \frac{k_{z,2}^{(x)}}{k_{z,1}} \right)\varepsilon_{ryy}k_0^2 = k_x^2 \left(\frac{k_{z,2}^{(x)}k_{z,2}^{(z)}}{\left(k_{z,1}\right)^2} - 1 \right)\left(\frac{1}{\varepsilon_{ryy}} - \frac{1}{\varepsilon_{rxx}} \right)
\tag{3.17}
$$

Whenever $\varepsilon_{ryy} = \varepsilon_{rxx} = \varepsilon_{r\|}$, the material may support two different modes with characteristic equations given by $\varepsilon_{r\|} + k_{z,2}^{(z)}k_{z,1}^{-1} = 0$ and $\mu_r k_{z,1} + k_{z,2}^{(x)} = 0$ for electric and magnetic polaritons, respectively.

Although TR already occurs at a single interface, practical situations more often consider thin films. In thin films, TR, CR, and guided modes contribute significantly to the radiation loss and consequently the energy loss suffered by the electron. As a matter of interest, in isotropic metallic thin films below the plasma frequency,

CR cannot take place. However, in this energy range, surface plasmons contribute significantly to the EELS and to the radiation spectra. It is known that far-field optical radiation cannot excite bulk plasmons, and hence, longitudinal currents and bulk plasmons in a reciprocal picture cannot contribute to the far-field radiation either. However, the situation changes when an emitter such as a relativistic electron is placed inside matter [26]. In this case, EELS experiments have already demonstrated a significant signal due to the excitation of bulk plasmons with moving electrons [70], which has applications in mapping different elements of composites. In the following, we show how to calculate the MREEL spectra, including those of thin films.

We consider here an anisotropic thin film positioned at $|z| < d$. The only difference between this problem and the problem with a single interface is that for this case, the vector potential should be introduced for three domains, specified as $z \geq 0$, $|z| \leq d$, and $z \leq 0$. For this purpose, we construct the solution as

$$A_\alpha(x, y, z; \omega) = \int\limits_{-\infty}^{+\infty} dk_x \int\limits_{-\infty}^{+\infty} dk_y$$

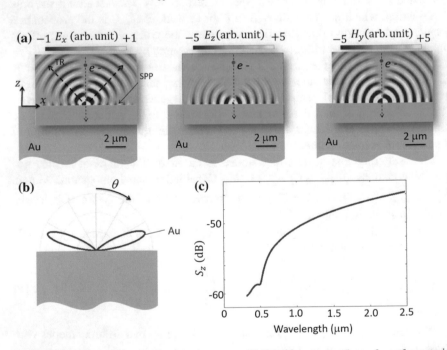

Fig. 3.4 A moving electron with a kinetic energy of 200 keV traversing the surface of a material normal to the surface plane. **a** Calculated scattered-field amplitudes for a vacuum/gold interface at a photon energy of $E = 1.5$ eV. **b** Radiation pattern of the emitted photons at an energy of $E = 0.8$ eV for an Au surface. **c** Radiation power spectra (z-component of the Poynting vector S_z) in the $+z$ direction for the TR from a Au surface

$$\begin{cases} \tilde{A}_1^\alpha e^{-i\kappa_z^{(1)}(z-d)} + \delta_{\alpha z} e^{i\frac{\omega}{v_{el}}z} \dfrac{-e\mu_0}{2\pi} \dfrac{1}{k_0^2-k_x^2-k_y^2-(\omega/v_{el})^2} & \text{for } z \geq d \\[2mm] \tilde{A}_{21}^\alpha e^{i\kappa_z^{(2,\alpha)}(z-d)} + \tilde{A}_{22}^\alpha e^{-i\kappa_z^{(2,\alpha)}(z+d)} \\[1mm] \quad + \delta_{\alpha z} e^{i\frac{\omega}{v_{el}}z} \dfrac{-e\mu_0\mu_r}{2\pi} \dfrac{1}{\varepsilon_{rzz}\mu_r k_0^2-k_x^2-k_y^2-(\omega/v_{el})^2} & \text{for } |z| \leq d \\[2mm] \tilde{A}_3^\alpha e^{+i\kappa_z^{(1)}(z+d)} + \delta_{\alpha z} e^{i\frac{\omega}{v_{el}}z} \dfrac{-e\mu_0}{2\pi} \dfrac{1}{k_0^2-k_x^2-k_y^2-(\omega/v_{el})^2} & \text{for } z \leq -d \end{cases} \tag{3.18}$$

where both the scattered and excitation potentials have been considered. Here, $\left(\kappa_z^{(1)}\right)^2 = k_0^2 - \beta^2$, $\left(\kappa_z^{(\alpha,2)}\right)^2 = \varepsilon_{r\alpha\alpha}\mu_r k_0^2 - \beta^2$. Only two components of the vector potential are, in principle, sufficient to satisfy the boundary conditions. These choices are arbitrary as far as they do not generate an underdetermined system of equations. By the choice of $\vec{A} = (A_x, 0, A_z)$ and upon satisfying the boundary conditions, the matrix components for the unknown values are provided as $\left[M_{ij}\right]\left[\tilde{A}\right]^T = \left[\tilde{E}\right]^T$, where $\left[\tilde{A}\right] = \left[\tilde{A}_1^x \; \tilde{A}_{21}^x \; \tilde{A}_{22}^x \; \tilde{A}_3^x \; \tilde{A}_1^z \; \tilde{A}_{21}^z \; \tilde{A}_{22}^z \; \tilde{A}_3^z\right]$, $\left[\tilde{E}\right] = \left[\tilde{E}_1, \tilde{E}_2, \ldots, \tilde{E}_8\right]$ and

$$[M_{ij}] =$$

$$\begin{bmatrix} m_{11}^x & m_{12}^x & m_{12}^x e^{-2i\kappa_z^{(2,x)}d} & 0 & m_{11}^z & m_{12}^z & -m_{12}^z e^{-2i\kappa_z^{(2,z)}d} & 0 \\ 0 & m_{12}^x e^{-2i\kappa_z^{(2,x)}d} & m_{12}^x & m_{11}^x & 0 & m_{12}^z e^{-2i\kappa_z^{(2,z)}d} & -m_{12}^z & -m_{11}^z \\ m_{31}^x & m_{32}^x & m_{32}^x e^{-2i\kappa_z^{(2,x)}d} & 0 & m_{31}^z & m_{32}^z & -m_{32}^z e^{-2i\kappa_z^{(2,z)}d} & 0 \\ 0 & m_{32}^x e^{-2i\kappa_z^{(2,x)}d} & m_{32}^x & m_{31}^x & 0 & m_{32}^z e^{-2i\kappa_z^{(2,z)}d} & -m_{32}^z & -m_{31}^z \\ 0 & 0 & 0 & 0 & m_{51}^z & m_{52}^z & m_{52}^z e^{-2i\kappa_z^{(2,z)}d} & 0 \\ 0 & 0 & 0 & 0 & 0 & m_{52}^z e^{-2i\kappa_z^{(2,z)}d} & m_{52}^z & m_{51}^z \\ m_{71}^x & m_{72}^x & -m_{72}^x e^{-2i\kappa_z^{(2,x)}d} & 0 & m_{71}^z & m_{72}^z & m_{72}^z e^{-2i\kappa_z^{(2,z)}d} & 0 \\ 0 & m_{72}^x e^{-2i\kappa_z^{(2,x)}d} & -m_{72}^x & -m_{71}^x & 0 & m_{72}^z e^{-2i\kappa_z^{(2,z)}d} & m_{72}^z & m_{71}^z \end{bmatrix}$$

$$\tag{3.19}$$

and the matrix elements are given by $m_{11}^x = k_0^2 - k_x^2$, $m_{12}^x = -\left(\varepsilon_{rxx}\mu_r k_0^2 - k_x^2\right)/\varepsilon_{rxx}\mu_r$, $m_{11}^z = k_x\kappa_z^{(1)}$, $m_{12}^z = k_x k_z^{(2,z)}(\varepsilon_{rxx}\mu_r)^{-1}$, $m_{31}^x = -k_x k_y$, $m_{32}^x = k_x k_y\left(\varepsilon_{ryx}\mu_r\right)^{-1}$, $m_{31}^z = k_y\kappa_z^{(d)}$, $m_{51}^z = ik_y$, $m_{52}^z = -ik_y\mu_r^{-1}$, $m_{71}^x = -i\kappa_z^{(d)}$, $m_{72}^x = -\mu_r^{-1} i\kappa_z^{(2,x)}$, and $m_{71}^z = -ik_x$, $m_{72}^z = ik_x\mu_r^{-1}$. The elements of $\tilde{E}_i$ are given by

$$E_1 = \left(\frac{-1}{\varepsilon_{rxx}\mu_r\left(\varepsilon_{rzz}\mu_r k_0^2 - k_x^2 - k_y^2 - (\omega/v_{el})^2\right)} + \frac{1}{k_0^2 - k_x^2 - k_y^2 - (\omega/v_{el})^2}\right)\frac{-e\mu_0}{2\pi}k_x\frac{\omega}{v_{el}}e^{i\frac{\omega}{v_{el}}d}$$

$$E_2 = e^{-i\frac{\omega}{v_{el}}2d}E_1$$

$$E_3 = \left(\frac{-1}{\varepsilon_{ryy}\mu_r\left(\varepsilon_{rzz}\mu_r k_0^2 - k_x^2 - k_y^2 - (\omega/v_{el})^2\right)} + \frac{1}{\left(k_0^2 - k_x^2 - k_y^2 - (\omega/v_{el})^2\right)}\right)\frac{-e\mu_0}{2\pi}k_y\frac{\omega}{v_{el}}e^{i\frac{\omega}{v_{el}}d}$$

$$E_4 = e^{-i\frac{\omega}{v_{el}}2d}E_3$$

$$E_5 = \left(\frac{1}{\mu_r\left(\varepsilon_{rzz}\mu_r k_0^2 - k_x^2 - k_y^2 - (\omega/v_{\mathrm{el}})^2\right)} - \frac{1}{\left(k_0^2 - k_x^2 - k_y^2 - (\omega/v_{\mathrm{el}})^2\right)} \right) \frac{-e\mu_0}{2\pi} ik_y e^{i\frac{\omega}{v_{\mathrm{el}}}d}$$

$$E_6 = e^{-i\frac{\omega}{v_{\mathrm{el}}}2d} E_5$$

$$E_7 = \left(\frac{-1}{\mu_r\left(\varepsilon_{rzz}\mu_r k_0^2 - k_x^2 - k_y^2 - (\omega/v_{\mathrm{el}})^2\right)} + \frac{1}{\left(k_0^2 - k_x^2 - k_y^2 - (\omega/v_{\mathrm{el}})^2\right)} \right) \frac{-e\mu_0}{2\pi} ik_x e^{i\frac{\omega}{v_{\mathrm{el}}}d}$$

$$E_8 = e^{-i\frac{\omega}{v_{\mathrm{el}}}2d} E_7 \tag{3.20}$$

As before, the MREEL spectrum in diffraction is defined by (3.9). To calculate the MREEL spectrum, it is then sufficient to calculate the z-component of the electric field using (3.5). The bulk spectrum is calculated using (3.11) as

$$\Gamma_{\mathrm{Bulk}}^{\mathrm{EELS}}(\omega; k_x, k_y) = \frac{e^2 d}{4\pi^2 \hbar \omega^2 \varepsilon_0} \mathrm{Im}\left\{ \frac{1}{\varepsilon_{rzz}} \frac{\left(\varepsilon_{rzz}\mu_r k_0^2 - (\omega/v_{\mathrm{el}})^2\right)}{\varepsilon_{rzz}\mu_r k_0^2 - k_x^2 - k_y^2 - (\omega/v_{\mathrm{el}})^2} \right\} \tag{3.21}$$

Additionally, using the electric field in the regions above and below the thin film, we calculate their contribution to the EEL spectrum as

$$\Gamma_{\mathrm{rad}}^{\mathrm{EELS}}(k_x, k_y; \omega) = \frac{e}{2\pi\omega^2\varepsilon_0\mu_0} \mathrm{Re}\left\{ \frac{e^{-i(\omega/v_{\mathrm{el}})d}}{k_z^{(d)} + \omega/v_{\mathrm{el}}} \left[\left(k_0^2 - \left(\kappa_z^{(1)}\right)^2\right)\left[\tilde{A}_1^z + \tilde{A}_3^z\right] + k_x\kappa_z^{(1)}\left[\tilde{A}_1^x - \tilde{A}_3^x\right] \right] \right\} \tag{3.22}$$

Additionally, considering the scattered electric field within the thin film, we calculate the contribution of the guided waves to the EELS as

$$\begin{aligned} \Gamma_{\mathrm{Guided}}^{\mathrm{EELS}}(k_x, k_y; \omega) = \frac{e\,2d}{2\pi\omega^2\varepsilon_0\mu_0} \mathrm{Im}\bigg\{ & \frac{\varepsilon_{r,zz}\mu_r k_0^2 - \left(\kappa_z^{(2,z)}\right)^2}{\varepsilon_{r,zz}\mu_r} e^{-i\kappa_z^{(2,z)}d} \\ & \times \left[+\tilde{A}_{21}^z \mathrm{sinc}\left(\left(\kappa_z^{(2,z)} - \omega/v_{\mathrm{el}}\right)d\right) + \tilde{A}_{22}^z \mathrm{sinc}\left(\left(\kappa_z^{(2,z)} + \omega/v_{\mathrm{el}}\right)d\right) \right] \\ & + \left(\frac{k_x\kappa_z^{(2,x)}}{\varepsilon_{r,zz}\mu_r} \right) e^{-i\kappa_z^{(2,x)}d} \\ & \times \left[-\tilde{A}_{21}^x \mathrm{sinc}\left(\left(\kappa_z^{(2,x)} - \omega/v_{\mathrm{el}}\right)d\right) + \tilde{A}_{22}^x \mathrm{sinc}\left(\left(\kappa_z^{(2,z)} + \omega/v_{\mathrm{el}}\right)d\right) \right] \bigg\} \end{aligned} \tag{3.23}$$

where $\mathrm{sinc}(X) = \sin(X)/X$. The entire EEL spectrum is computed by summing the individual terms given by (3.21)–(3.23).

The inclusion of a second interface and of multilayered structures has another consequence, which is the occurrence of interferences within the thin film, because of the multiple reflections of the emitted light from the boundaries (Fig. 3.5b). This interference phenomenon significantly affects the overall EELS signal and the TR emission and has applications in measuring the electron beam energy with high accuracy [71]. Dielectric and metallic thin films can also support guided modes, whereas the field profile in the latter case can be decomposed into symmetric and antisymmetric modal configurations formed by the coupled surface plasmon polaritons at the interfaces. The hybridization of the surface plasmon modes for pure thin films is perfectly captured in the MREEL spectra [72], for example, for aluminium

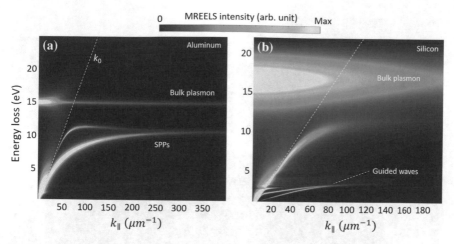

Fig. 3.5 MREELS in the interaction of electrons with a kinetic energy of 200 keV with **a** an Al film with a thickness of 12 nm and **b** a silicon film with a thickness of 200 nm

(Fig. 3.5a). Basically, the excitations of symmetric and antisymmetric SPPs, bulk plasmons, and transverse plasmons are distinguished. The situation for silicon is, however, different. At lower energies where the material response is purely dielectric, several dispersion bands are formed, thanks to the excitation of guided modes of various symmetries. At energies above $E = 4.22$ eV, the permittivity becomes negative, and surface polaritons can be excited up to an energy of $E = 11.5$ eV. At $E = 16.3$ eV, the real part of the permittivity crosses zero, and as a result, a strong bulk polariton excitation becomes observable in the MREEL spectrum.

For hyperbolic metamaterials such as h-BN, there exists a more complicated classification of optical modes [62, 73]. In general, six different modal classes are possible, namely, symmetric and antisymmetric transverse magnetic with respect to the x, y, and z axes (TM_x, TM_y, and TM_z, respectively), where not all of them can be excited by moving electrons interacting with thin films composed of hyperbolic materials. Moreover, as a result of the interaction of electrons with an h-BN thin film, enhanced ultrabroadband TR will be generated (Fig. 3.6a, b), which renders itself in ultrabroadband EELS resonances as well. In the MREELS energy-momentum map, this radiation is apparent as strong MREELS intensity within the light cone (Fig. 3.6c, d). In fact, hyperbolic metamaterials have already been demonstrated to be able to enhance and control the emission from nanoemitters in an efficient and directional way [74]. However, the possibility of controlling the directionality of the TR with hyperbolic materials has not yet been fully investigated. As previously mentioned, EELS can map the photonic local density of states projected along the electron trajectory [4]. Exactly this projection along with the momentum conservation picture serves as a criterion for the EELS to be able to map only certain modes of the structure [21]. Nevertheless, the mentioned criterion is advantageous in designing coherent radiation sources using synchronization, as in Smith–Purcell free-electron lasers [75, 76] (see Sect. 3.5). Here, we only mention that when using EELS, only those modes

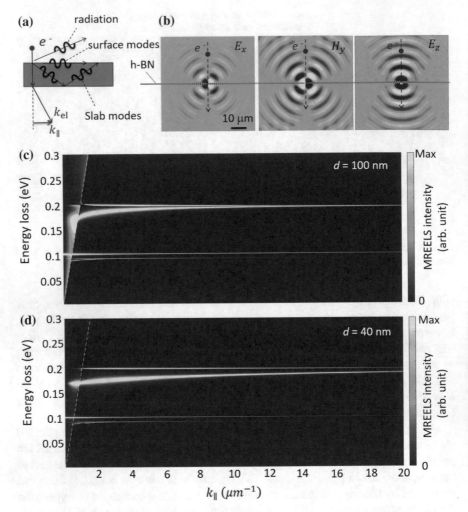

Fig. 3.6 Interaction of an electron at a kinetic energy of 200 keV with an h-BN thin film. **a** During the interaction, both the radiating and propagating optical modes are excited, as a result of which the electron receives a recoil and also loses energy. **b** Fourier-transformed scattered electromagnetic fields excited during the interaction of the electron with an h_BN with a thickness of $2d = 80$ nm at a photon energy of $E = 0.18$ eV. MREELS maps resolved by the interactions of electrons with h-BN thin films with thicknesses of **c** $2d = 200$ nm and **d** $2d = 80$ nm

of the structure that support an electric field component along the electron trajectory are excitable. In this regard, the TM_y modes are not excited, as these modes sustain only E_y, H_x, and H_z field components. The dispersion of the excitable modes, however, can be captured using MREELS. Since a hyperbolic material has a combination of both metallic and dielectric responses, a rich class of optical excitations is expected. However, due to the thicknesses of the films considered here and in most practical situations, only polaritonic responses are observed—the photonic modes

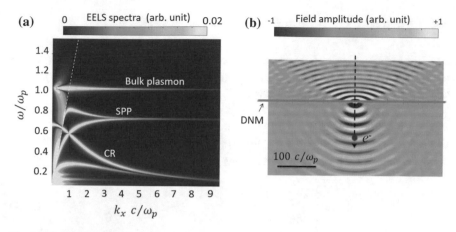

Fig. 3.7 Interaction of a moving electron with a thin film made of an LHM (reprinted by permission from [3]; © 2017 IOP Publishing Ltd). **a** Momentum-resolved EEL spectra for an electron at a kinetic energy of 200 keV interacting with an LHM film with a thickness of $D = 2c/\omega_p$. **b** Profile of the z-component of the electric field at $\omega = 0.62\omega_p$

can be excited at larger thicknesses. We finally briefly mention that our retarded analytical treatment is consistent with the non-retarded calculations reported in [77], within the upper reststrahlen band. The interaction of extremely concise wave packets of relativistic electrons in space-time with thin samples and interfaces takes place within only a few attoseconds. This short interaction time provides a large bandwidth for the spectroscopic investigations of optical modes and obviously leads to ultrashort light emission, which has recently been studied by taking the Fourier transform of time-harmonic solutions of Maxwell's equations over a large spectral bandwidth [78].

Finally, we mention here that there still exists a great potential for controlling the emission from electron-driven photon sources using metamaterials. The above-mentioned example included a natural hyperbolic metamaterial. An even more interesting phenomenon is to be exploited by using LHM. As a prototype, an LHM film with both permittivity and permeability modelled by a Drude function as $\varepsilon_r = \mu_r = 1 - \omega_p^2/\omega(\omega - i\gamma_p)$ is considered. Figure 3.7a shows the calculated MREEL spectra, considering an electron at a kinetic energy of 200 keV interacting with a film with a thickness of $D = 2c/\omega_p$. The exotic dispersions of the possible excitations are well captured in this picture: 1—inverted CR emission, 2—SPP modes that are decomposed into Symmetric and antisymmetric modes, 3—bulk plasmon excitation exactly at the plasma energy, and 4—a plethora of guided mode excitations occurring at energies below the CR dispersion. The different crossings that occur for CR and SPPs may be an interesting option for investigating weak and strong couplings in metamaterials [79, 80]. The TR emission from such a structure Demonstrates the interferences between the inverted CR and TR, which is understood by the splitting of the field profile at $z > D/2$, in contrast to the transmitted field (Fig. 3.7b).

3.4　Electron Travelling Parallel to the Interface

We consider here an electron travelling parallel to the surface at a distance d from the surface (Fig. 3.9a) (aloof excitation). For this system, the current density in the frequency domain is given by $J_x(\vec{r}, \omega) = q(2\pi)^{-1}\delta(z - d)\exp(\omega x v_{el}^{-1})\int_{-\infty}^{+\infty}\exp(-ik_y y)dk_y$. We expand the vector potential as $A_\alpha(\vec{r}, \omega) = (2\pi)^{-1}\int_{-\infty}^{+\infty}\tilde{A}_\alpha(k_x = \omega v_{el}^{-1}, k_y, z; \omega)\exp(-i\omega x v_{el}^{-1})\exp(-ik_y y)dk_y$. Moreover, $\tilde{A}_\alpha(k_x = \omega v_{el}^{-1}, k_y, z; \omega)$ is constructed as

$$\tilde{A}_\alpha = \begin{cases} \tilde{A}_1^\alpha e^{-ik_{z,1}z} + \delta_{\alpha x}\dfrac{\mu_0 q}{2ik_{z,1}}e^{-ik_{z,1}|z-d|} & \forall\, z \geq 0 \\ \tilde{A}_2^\alpha e^{+ik_{z,2}^{(\alpha)}z} & \forall\, z \leq 0 \end{cases} \tag{3.24}$$

with $(k_{z,1})^2 = k_0^2 - (\omega v_{el}^{-1})^2 - k_y^2$ and $(k_{z,2}^{(\alpha)})^2 = \varepsilon_{r\alpha\alpha}\mu_r k_0^2 - (\omega v_{el}^{-1})^2 - k_y^2$. The unknown coefficients $\tilde{A}_{1,2}^\alpha$ can be solved by $\sum_{\alpha\in\{x, z\}}C_\alpha^\beta\tilde{A}_2^\alpha = I^\beta$, where C_α^β are the same as those obtained above for the penetrating trajectory (see (3.12a, b) and discussions therein). The excitation components are, however, different and are obtained as $I^1 = \mu_0\mu_r q(ik_{z,1})^{-1}(k_0^2 - (\omega/v_{el})^2)e^{-ik_{z,1}d}$ and $I^2 = \mu_0\mu_r q(ik_{z,1})^{-1}k_x e^{-ik_{z,1}d}$, respectively. It has already been noted that the z-component of the wave vector at the region above the interface can be obtained as $k_{z,1} = i\sqrt{\omega^2 v_{el}^{-2}\gamma^{-2} + k_y^2}$, where γ is the Lorentz factor. In other words, the generated light fields are in the form of evanescent waves in this region. However, if the material supports the excitation of interface optical waves in the form of polaritons, the aloof trajectory will result in an efficient coupling of the electron waves to the polaritons. Depending on the electron velocity, and thanks to the momentum matching condition specified in (3.2), this coupling can only happen in selected energy ranges (see Fig. 3.8 for the electron travelling parallel to the vacuum/silver interface). Depending on the interaction length, the electron continuously loses energy by propagation parallel to the interface. The rate of energy loss for this case can be obtained as

$$\frac{d}{dx}\Gamma(k_y, \omega) = \frac{-e}{8\pi^2\hbar k_0^2}\mathrm{Im}\left(\tilde{A}_1^x\omega^2 v_{el}^{-2}\gamma^{-2}e^{-ik_{z,1}d} + \tilde{A}_1^z k_{z,1}\omega v_{el}^{-1}e^{-ik_{z,1}d}\right)$$

$$= \frac{-e}{8\pi^2\hbar}\beta^{-2}\gamma^{-2}\left[\mathrm{Im}\left(\tilde{A}_1^x e^{-ik_{z,1}d}\right) + \gamma\sqrt{1 + \gamma^2\tan^2\varphi_y}\,\mathrm{Re}\left(\tilde{A}_1^z e^{-ik_{z,1}d}\right)\right] \tag{3.25}$$

For this case, the energy loss rate depends explicitly on the electron impact position. Figure 3.8c shows the energy loss rate for an electron at various kinetic energies travelling at a distance of 10 nm from a semi-infinite Ag medium. The polariton waves that are excited by the travelling electron propagate at certain angles with respect to the trajectory line of the electron. This angle can be calculated as $\varphi_y = \tan^{-1}(k_y/k_x)$,

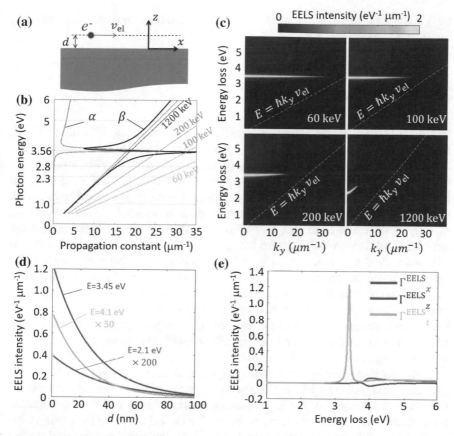

Fig. 3.8 Electron beam travelling parallel to a surface. **a** Schematic of the system composed of an electron with a velocity of $\vec{v} = v_{el}\hat{x}$ and an impact parameter ($y = 0$, $z = d$) interacting with a Ag/vacuum interface located at $z = 0$. **b** Propagation constant of the SPPs at the Ag/vacuum interface. β is the phase constant, and α is the attenuation constant. Depicted as the dash-dotted line is the momentum conservation (3.2) at different electron energies. **c** Momentum-resolved EEL spectra of a Ag/vacuum interface for several electron energies. **d** EELS intensity versus the distance d from the Ag surface at the energy losses depicted for each plot. **e** EELS intensity for the contribution of each term in (3.15)

where $k_x = \omega v_{el}^{-1}$. This angle will always remain below 90 degrees, or in other words, $k_y < k_x$. The dashed lines in Fig. 3.8c mark the dispersion line associated with the maximum amount of $E = \hbar k_y v_{el} = \hbar\omega$. As noted, the maximum intensity for the energy-loss peak always remains within the light cone associated with this dispersion line. This means that the probability that an electron loses the associated photon energy at momentum k_y at regions outside the mentioned light cone is negligible. As expected, the EELS intensity decreases by increasing the impact parameter d at a rate related to the electron velocity and the damping ratio of the interface polaritons (Fig. 3.8d). Interestingly, the contribution of the z-component of the vector potential

(A_z) to the total EELS signal in (3.15) is more prominent than the contribution of the A_x component (Fig. 3.8e).

3.5   Smith–Purcell Effect

In 1953 [81], even before the detection of TR by Goldsmith and Jelly in 1959 [82], Smith and Purcell investigated the emission properties of an electron beam propagating parallel to a metal diffraction grating. Based on geometrical-optical assumptions, they found that there is a simple relation between the wavelength of the emitted light and the velocity of the electron, namely, $\lambda = d(\beta^{-1} - \cos(\theta))$, where d is the period of the grating, $\beta = v_{el}/c$, and θ is the angle between the emitted light ray and the velocity of the electron. In fact, this relation satisfies the momentum conservation criterion mentioned in (3.2) by considering $k = k_0 + 2m\pi/d$, where m is the diffraction order. First-harmonic Smith–Purcell radiation occurs whenever $m = 1$ is the dominant response. Smith-Purcell radiation has been used so far to design various forms of free-electron lasers [83] and, in particular, THz radiation sources [84]. It has also been demonstrated that, in addition to using ultra-relativistic electron beams, high-quality electron sources of the sort utilized in scanning electron microscopes can be used to generate coherent superradiant Smith-Purcell radiation within the far-infrared spectral range [85].

In the work reported by Urata et al. [85], a one-dimensional linear grating with a periodicity ranging from 128 to 308 µm has been considered. Normally, to increase the coupling efficiency of the beam with the grating, distributed feedback grating elements are used. In this work, the non-radiative optical modes of the grating were used to enhance the coupling efficiency. In a better configuration, a cylindrical grating might be used as an effective medium for electron-light interactions. Such a medium can be fabricated by means of multilayered nanofabrication technique, resulting in periodicities that can be used to generate Smith-Purcell radiation within the visible range (Fig. 3.9a) [86]. As expected from the theory of the Smith-Purcell effect, the spectral positioning of the optical emission peaks will vary upon changing the kinetic energy of the electron beam (Fig. 3.9b). Nevertheless, two peaks are observed, which is due to the coupling of the electrons to the optical modes of different azimuthal symmetries in the cylindrical configuration ($m = 0$ and $m = 1$). The emission intensity can also be controlled by simply controlling the electron beam current (Fig. 3.9c). Nevertheless, the emission was not expected to be in a coherent state, even though high-quality low-emittance electron beams, as in [85], were used [86].

Although Smith–Purcell emission is generally described by the grating diffraction orders, practical situations mostly consider a finite number of grating elements. The effect of the number of grating elements on the probability of photon generation and EELS has been considered in [87], which demonstrates a transition from individual resonances to an interference pattern in the far-field zone upon increasing the number of elements (Fig. 3.10a).

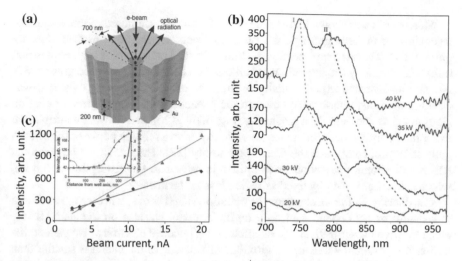

Fig. 3.9 A Smith-Purcell light source in the visible range. **a** Schematic of an electron-driven Smith-Purcell light source generated by a multilayered Au/SiO$_2$ structure with a periodicity of 400 nm. **b** Dependence of the emitted optical wavelength on the kinetic energy of the electron. **c** Dependence of the optical intensity on the beam current for both peaks observed in the spectrum. The inset shows the calculations for the optical intensity for peak I and the photon generation probability for the same peak versus the position of the beam with respect to the symmetry axis of the light well [86]. Reprinted by slight modifications from [86]; © 2009 American Physical Society

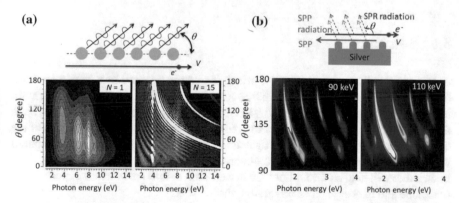

Fig. 3.10 Smith-Purcell radiation in the interaction of moving electrons with optical gratings. **a** Simulated probability of photon generation from the interaction of a relativistic electron at a kinetic energy of 200 keV with chains of Al spheres composed of $N = 1$ and $N = 15$ elements. **b** Measured CL spectra from the interaction of electron beams at different energies with a silver grating. Reprinted by permission from [87]; © 2000 American Physical Society. Reprinted by permission from [20]; © 2015 American Physical Society

Moreover, it was shown that for the case of a grating composed of silica spheres, in contrast to metallic gratings, the EELS and photon generation probabilities are quite similar. The interplay between the Smith–Purcell radiation and surface plasmon radiation caused by the interaction of an electron beam and a metallic grating was recently investigated experimentally in the case of a silver grating, where it was demonstrated that interference between the plasmons and Smith–Purcell radiation leads to a characteristic Fano resonance (Fig. 3.10b) [20]. In a relevant contribution, using the Poynting theorem, it was possible to derive a relation between the photon generation probability and the EELS probability [88]. The difference between the EELS and photon generation probabilities is related to the absorption spectrum, which is itself governed by the dissipation loss in the matter.

Considering all the mechanisms of radiation stated above, whether the emission is coherent or not depends not only on the material but also on the shape of the electron wave packet [89]. As an example, for a bunched electron wave packet, the emission is coherent when the longitudinal broadening of the bunch is smaller than the wavelength of the emission [90]. In general, an ensemble of emitters positioned in such a bunch can lead to a coherent spontaneous emission known as superradiance [91]. Interestingly, this phenomenon can be inversed, as coherent optical radiation interacting with an electron wave function can cause the wave function to evolve into bunches [92, 93], even for a single-electron wave packet. The inverse Smith-Purcell effect will be discussed in more detail in Chap. 8.

3.6   Summary

In this chapter, we studied the first principles of the interaction between electron beams and electromagnetic waves in free space and inside materials. The momentum and energy conservation required for initiating electron-photon interactions has been described and applied to understand Cherenkov radiation. Several mechanisms of the electron beam radiation were investigated in more detail. In particular, we compared the radiated power and the electron-energy loss spectra in various systems. This knowledge will help us throughout the dissertation to interpret various experimental outcomes and to design electron–driven photon sources.

References

1. F.J.G. de Abajo, Optical excitations in electron microscopy (in English). Rev. Mod. Phys. **82**(1), 209–275 (2010). https://doi.org/10.1103/revmodphys.82.209
2. R.H. Ritchie, A. Howie, Inelastic-scattering probabilities in scanning-transmission electron-microscopy (in English). Philos. Mag. A **58**(5), 753–767 (1988). [Online]. Available: < Go to ISI > ://WOS:A1988Q932100005
3. N. Talebi, Interaction of electron beams with optical nanostructures and metamaterials: from coherent photon sources towards shaping the wave function. J. Opt. UK **19**(10), 103001 (2017).

https://doi.org/10.1088/2040-8986/aa8041

4. F.J.G. de Abajo, M. Kociak, Probing the photonic local density of states with electron energy loss spectroscopy (in English). Phys. Rev. Lett. **100**(10), 106804 (2008). https://doi.org/10.1103/physrevlett.100.106804

5. B. Ögüt, N. Talebi, R. Vogelgesang, W. Sigle, P.A. van Aken, Toroidal Plasmonic Eigenmodes in Oligomer Nanocavities for the visible. Nano Lett. **12**(10), 5239–5244 (2012). https://doi.org/10.1021/nl302418n

6. F.P. Schmidt, H. Ditlbacher, U. Hohenester, A. Hohenau, F. Hofer, J.R. Krenn, Universal dispersion of surface plasmons in flat nanostructures (in English). Nat. Commun. **5**, 3604 (2014). https://doi.org/10.1038/ncomms4604

7. S.J. Barrow, D. Rossouw, A.M. Funston, G.A. Botton, P. Mulvaney, Mapping bright and dark modes in gold nanoparticle chains using electron energy loss spectroscopy (in English). Nano Lett. **14**(7), 3799–3808 (2014). https://doi.org/10.1021/nl5009053

8. F.P. Schmidt, H. Ditlbacher, U. Hohenester, A. Hohenau, F. Hofer, J.R. Krenn, Dark plasmonic breathing modes in silver nanodisks (in English). Nano Lett. **12**(11), 5780–5783 (2012). https://doi.org/10.1021/nl3030938

9. J. Nelayah et al., Mapping surface plasmons on a single metallic nanoparticle (in English). Nat. Phys. **3**(5), 348–353 (2007). https://doi.org/10.1038/nphys575

10. D. DeJarnette, D.K. Roper, Electron energy loss spectroscopy of gold nanoparticles on graphene (in English). J. Appl. Phys. **116**(5), 054313 (2014). https://doi.org/10.1063/1.4892620

11. O. Nicoletti, M. Wubs, N.A. Mortensen, W. Sigle, P.A. van Aken, P.A. Midgley, Surface plasmon modes of a single silver nanorod: an electron energy loss study (in English). Opt. Express **19**(16), 15371–15379 (2011). https://doi.org/10.1364/Oe.19.015371

12. G. Boudarham, M. Kociak, Modal decompositions of the local electromagnetic density of states and spatially resolved electron energy loss probability in terms of geometric modes (in English). Phys. Rev. B **85**(24), 245447 (2012). https://doi.org/10.1103/physrevb.85.245447

13. N. Talebi, B. Ögüt, W. Sigle, R. Vogelgesang, P.A. van Aken, On the symmetry and topology of plasmonic eigenmodes in heptamer and hexamer nanocavities. Appl. Phys. A **116**(3), 947–954 (2014). https://doi.org/10.1007/s00339-014-8532-y

14. B. Ogut, R. Vogelgesang, W. Sigle, N. Talebi, C.T. Koch, P.A. van Aken, Hybridized metal slit eigenmodes as an illustration of Babinet's principle (in English). ACS Nano **5**(8), 6701–6706 (2011). https://doi.org/10.1021/nn2022414

15. R. Walther et al., Coupling of surface-plasmon-polariton-hybridized cavity modes between submicron slits in a thin gold film (in English). Acs Photonics **3**(5), 836–843 (2016). https://doi.org/10.1021/acsphotonics.6b00045

16. A. Salomon, Y. Prior, M. Fedoruk, J. Feldmann, R. Kolkowski, J. Zyss, Plasmonic coupling between metallic nanocavities (in English). J. Opt. UK **16**(11), 114012 (2014). https://doi.org/10.1088/2040-8978/16/11/114012

17. L. Gu et al., Resonant wedge-plasmon modes in single-crystalline gold nanoplatelets (in English). Phys. Rev. B **83**(19), 195433 (2011). https://doi.org/10.1103/physrevb.83.195433

18. X.B. Xu et al., Tunable nanoscale confinement of energy and resonant edge effect in triangular gold nanoprisms (in English). J. Phys. Chem. C **117**(34), 17748–17756 (2013). https://doi.org/10.1021/jp4051929

19. E.P. Bellido, A. Manjavacas, Y. Zhang, Y. Cao, P. Nordlander, G.A. Botton, Electron energy-loss spectroscopy of multipolar edge and cavity modes in silver nanosquares (in English). Acs Photonics **3**(3), 428–433 (2016). https://doi.org/10.1021/acsphotonics.5b00594

20. N. Yamamoto, F.J.G. de Abajo, V. Myroshnychenko, Interference of surface plasmons and Smith-Purcell emission probed by angle-resolved cathodoluminescence spectroscopy (in English). Phys. Rev. B **91**(12), 125144 (2015). https://doi.org/10.1103/physrevb.91.125144

21. N. Talebi et al., Excitation of mesoscopic plasmonic tapers by relativistic electrons: phase matching versus eigenmode resonances (in English). ACS Nano **9**(7), 7641–7648 (2015). https://doi.org/10.1021/acsnano.5b03024

22. S.R. Guo et al., Reflection and phase matching in plasmonic gold tapers (in English). Nano Lett. **16**(10), 6137–6144 (2016). https://doi.org/10.1021/acs.nanolett.6b02353

23. T. Coenen, S.V. den Hoedt, A. Polman, A new cathodoluminescence system for nanoscale optics, materials science, and geology. Microsc. Today **24**(3), 12–19 (2016). https://doi.org/10.1017/S1551929516000377

24. D.R. Glenn et al., Correlative light and electron microscopy using cathodoluminescence from nanoparticles with distinguishable colours (in English). Sci. Rep. **2**, 865 (2012). https://doi.org/10.1038/srep00865

25. M. Kociak et al., Seeing and measuring in colours: electron microscopy and spectroscopies applied to nano-optics (in English). C. R. Phys. **15**(2–3), 158–175 (2014). https://doi.org/10.1016/j.crhy.2013.10.003

26. F.J.G. de Abajo et al., Plasmonic and new plasmonic materials: general discussion (in English). Faraday Discuss. **178**, 123–149 (2015). https://doi.org/10.1039/c5fd90022k

27. B. Barwick, D. J. Flannigan, A. H. Zewail (2009) Photon-induced near-field electron microscopy. Nature 462, 902. 12/17/online 2009, https://doi.org/10.1038/nature08662; https://www.nature.com/articles/nature08662#supplementary-information

28. B. Barwick, A.H. Zewail, Photonics and plasmonics in 4D ultrafast electron microscopy (in English). Acs Photonics **2**(10), 1391–1402 (2015). https://doi.org/10.1021/acsphotonics.5b00427

29. S.T. Park, M.M. Lin, A.H. Zewail, Photon-induced near-field electron microscopy (PINEM): theoretical and experimental (in English). New J. Phys. **12**, 123028 (2010). https://doi.org/10.1088/1367-2630/12/12/123028

30. Y.M. Liu, X. Zhang, Metamaterials: a new frontier of science and technology (in English). Chem. Soc. Rev. **40**(5), 2494–2507 (2011). https://doi.org/10.1039/c0cs00184h

31. N. Engheta, R.W. Ziolkowski, A positive future for double-negative metamaterials (in English). IEEE Trans. Microw. Theor. **53**(4), 1535–1556 (2005). https://doi.org/10.1109/Tmtt.2005.845188

32. P. L. Kapitza, P. A. M. Dirac, The reflection of electrons from standing light waves (in English). Proc. Camb. Philos. Soc. **29**, 297–300 (1993) [Online]. Available: <Go to ISI>://WOS:000200163900030

33. H. Batelaan, Colloquium: Illuminating the Kapitza-Dirac effect with electron matter optics (in English). Rev. Mod. Phys. **79**(3), 929–941 (2007). https://doi.org/10.1103/revmodphys.79.929

34. S.J. Wu, Y.J. Wang, Q. Diot, M. Prentiss, Splitting matter waves using an optimized standing-wave light-pulse sequence (in English). Phys. Rev. A **71**(4), 043602 (2005). https://doi.org/10.1103/physreva.71.043602

35. E.M. Rasel, M.K. Oberthaler, H. Batelaan, J. Schmiedmayer, A. Zeilinger, Atom wave interferometry with diffraction gratings of light (in English). Phys. Rev. Lett. **75**(14), 2633–2637 (1995). https://doi.org/10.1103/PhysRevLett.75.2633

36. A.G. Hayrapetyan, K.K. Grigoryan, J.B. Gotte, R.G. Petrosyan, Kapitza-Dirac effect with traveling waves (in English). New J. Phys. **17**, 082002 (2015). https://doi.org/10.1088/1367-2630/17/8/082002

37. D.L. Freimund, K. Aflatooni, H. Batelaan, Observation of the Kapitza-Dirac effect (in English). Nature **413**(6852), 142–143 (2001). https://doi.org/10.1038/35093065

38. A. Friedman, A. Gover, G. Kurizki, S. Ruschin, A. Yariv, Spontaneous and stimulated-emission from quasifree electrons (in English). Rev. Mod. Phys. **60**(2), 471–535 (1988). https://doi.org/10.1103/RevModPhys.60.471

39. A. Gover, P. Sprangle, A unified theory of magnetic Bremsstrahlung, electrostatic Bremsstrahlung, Compton-Raman Scattering, and Cerenkov-Smith-Purcell free-electron lasers (in English). IEEE J. Quantum Electron **17**(7), 1196–1215 (1981). https://doi.org/10.1109/Jqe.1981.1071257

40. P. Bucksbaum, T. Moller, K. Ueda, Frontiers of free-electron laser science (in English). J. Phys. B At. Mol. Opt. **46**(16), 160201 (2013). https://doi.org/10.1088/0953-4075/46/16/160201

41. R. Falcone, M. Dunne, H. Chapman, M. Yabashi, K. Ueda, Frontiers of free-electron laser science II (in English). J. Phys. B At. Mol. Opt. **49**(18), 180201 (2016). https://doi.org/10.1088/0953-4075/49/18/180201

42. Z. R. Huang, K. J. Kim, Review of x-ray free-electron laser theory (in English). Phys. Rev. Spec. Top. Ac. **10**(3), 034801 (2007). https://doi.org/10.1103/physrevstab.10.034801
43. B.W.J. McNeil, N.R. Thompson, X-ray free-electron lasers (in English). Nat. Photonics **4**(12), 814–821 (2010). https://doi.org/10.1038/nphoton.2010.239
44. D.B. Melrose, K.G. Ronnmark, R.G. Hewitt, Terrestrial kilometric radiation—the cyclotron theory (in English). J. Geophys. Res. Space **87**(Na7), 5140–5150 (1982). https://doi.org/10.1029/Ja087ia07p05140
45. D.M. Asner et al., Single-electron detection and spectroscopy via relativistic cyclotron radiation (in English). Phys. Rev. Lett. **114**(16), 162501 (2015). https://doi.org/10.1103/physrevlett.114.162501
46. V. L. Ginzburg, Radiation of uniformly moving sources (Vavilov-Cherenkov effect, transition radiation, and other phenomena) (in Russian). Usp Fiz Nauk+ **166**(10), 1033–1042 (1996). [Online]. Available: <Go to ISI>://WOS:A1996VV43300001
47. R. Garciamolina, A. Grasmarti, A. Howie, R.H. Ritchie, Retardation effects in the interaction of charged-particle beams with bounded condensed media (in English). J. Phys. C Solid State **18**(27), 5335–5345 (1985). https://doi.org/10.1088/0022-3719/18/27/019
48. E. Kröger, Calculations of the energy losses of fast electrons in thin foils with retardation. Zeitschrift für Physik A Hadrons and Nuclei, Journal Article **216**(2), 115–135 (1968). https://doi.org/10.1007/bf01390952
49. P. A. Cherenkov, The spectrum of visible radiation produced by fast electrons (in English). C. R. Acad. Sci. Urss **20**, 651–655 (1938). [Online]. Available: <Go to ISI>://WOS:000201891900170
50. P. A. Cherenkov, Absolute output of radiation caused by electrons moving within a medium with super-light velocity (in English). C. R. Acad. Sci. Urss **21**, 116–121 (1938). [Online]. Available: <Go to ISI>://WOS:000201892000033
51. P. A. Cherenkov, Spatial distribution of visible radiation produced by fast electrons (in English). C. R. Acad. Sci. Urss **21**, 319–321 (1938). [Online]. Available: <Go to ISI>://WOS:000201892000086
52. I. Frank, I. Tamm, Coherent visible radiation of fast electrons passing through matter (in English). C. R. Acad. Sci. Urss **14**, 109–114 (1937). [Online]. Available: <Go to ISI>://WOS:000201973400025
53. W. Li, C.-X. Yu, S.-B. Liu, Quantum theory of Cherenkov radiation in an anisotropic absorbing media, vol. 7501, pp. 750108 (2009). [Online]. Available: http://dx.doi.org/10.1117/12.847455
54. I. Kaminer et al., Quantum Čerenkov radiation: spectral cutoffs and the role of spin and orbital angular momentum. Phys. Rev. X **6**(1), 011006 (2016). [Online]. Available: http://link.aps.org/doi/10.1103/PhysRevX.6.011006
55. R. Matloob, A. Ghaffari, Cerenkov radiation in a causal permeable medium (in English). Phys. Rev. A **70**(5), 052116 (2004). https://doi.org/10.1103/physreva.70.052116
56. S.G. Chefranov, Relativistic generalization of the Landau criterion as a new foundation of the Vavilov-Cherenkov radiation theory (in English). Phys. Rev. Lett. **93**(25), 254801 (2014). https://doi.org/10.1103/physrevlett.93.254801
57. K. Tanha, A.M. Pashazadeh, B.W. Pogue, Review of biomedical Cerenkov luminescence imaging applications (in English). Biomed. Opt. Express **6**(8), 3053–3065 (2015). https://doi.org/10.1364/Boe.6.003053
58. M. Buchanan, Thesis: Minkowski, Abraham and the photon momentum. Nat. Phys. **3**(2), 73–73 02//print 2007. [Online]. Available: http://dx.doi.org/10.1038/nphys519
59. C. V. Festenberg, Energy loss measurements on III–V compounds. Zeitschrift für Physik, journal article **227**(5), 453–481 (1969). https://doi.org/10.1007/bf01394892
60. C.H. Chen, J. Silcox, R. Vincent, Electron-energy losses in silicon—bulk and surface Plasmons and Cerenkov Radiation (in English). Phys. Rev. B **12**(1), 64–71 (1975). https://doi.org/10.1103/Physrevb.12.64
61. M. Stoger-Pollach et al., Cerenkov losses: a limit for bandgap determination and Kramers-Kronig analysis (in English). Micron **37**(5), 396–402 (2006). https://doi.org/10.1016/j.micron.2006.01.001

62. N. Talebi, Optical modes in slab waveguides with magnetoelectric effect (in English). J. Opt. UK **18**(5), 055607 (2016). https://doi.org/10.1088/2040-8978/18/5/055607

63. P. Schattschneider, *Fundamentals of Inelastic Electron Scattering* (Springer-Verlag, Wien, Austria, 1986)

64. P. Li et al., Hyperbolic phonon-polaritons in boron nitride for near-field optical imaging and focusing. Nat. Commun. **6**, 7507 (2015). https://doi.org/10.1038/ncomms8507; https://www.nature.com/articles/ncomms8507#supplementary-information

65. V. Ginsburg, I. Frank, Radiation of a uniformly moving electron due to its transition from one medium into another (in Russian). Zh Eksp Teor Fiz+ **16**(1), 15–28 (1946). [Online]. Available: <Go to ISI>://WOS:A1946YB70900002

66. V.L. Ginzburg, Transition radiation and transition scattering (in English). Phys. Scripta **T2**, 182–191 (1982). https://doi.org/10.1088/0031-8949/1982/T2a/024

67. V.L. Ginzburg, V.N. Tsytovich, Several problems of the theory of transition radiation and transition scattering (in English). Phys. Rep. **49**(1), 1–89 (1979). https://doi.org/10.1016/0370-1573(79)90052-8

68. F.J.G. de Abajo, A. Rivacoba, N. Zabala, N. Yamamoto, Boundary effects in Cherenkov radiation (in English). Phys. Rev. B **69**(15), 155420 (2004). https://doi.org/10.1103/physrevb.69.155420

69. A. Losquin, M. Kociak, Link between cathodoluminescence and electron energy loss spectroscopy and the radiative and full electromagnetic local density of states (in English). Acs Photonics **2**(11), 1619–1627 (2015). https://doi.org/10.1021/acsphotonics.5b00416

70. R.F. Egerton, Electron energy-loss spectroscopy in the TEM (in English). Rep. Prog Phys. **72**(1), 016502 (2009). https://doi.org/10.1088/0034-4885/72/1/016502

71. L. Wartski, S. Roland, J. Lasalle, M. Bolore, G. Filippi, Interference phenomenon in optical transition radiation and its application to particle beam diagnostics and multiple-scattering measurements (in English). J. Appl. Phys. **46**(8), 3644–3653 (1975). https://doi.org/10.1063/1.322092

72. R. Vincent, J. Silcox, Dispersion of radiative surface plasmons in aluminum films by electron-scattering (in English). Phys. Rev. Lett. **31**(25), 1487–1490 (1973). https://doi.org/10.1103/PhysRevLett.31.1487

73. N. Talebi, C. Ozsoy-Keskinbora, H.M. Benia, K. Kern, C.T. Koch, P.A. van Aken, Wedge Dyakonov waves and Dyakonov plasmons in topological insulator Bi_2Se_3 probed by electron beams. ACS Nano **10**(7), 6988–6994 (2016). https://doi.org/10.1021/acsnano.6b02968

74. C.L. Cortes, W. Newman, S. Molesky, Z. Jacob, Quantum nanophotonics using hyperbolic metamaterials (in English). J. Opt. UK **14**(6), 063001 (2012). https://doi.org/10.1088/2040-8978/14/6/063001

75. L. Schachter, A. Ron, Smith-Purcell free-electron laser (in English). Phys. Rev. A **40**(2), 876–896 (1989). https://doi.org/10.1103/PhysRevA.40.876

76. M.H. Wang, X.G. Xiao, J.Y. Chen, Y.Y. Wei, Study on a novel Smith-Purcell free-electron laser (in English). Phys. Lett. A **345**(4–6), 423–427 (2005). https://doi.org/10.1016/j.physleta.2005.07.020

77. A.A. Govyadinov et al., Probing low-energy hyperbolic polaritons in van der Waals crystals with an electron microscope (in English). Nat. Commun. **8**, 95 (2017). https://doi.org/10.1038/s41467-017-00056-y

78. B. J. M. Brenny, A. Polman, F. J. García de Abajo, Femtosecond plasmon and photon wave packets excited by a high-energy electron on a metal or dielectric surface. Phys. Rev. B **94**(15), 155412 (2016). [Online]. Available: http://link.aps.org/doi/10.1103/PhysRevB.94.155412

79. L. Novotny, Strong coupling, energy splitting, and level crossings: a classical perspective (in English). Am. J. Phys. **78**(11), 1199–1202 (2010). https://doi.org/10.1119/1.3471177

80. S.R.K. Rodriguez, Classical and quantum distinctions between weak and strong coupling (in English). Eur. J. Phys. **37**(2), 025802 (2016). https://doi.org/10.1088/0143-0807/37/2/025802

81. S.J. Smith, E.M. Purcell, Visible light from localized surface charges moving across a grating (in English). Phys. Rev. **92**(4), 1069–1069 (1953). https://doi.org/10.1103/PhysRev.92.1069

82. P. Goldsmith, J.V. Jelley, Optical transition radiation from protons entering metal surfaces (in English). Philos. Mag. **4**(43), 836–844 (1959). https://doi.org/10.1080/14786435908238241
83. J.M. Wachtel, Free-electron lasers using the Smith-Purcell Effect (in English). J. Appl. Phys. **50**(1), 49–56 (1979). https://doi.org/10.1063/1.325642
84. R.P. Leavitt, D.E. Wortman, C.A. Morrison, Orotron—free-electron laser using the Smith-Purcell Effect (in English). Appl. Phys. Lett. **35**(5), 363–365 (1979). https://doi.org/10.1063/1.91151
85. J. Urata, M. Goldstein, M.F. Kimmitt, A. Naumov, C. Platt, J.E. Walsh, Superradiant Smith-Purcell emission (in English). Phys. Rev. Lett. **80**(3), 516–519 (1998). https://doi.org/10.1103/PhysRevLett.80.516
86. G. Adamo et al., Light well: a tunable free-electron light source on a chip (in English). Phys. Rev. Lett. **103**(11), 113901 (2009). https://doi.org/10.1103/physrevlett.103.113901
87. F.J.G. de Abajo, Smith-Purcell radiation emission in aligned nanoparticles (in English). Phys. Rev. E **61**(5), 5743–5752 (2000). https://doi.org/10.1103/PhysRevE.61.5743
88. N. Talebi, A directional, ultrafast and integrated few-photon source utilizing the interaction of electron beams and plasmonic nanoantennas. New J. Phys. **16**(5), 053021 (2014). https://doi.org/10.1088/1367-2630/16/5/053021
89. A. Gover, Y. Pan, Stimulated radiation interaction of a single electron quantum wavepacket. ArXiv e-prints **1702**. [Online]. Available: http://adsabs.harvard.edu/abs/2017arXiv170206394G
90. H.L. Andrews, C.H. Boulware, C.A. Brau, J.D. Jarvis, Superradiant emission of Smith-Purcell radiation (in English). Phys. Rev. Spec. Top. Ac. **8**(11), 110702 (2005). https://doi.org/10.1103/physrevstab.8.110702
91. R.H. Dicke, Coherence in spontaneous radiation processes (in English). Phys. Rev. **93**(1), 99–110 (1954). https://doi.org/10.1103/Physrev.93.99
92. R.M. Phillips, History of the Ubitron (in English). Nucl. Instrum. Meth. A **272**(1–2), 1–9 (1988). https://doi.org/10.1016/0168-9002(88)90185-4
93. J.M.J. Madey, Stimulated emission of bremsstrahlung in a periodic magnetic field (in English). J. Appl. Phys. **42**(5), 1906–1913 (1971). https://doi.org/10.1063/1.1660466

Chapter 4
Electron–Induced Domain

Abstract EELS and CL have been introduced as efficient tools for probing nanooptical excitations in single nanostructures, with a nanometre spatial resolution and meV energy resolution. Thanks to the ultrafast interaction of localized relativistic electrons with the optical modes of nanostructures in TEMs, electron beams appear as an ultra-broadband probe of sample resonances. The inelastic interaction of a swift electron with nanostructures can be understood using a useful classical approach, as discussed in the previous chapter, that has been proven to be identical to the quantum-mechanical treatment when averaging over electron impact parameters weighted by the spot intensity (Ritchie and Howie in Philos Mag A 58(5):753–767, 1988 [3]).

When interpreting electron energy-loss spectra as the probability of an electron losing an amount of energy equal to $\hbar\omega$, the loss probability is given as

$$\Gamma^{\text{EELS}}(\omega) = \frac{e}{\pi\hbar\omega} \int dt \, \text{Re}\left\{ e^{-i\omega t} \vec{v}_{\text{el}} \cdot \vec{E}^{\text{ind}}\left[\vec{r}_{\text{el}}(t), \omega \right] \right\} \qquad (4.1)$$

where $\vec{v}_{\text{el}}$ is the electron velocity, $\vec{r}_{\text{el}}$ is the electron trajectory, e is the electron charge, and $\hbar$ is the reduced Planck constant [4]. A complete treatment then requires knowledge of the electron trajectory, as well as the induced electric field ($\vec{E}^{\text{ind}}$) along the electron trajectory. Assuming a uniform electron trajectory along the z-axis as $\vec{r}_{\text{el}}(t) = (x_0, y_0, z = v_{\text{el}}t)$ (nonrecoil approximation), (4.1) can be further simplified as

$$\Gamma^{\text{EELS}}(x_0, y_0, \omega) = \frac{e}{\pi\hbar\omega} \text{Re} \int_{-\infty}^{+\infty} dz \, \tilde{E}_z(x_0, y_0, z; \omega) \, e^{i\frac{\omega}{v_{\text{el}}} z}$$

$$= \frac{e}{\pi\hbar\omega} \text{Re}\, \tilde{E}_z(x_0, y_0, k_z = \omega/v_{\text{el}}; \omega) \qquad (4.2)$$

© Springer Nature Switzerland AG 2019
N. Talebi, *Near-Field-Mediated Photon–Electron Interactions*,
Springer Series in Optical Sciences 228,
https://doi.org/10.1007/978-3-030-33816-9_4

which clearly indicates the role of momentum conservation during the interaction of electrons with near-field distributions of nanostructures as $\hbar k_z = \hbar \omega / v_{\mathrm{el}}$. Interestingly, the loss probability is directly related to the induced electric field, which renders EELS a powerful technique to directly map the electric field projected along the electron trajectory [5]. One might find another useful derivation of the loss probability by treating EELS as the *rate of energy leaving the electron beam* as [6]

$$\Gamma^{\mathrm{EELS}}(\omega) = \frac{-1}{\pi \hbar \omega} \mathrm{Re} \iiint \vec{E}(\vec{r}, \omega) \cdot \vec{J}^*(\vec{r}, \omega) \, \mathrm{d}^3 r \qquad (4.3)$$

where $\vec{J}(\vec{r}, \omega)$, is the current density function in the frequency domain. Equation (4.3) is further simplified to (4.2) by using the nonrecoil approximation as $\vec{J}(\vec{r}, \omega) = \Im \vec{J}(\vec{r}, t) = -e\delta(x - x_0)\delta(y - y_0) \exp(i\omega z / v_{\mathrm{el}})$ [2]. Introducing (4.3) has the advantage that the so-called Poynting theorem can be used to link the EELS outcome to that of the photon-generation probability, where the latter is obtained by integrating the Poynting vector as $\Gamma^{PG}(\omega) = \frac{1}{\pi \hbar \omega} \iint \mathrm{Re}\left\{ \vec{E}(\vec{r}, \omega) \times \vec{H}^*(\vec{r}, \omega) \right\} \cdot \vec{\mathrm{d}s}$. The difference between the photon generation probability and loss probability is the so-called absorption spectrum [2]. Note that $\Gamma^{PG}(\omega)$ is directly related to the CL spectra whenever the far-field radiation is considered, for which $\vec{H}(\vec{r}, \omega) = (\omega \mu_0)^{-1} \vec{k}_0 \times \vec{E}(\vec{r}, \omega)$, where μ_0 is the permeability of free space [4]. In this regard, CL is complementary to EELS in studying the near-field zone of nanophotonic excitations and can be used to map radiative modes [7, 8]. CL detection systems incorporated in a TEM can be used to compare EELS and CL [9]. Advanced CL systems have recently been introduced in SEMs with the ability to perform angle-resolved CL mapping and CL polarimetry [10–12].

A great improvement in interpreting the outcomes of low-loss EELS has been achieved by linking EELS to Green's function and the photonic local density of state (PLDOS) [13]. In this context, EELS is related to the PLDOS projected along the electron trajectory. In fact, (4.2) demonstrates that EELS can be used to directly detect the electric field component projected along the electron trajectory. Pioneered by the work of Nelayah et al. [14], EELS is applied to many nanophotonics systems to understand the near-field distribution and resonant energies of optical modes, especially localized plasmons and plasmon polaritons (see Fig. 4.1) [15–24]. The breathing modes of nanotriangles have emerged as an interesting nonradiative mode, at least when the size of the structure is small enough [20]. Zhu and coworkers probed the PLDOS of a void structure in an Al matrix and decomposed the full PLDOS into edge and surface plasmons [24]. Using EELS, it was demonstrated that long-range plasmon polaritons propagating in an ultrathin nanowire are robust to the discontinuities induced by bends [22]. Guzzinati et al. showed that a properly shaped electron beam can be used to probe symmetrically decomposed modes of plasmonic nanostructures [25]. Acquiring the momentum-resolved energy-loss spectra of thin aluminium nanodiscs, Shekhar et al. proved that there is a substantial difference between thin films and nanodiscs due to the mode confinement in nanodiscs [23].

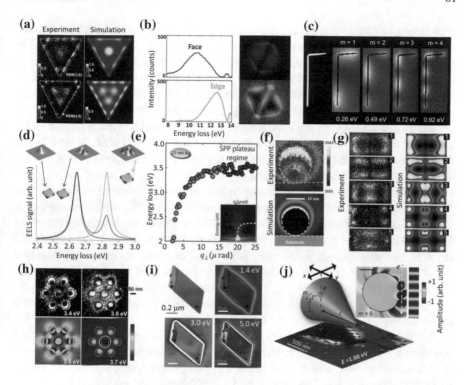

Fig. 4.1 EELS investigation of optical excitations in several nanostructures: **a** plasmonic edge and breathing modes (from [26]); **b** edge and face modes of topologically enclosed void structures in an Al matrix (from [27]); **c** long-range plasmons in bent silver nanowires (from [28]); **d** dipolar and quadrupolar optical modes of a gold cubic nanostructure decomposed with shaped electron beams (from [16]); **e** plasmonic modes of thin silver discs acquired using momentum-resolved EELS (from [29]); **f** localized plasmons of nanospheres in the presence of a substrate (from [11]); **g** coupled edge and gap plasmons in adjacent silver nanoparticles (from [30]); **h** toroidal, azimuthal, and transverse plasmonic modes in void oligomers (from [31]); **i** wedge hyperbolic polaritons and Dyakonov waves in Bi_2Se_3 nanoparticles (from [32]); **j** plasmonic modes with higher angular momentum orders in mesoscopic gold tapers (from [2]). Reprinted by permission from [1]

The effect of the substrate on the spatial distribution of the localized plasmons has also been extensively studied using EELS [21].

Bellido et al. investigated the coupling between adjacent nanoparticles in close proximity to each other, which causes the formation of gap plasmons [19]. The electron beam excitation of exotic toroidal moments in oligomer nanocavities has also been investigated and proven to contribute negligibly to the far-field radiation in the direction normal to the surface [18]. The formation of wedge hyperbolic polaritons, which are long-range excitations of hyperbolic polaritons propagating along wedges, via the coupling between two adjacent edge polaritons has also been studied in Bi_2Se_3 nanoflakes [15]. Finally, considering rotationally symmetric plasmon polaritons of mesoscopic gold tapers, it was shown that phase matching between the near-field zones of the electron and excited plasmons can be captured as individual resonances in EELS energy-distance maps [16, 33]. Indeed, it should be recalled that the momentum selection rule $k_z = \omega/v_{el}$, as noted in (4.2), restricts EELS to map only those photonic states that can afford the necessary momentum. In mesoscopic samples, phase-matching (synchronization) between the electron's self-field and near-field distributions can be manifested in EELS as *new resonances* in structures that support optical modes with higher-order angular momentum [16] or in optical gratings [34] (see Fig. 4.1j). Especially for a grating with a period L, the phase-matching condition will result in $\omega/v_{el} = k_0 \sin\theta + 2m\pi/L$, where k_0 and θ are respectively the emitted photon wave number and its direction of propagation with respect to the axis of the grating and $m = 0, 1, 2, \dots$ is the diffraction order of the grating. This geometrical condition almost precisely describes the criterion for the emission of coherent Smith-Purcell radiation [34, 35], which can interfere with the electron-induced plasmon radiation [36]. Note that the Smith-Purcell effect is widely accepted as the radiation from the electron beam itself. In the experiment involved, essentially a single electron is expected to arrive at each given time in a TEM; despite this fact, however, the Smith-Purcell radiation is a coherent emission process. This concept is also correct for transition radiation, which occurs due to the fast annihilation of the created dipole from the interaction of a swift electron with its image charge at the surface of a metal.

Any energy loss experienced by an electron should affect its kinetic energy and consequently its momentum and trajectory. In the remainder of this chapter, we discuss the recoil experienced by the electron by traversing the near-field distribution of nanostructures in an aloof trajectory, i.e., avoiding a penetrating trajectory into the material. Hence, the recoil experienced is due to only the inelastic interaction related to the optical field distributions and not the electron-electron or electron-ion interactions. In the following, we provide a classical viewpoint of the problem, whereas the quantum mechanical recoil and interference effects between various quantum paths to achieve such a recoil will be discussed in Chap. 8.

4.1 Electron Recoil

The applicability of EELS for mapping the photonic local density of states is greatly based on the nonrecoil approximation. It is interesting, however, to note that as the electron passes within the vicinity of nanostructures, even in an aloof experiment, the electromagnetic excitations will act on the electron and change its trajectory. Thus, the electron should be subjected to longitudinal as well as transverse recoils because of the Coulomb potential or—in the retarded picture—the Lorentz force. The advantage of the non-recoil approximation, however, is that the change in momentum (both transverse and longitudinal) during the interaction is insignificant to the final probability and spectral distribution of the interaction, as will be discussed later. Moreover, the correctness of the non-recoil approximation for EELS is better justified by the weak dependence of the loss spectrum on the incident energy across the entire range of the loss spectrum.

In a combined laser and electron microscope setup, when the structures are illuminated by an external laser light, this recoil can be better observed. However, to the best of our knowledge, such an experiment has not yet been reported for the electron-induced polarizations in a single structure in the absence of laser excitation.

In addition to the experiments, numerical studies should also be employed to obtain a better understanding of the interaction of electrons with optical near-fields, both in the presence and in the absence of external laser excitations, particularly beyond the nonrecoil approximation. In a classical approach, one might combine the Lorentz and Maxwell equations in a self-consistent way, as routinely employed in particle-in-cell (PIC) numerical procedures [37, 38]. Indeed, the success of the PIC method in simulating many physical processes in plasma physics [39, 40], free-electron lasers [41], accelerators [42, 43], and general systems of interacting electromagnetic fields and charged particles has initiated research on self-consistent simulation approaches.

We combined a finite-difference time-domain (FDTD) electromagnetic solver [44, 45] with a Lorentz equation solver to simulate the interaction of a single relativistic electron with the electron-induced excitations in nanoparticles. Apparently, the system should be treated in a self-consistent way, as the electron first induces near-field polarization and is then scattered by the self-induced polarization. The electromagnetic field components are assigned to fixed grid points of the FDTD simulation domain (see Fig. 4.2a), whereas the particles are tracked in the continuous domain. The combined equations can be written as

$$\vec{\nabla} \times \vec{E}(\vec{r}, t) = -\frac{\partial \vec{B}(\vec{r}, t)}{\partial t}$$

$$\vec{\nabla} \times \vec{H}(\vec{r}, t) = \frac{\partial \vec{D}(\vec{r}, t)}{\partial t} + \vec{J}_{\text{el}}(\vec{r}, t) \tag{4.4a}$$

for the electromagnetic fields and

$$\frac{d}{dt} \gamma m \vec{v}_{\text{el}} = \int \rho(\vec{r}, t) \left(\vec{E}(\vec{r}, t) + \vec{v}_{\text{el}} \times \vec{B}(\vec{r}, t) \right) d^3 r$$

$$\frac{d \vec{r}_{\text{el}}(t)}{dt} = \vec{v}_{\text{el}} \tag{4.4b}$$

where $\vec{E}$, $\vec{D}$, $\vec{H}$, and $\vec{B}$ are the electric field, magnetic field, displacement vector, and magnetic flux density, respectively. γ is the Lorentz factor, m is the electron mass, $\vec{r}_e(t)$ is the electron trajectory, and $\rho(\vec{r}, t)$ is the electron charge density. The current distribution at the boosted frame is computed as $J'^{\alpha} = \frac{\partial x'^{\alpha}}{\partial x^{\alpha}} J^{\alpha}$, in which J^{α} is the four-vector current density distribution in the laboratory frame. In theory, the charge distribution of an electron is perfectly approximated by the Dirac delta function ($\rho(\vec{r}, t) = -e\delta(\vec{r} - \vec{r}_e(t))$), as understood from the Coulomb potential. In practice, however, an electron toy model is often introduced [38, 46], or the charge

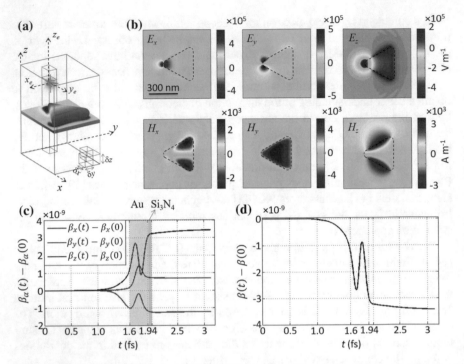

Fig. 4.2 Electron at a kinetic energy of 200 keV interacting with a single gold nanotriangle prism 10 nm away from the edge of the structure. **a** Simulation setup including an FDTD electromagnetic field solver and a particle-in-cell tracker. **b** The spatial distribution of the induced time-dependent electromagnetic field components at $t = 1.8$ fs in the plane 5 nm above the gold prism. Modulation of **c** individual and **d** total electron velocity components versus time. Reprinted by permission from [1]

distribution in the continuous particle space is mapped to the grids of the electromagnetic solver using an extrapolation technique. We used a symmetric Gaussian charge distribution as $-q(2\pi\ W)^{-3}\exp\left(-0.5|\vec{r} - \vec{r}_e(t)|^2/W^2\right)$, where W is the broadening of the charge distribution. The electromagnetic field in the laboratory frame in free space is given by Jackson [47]

$$E'_{z,\,\text{inc}}(\rho', z'; t) = \frac{-q\gamma(z' - \beta ct')}{\varepsilon_0\left(\sqrt{2\pi}\right)^3 W\left(\rho'^2 + \gamma^2(z' - \beta ct')^2\right)}\exp\left(-\frac{1}{2}\rho'^2 + \gamma^2(z' - \beta ct')^2/W^2\right)$$

$$+ \frac{-q\gamma(z' - \beta ct')}{4\pi\varepsilon_0\left(\rho'^2 + \gamma^2(z' - \beta ct')^2\right)^{\frac{3}{2}}}\text{erf}\left(\left(\rho'^2 + \gamma^2(z' - \beta ct')^2\right)^{\frac{1}{2}}/\sqrt{2}W\right)^{\frac{1}{2}}$$

$$(4.5a)$$

and

$$E'_{\rho', \text{inc}}(\rho', z'; t)$$

$$= \frac{-q\gamma\rho'}{\varepsilon_0\left(\sqrt{2\pi}\right)^3 W\left(\rho'^2 + \gamma^2(z' - \beta ct')^2\right)} \exp\left(-\frac{1}{2}\frac{\rho'^2 + \gamma^2(z' - \beta ct')^2}{W^2}\right)$$

$$+ \frac{-q\gamma\rho'}{4\pi\varepsilon_0\left(\rho'^2 + \gamma^2(z' - \beta ct')^2\right)^{\frac{3}{2}}} \text{erf}\left(\frac{\left(\rho'^2 + \gamma^2(z' - \beta ct')^2\right)^{\frac{1}{2}}}{\sqrt{2}W}\right) \tag{4.5b}$$

where propagation along the z-axis has been considered. erf() is the error function, $(\rho', z') = (x', y', z')$ is the boosted coordinate system, and $\vec{\beta} = (0, 0, v_e/c)$. The self-interaction, which can be modelled by $\iiint \vec{E}'_{\text{inc}}(\vec{r}', t') \cdot \vec{J}'(\vec{r}', t') \mathrm{d}^3 r'$, is then identically zero, thanks to the symmetry of the field components. In this way, the total field rather than the scattered fields can be used in (4.4). However, the numerical implementation of (4.5) inside an electromagnetic solver introduces additional errors. To cancel the interaction of the electron with its own field, at each time loop, we calculate the total field using (4.4a) and use $\vec{H}^{\text{sca}} = \vec{H}^t - \vec{H}_{\text{inc}}$ and $\vec{E}^{\text{sca}} = \vec{E}^t - \vec{E}_{\text{inc}}$. For an electron interacting with a triangular gold nanoprism with a thickness of 50 nm (see Fig. 4.2a), the induced electromagnetic field components at a given time are shown in Fig. 4.2b. The modulations of the electron velocity in both the transverse and longitudinal directions and the total change of the velocity are shown in Fig. 4.2c, d, respectively. Interestingly, the change in the electron velocity is initiated even before the electron reaches the structure, when it enters the near-field domain. The near-field domain is defined by the region in the vicinity of a given nanostructure at which the momentum of light is larger than the free space momentum. This associated short-range decay of the optical near field results in an inelastic interaction with the electron. Surprisingly, however, the modulation in the electron velocity is incredibly low, and the final electron velocity is reduced by only $\delta v_{\text{el}} = 3.45 \times 10^{-9}c$, where c is the speed of light in a vacuum. These findings further confirm the appropriateness of the nonrecoil approximation in treating the energy-loss probability and CL spectra. Two points, however, should be noted here: (i) The final velocity of the electron and the lateral recoil it experiences both depend on the impact parameter. In general, it is expected that the electron undergoes less recoil at larger impacts. In other words, in the structure considered above, an impact position at a distance of 10 nm away from the structure results in negligible recoil. (ii) In EELS, we detect the probability of an electron losing a certain amount of photon energy. The probability amplitudes indeed depend on the impact parameter but not on the resonant energies. Hence, we conclude here that it is the form of the effective interaction potential, not the impact parameter that affects the resonances. As we previously observed, the probability spectra depend on the frequency-dependent electric-field component. We may relate the change in the longitudinal momentum to the magnetic vector potential as $m_0\delta\tilde{v}_{\text{el}}(\omega) = q A_z(\vec{r}_{\text{el}}, \omega)$ and hence to the electric field as $E_z(r_{\text{el}}, \omega) = i\omega_{ph}m_0\delta\tilde{v}_{\text{el}}(\omega)/q$. Using (4.2), we may then conclude that it is the frequency dependence of the velocity modulation, not $\delta v_{\text{el}} = v_{\text{el}}(t \to +\infty) - v_{\text{el}}(t \to 0)$, that is related to the EEL spectra.

Although the electron recoil experienced is apparently negligible, electron energy loss detectors can nowadays precisely determine the electron energy loss probabilities up to the limit of a few meV per electron [48, 49]. In systems where a phase-matching (synchronicity) condition between the electron and photons is obtained (see Fig. 4.1j), the overall recoil that an electron experiences can be additively manipulated. This can be achieved by an optical grating as well (see Fig. 4.3a). We consider each element of a grating to be composed of two gold nanowires with a 20 nm gap between them in order to enhance the interaction and to maintain a symmetric excitation. The synchronicity condition in this case is exactly analogous to that of the Smith-Purcell effect. The criterion for Smith-Purcell radiation in free space is given by $\omega/v_{el} = k_0 \sin\theta + 2m\pi/L$, as described in Sect. 4.1. However, we note that the power that is transmitted into the substrate is more pronounced than the power radiated in other directions. Additionally, a planar waveguide incorporated inside the substrate, such as a thin film of HfO$_2$, can be used to guide the electromagnetic radiation along the

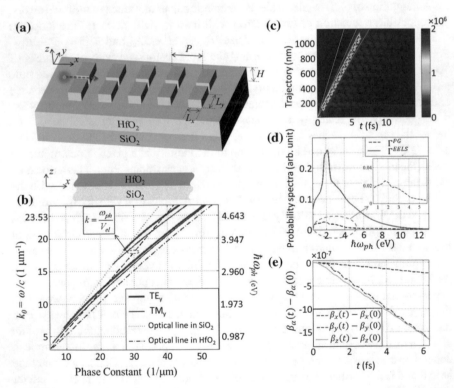

Fig. 4.3 Guided Smith-Purcell radiation. **a** Interaction of an electron at the velocity of $v_e = 0.598c\,\hat{x}$ with a grating positioned upon a planar waveguide. The grating is composed of gold nanorods with $P = 185$ nm, $L_x = 50$ nm, $L_y = 147$ nm, and $H = 180$ nm, and the gaps between two nanorods are 20 nm. **b** Dispersion of the propagating modes of the waveguide, which can be used to meet the synchronicity condition. **c** Induced x-component of the electric field versus time and electron trajectory. **d** EELS and photon-generated probability spectra. **e** Relative velocity of the electron versus time. Reprinted by permission from [1]

symmetry axis of the grating. The Smith-Purcell radiation can hence be coupled into the propagating modes of the waveguide, which are themselves decomposed, as usual, into the TE and TM waves (see Fig. 4.3b). The time of travel of the electron between two adjacent grating elements is given by $\delta t_{el} = P/v_{el}$, while for the emitted photons, $\delta t_{ph} = (\beta_r + 2m\pi/P)P/\omega_{ph}$, where P is the period of the grating, β_r is the phase constant of the propagating mode of the waveguide, and m is the diffraction order. To have constructive interference of the generated photons from the interaction of the electrons with the grating elements, the criterion $\delta t_{el} = \delta t_{ph}$ should be satisfied. We note, however, that in contrast to the Smith-Purcell effect in free space, we can even satisfy this criterion with $m = 0$ (see Fig. 4.3b), which greatly enhances the efficiency of the photon generation process. Figure 4.3c shows the induced x-component of the electric field versus time along the electron trajectory. The excited field from the interaction of the electron with the adjacent element is synchronous with the electron velocity and mostly propagating at velocities exceeding the velocity of light in free space. The calculated EEL spectrum sustains a double peak at energies of $E = 1.9$ and 2.1 eV, which correspond to the synchronicity condition ($\beta_r = \omega_{ph}/v_{el}$) with the lowest-order TM and TE modes, respectively (Fig. 4.3b, d). As the electron travels at a distance of 5 nm above the grating elements, the overall recoil experienced from the interaction of the electron with the grating is linearly accumulated and is 3 orders of magnitude higher than the single interaction demonstrated in Fig. 4.2. This is partly due to the incorporation of gap plasmons between adjacent nanorod antennas and partly because of the satisfaction of a synchronous excitation criterion. A comparison between the EELS and photon generation spectra, however, demonstrates that only a small percentage of the photons are coupled to the radiative and propagating photons. In fact, most of the energy is dissipated inside the metallic elements. Nevertheless, the radiation is mostly coupled to the propagating modes of the waveguide rather than to far-field radiation modes (see Fig. 4.4).

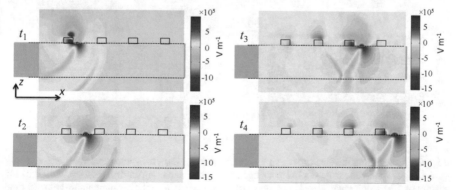

Fig. 4.4 Coupling of Smith-Purcell radiation to guided modes. Snapshots of the z-component of the electric field induced by the interaction between an electron with a velocity of $v_e = 0.598c\,\hat{x}$ and the structure shown in Fig. 4.4a ($t_1 < t_2 < t_3 < t_4$)

The experienced electron recoil can also be controlled using only two nanoanten-nas coupled to a ridge waveguide [2]. In such a configuration, the localized plasmons excited near the first interaction point are coupled to the plasmon polaritons of a ridge waveguide. A second nanoantenna is positioned along the electron trajectory, as shown in Fig. 4.5a, which supports the second interaction. The plasmon polari-tons of the first excitation point are then guided towards the second interaction point. They further interfere with the electron-induced plasmons of the second nanoantenna either constructively or destructively, depending on the distance Δx between the two nanoantennas. The simulated electric field versus time along the electron trajectory for $\Delta x = 327$ nm and $\Delta x = 130$ nm demonstrates this interference phenomenon (compare Fig. 4.5b, c). Figure 4.5d and e show the induced z-component of the elec-tric field at a given time for $\Delta x = 327$ nm and $\Delta x = 130$ nm, respectively. For the

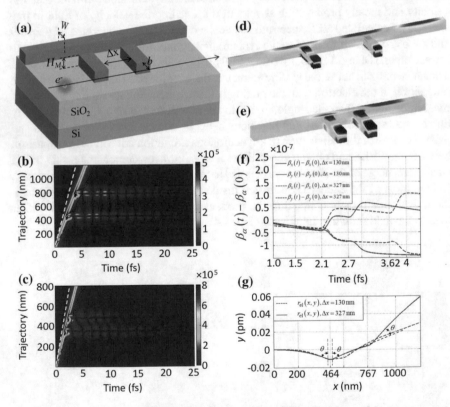

Fig. 4.5 Controlling the electron recoil with nanoantennas [50]. **a** Topology of the structure com-posed of gold nanoantennas coupled to a ridge waveguide interacting with an electron with a kinetic energy of 200 keV propagating along the x-axis 2 nm away from the nanoantennas. $W = 30$ nm, $H_M = 30$ nm, and the length of the nanoantennas is 147 nm. Induced electric field along the electron trajectory for **b** $\Delta x = 327$ nm and **c** $\Delta x = 130$ nm. z-component of the induced electric field for **d** $\Delta x = 327$ nm and **e** $\Delta x = 130$ nm. **f** The relative velocity of the electron ($\beta = v_{el}/c$) along the x- and y-directions. $\theta = 1.57 \times 10^{-8}$ rad is the scattering angle of an electron in interaction with a single nanoantenna. Reprinted by permission from [1]

former case, the two nanoantennas resonate out of phase. However, they excite the plasmon polaritons of the ridge wave guide in a constructive way. The overall transverse recoil experienced for this case is then doubled in comparison with the recoil experienced from a single nanoantenna. In contrast, for nanoantennas at a distance of $\Delta x = 130$ nm, the two nanoantennas resonate in phase. The plasmon polaritons that are excited near the first interaction point, however, are out of phase with the electron-induced plasmons of the second antenna and thus interfere destructively. The recoil that the electron experiences along the transverse direction (y-direction in this case) is then less than that of the previous case with $\Delta x = 327$ nm (see Fig. 4.5g, f).

The electromagnetic powers guided and reflected by the structure are approximately equal. A feedback element is thus required to improve the coupling efficiency of the generated photons into the guided modes. For the case of the feedback element, a split-ring resonator is chosen, as depicted in the inset of Fig. 4.6a, with $L_v = 110$ nm and $L_h = 100$ nm. The reason for this choice is that, first, the local field distribution along the trajectory is the same for both the nanoantenna and the vertical arms of the split-ring resonator, which provides a perfect coupling path for the photons produced in both elements. The second reason is to provide a guided path

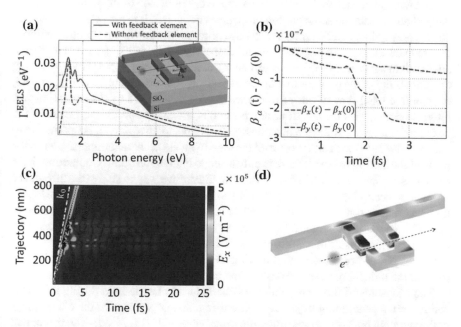

Fig. 4.6 Effect of a feedback element on the electron recoil and EELS. **a** EEL spectra for an electron moving adjacent to the structure with and without a split-ring resonator as a feedback element. The inset shows the structure including a split-ring resonator, the electron trajectory through the gap between the nanoantennas and the feedback element. **b** The relative velocity of the electron versus time projected along the x- and y-axes. **c** Induced time-dependent electric field along the electron trajectory. **d** Field profile for the induced z-component of the electric field at an energy of $E_{ph} = 1.3$ eV at a given time. Reprinted by modifications from [2] under the CC BY license

for the generated photons from the first interaction point to the second interaction point, which, on the other hand, can provide a means for constructive interference. This interference phenomenon can be seen from the computed EEL spectra, depicted in Fig. 4.6a. For the geometry of the structure to be designed correctly, the system should sustain the same resonance at 1.3 eV, which is fully satisfied by the introduced feedback element.

While there is no significant change in the x- and z-components of the velocity, the y-component of the velocity is reduced rather monotonically. The computed x-component of the induced electric field along the trajectory of the electron is depicted in Fig. 4.6c. There is a strong modulation of the generated optical field due to the presence of the resonator, which results in a generation of a series of optical pulses with a high damping rate. This modulation shows the efficiency of the feedback element, which results in the switching of the plasmon resonances between the nanoantennas and the feedback element, providing a channel between the two interaction points.

Figure 4.6d shows the z-component of the electric field within half of the time period of the excited plasmon. It is seen that the excited plasmon sustains a unidirectional flow of power along the $+x$-direction. The probability that the generated photon with an energy of 1.3 eV propagates to the right side of the wire is 4.6 times the probability that the same photon propagates to the left. It is important to note that there is no significant difference in this ratio in comparison with the structure without the feedback element. In fact, the introduction of the feedback element has resulted in an overall increase in the collection efficiency of the generated photons by decreasing the probability of the photons that are scattered into the surrounding medium. In fact, the unidirectional flow of power in the waveguide is achieved by incorporating two interaction points for the electrons and plasmons, along with providing a path for the generated photons to interfere in a constructive or destructive way. It is instructive to compare the mentioned situation with the case of an electron travelling normal to the plane of the circuit, in which there is only a single interaction point. Comparing these two cases, as demonstrated in Fig. 4.7a, is an illustrative way to study the differences between the coupling and interference phenomena. Figure 4.7b shows the EEL spectra for both penetrating and aloof electron trajectories. The prominent difference in the EEL spectra is a signature for the role of the interference paths experienced by the electron as a result of a multiple scattering effect (Fig. 4.7c). This interference results in power flow in one direction inside the waveguide, providing an efficient path for the transfer of energy from the first nanoantenna to the second nanoantenna.

The snapshots of the z-component of the induced electric field during the interaction for the penetrating trajectory demonstrate the propagation of the electromagnetic energy in both directions along the ridge waveguide (Fig. 4.7d). The calculated x-component of the Poynting vector along the ridge waveguide at given positions shows the unidirectional flow of the electromagnetic power for the parallel trajectory compared to the penetrating electron trajectory (Fig. 4.7e).

The abovementioned results show the possibility of observing and controlling the classical electron recoil along both the longitudinal and transverse directions. The transverse recoil can be experimentally observed by detecting the electron diffraction

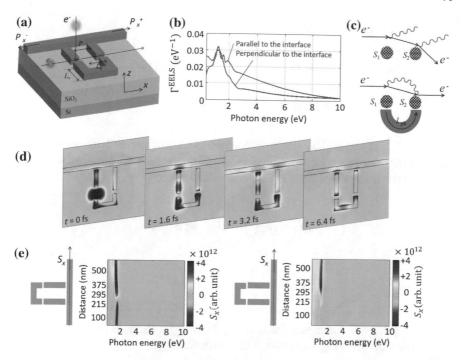

Fig. 4.7 Interference in EELS for an electron interacting with two nanoantennas compared to a single interaction. **a** Schematic of the structure including two trajectories for the electron. **b** EEL spectra calculated for both parallel and perpendicular trajectories. **c** Multiple scattering effect with two possible paths for electron-photon interactions when the feedback element is inserted. **d** Snapshots of the z-component of the induced electric field at given interaction times. **e** Calculated power flow along the waveguide for electron trajectories perpendicular (Left) and parallel (Right) to the symmetry axis of the ridge waveguide. Reprinted by modifications from [2] under the CC BY license

and can be used for mapping the vectorial field components as an example. The longitudinal recoils, however, lead to a change in the EELS signal, and consequently, interesting interference effects happen between the electron-light interaction paths.

4.2 Relation Between Photon-Generation and Electron Energy-Loss Probabilities

To formulate the conservation of the electromagnetic energy in the investigated system, Poynting's theorem is applied. We start with the Maxwell equations given by $\vec{\nabla} \times \vec{E}(\vec{r}, t) = -\partial \vec{B}(\vec{r}, t)/\partial t$ and $\vec{\nabla} \times \vec{H}(\vec{r}, t) = \partial \vec{D}(\vec{r}, t)/\partial t + \vec{J}(\vec{r}, t)$, in which $\vec{E}(\vec{r}, t)$, $\vec{H}(\vec{r}, t)$, $\vec{D}(\vec{r}, t)$, and $\vec{B}(\vec{r}, t)$ are the electric field, magnetic field, electric displacement, and magnetic flux density, respectively, computed at displacement vector $\vec{r}$ and time t. $\vec{J}(\vec{r}, t)$ is the current density function. We will consider a case of a dispersive medium with the constitutive relations

$$
\begin{aligned}
\vec{D}(\vec{r}, t) &= \int_{-\infty}^{+\infty} \hat{\varepsilon}(\vec{r}, t - \tau)\, \vec{E}(\vec{r}, \tau)\mathrm{d}\tau \\
\vec{B}(\vec{r}, t) &= \int_{-\infty}^{+\infty} \hat{\mu}(\vec{r}, t - \tau)\, \vec{H}(\vec{r}, \tau)\mathrm{d}\tau
\end{aligned}
\tag{4.6}
$$

Here, $\hat{\varepsilon}(\vec{r}, t)$ and $\hat{\mu}(\vec{r}, t)$ are the time-domain electric permittivity and magnetic permeability tensors, which can be obtained from the local dispersion relations of the materials using the Fourier transform as shown below:

$$
\hat{\varepsilon}(\vec{r}, t) = \frac{1}{2\pi} \int_{-\infty}^{+\infty} \hat{\varepsilon}(\vec{r}, \omega)\, \mathrm{e}^{i\omega t}\mathrm{d}\omega
$$

$$
\hat{\mu}(\vec{r}, t) = \frac{1}{2\pi} \int_{-\infty}^{+\infty} \hat{\mu}(\vec{r}, \omega)\, \mathrm{e}^{i\omega t}\mathrm{d}\omega
\tag{4.7}
$$

Using the usual procedure for deriving Poynting's theorem from the Maxwell equations, the following relation will be obtained:

$$
\vec{\nabla}\cdot\left(\vec{E}(\vec{r}, t) \times \vec{H}(\vec{r}, t)\right) + \vec{H}(\vec{r}, t)\cdot\frac{\partial \vec{B}(\vec{r}, t)}{\partial t}
$$
$$
+ \vec{E}(\vec{r}, t)\cdot\frac{\partial \vec{D}(\vec{r}, t)}{\partial t} = -\vec{E}(\vec{r}, t)\cdot\vec{J}(\vec{r}, t)
\tag{4.8}
$$

To attribute the spontaneous electromagnetic power to the power spectrum measured using spectroscopy techniques, we assume the field components to be periodic functions of time. This is a justified assumption, since the outputs of all the laser sources utilized in the experiments are periodic over a specified rate, e.g., considering the terahertz modulation of the femtosecond emitted pulses. Since we are dealing with periodic field components, we can expand all the field components with a Fourier series as

$$
\vec{A}(\vec{r}, t) = \sum_{n=-\infty}^{+\infty} \vec{A}_n(\vec{r}) \exp\left(i\frac{2n\pi}{T}t\right),
\tag{4.9}
$$

where $\vec{A}(\vec{r}, t)$ are the vector functions describing the field components and T is the temporal periodicity of the incident pulses. Note that the spatial vector components can be obtained using $\vec{A}_n(\vec{r}) = \frac{1}{T}\int_{-T/2}^{+T/2} \vec{A}(\vec{r}, t) \exp(-i\omega_n t)\mathrm{d}t$, where we introduced the sampling frequencies as $\omega_n = 2n\pi/T$. Using the introduced notations and using the real valued-ness of the electromagnetic fields as $\vec{E}_n(\vec{r}) = \vec{E}_{-n}^{*}(\vec{r})$, (4.8) can be written as below:

$$\nabla \cdot \left(\sum_{n=-\infty}^{+\infty} \sum_{m=-\infty}^{+\infty} \vec{\tilde{E}}_n^*(\vec{r}) \times \vec{\tilde{H}}_m(\vec{r}) \exp(i(\omega_m - \omega_n)t) \right)$$

$$+ \sum_{n=-\infty}^{+\infty} \sum_{m=-\infty}^{+\infty} \vec{\tilde{H}}_n^*(\vec{r}) \cdot i\omega_m \hat{\tilde{\mu}}(\omega_m) \cdot \vec{\tilde{H}}_m(\vec{r}) \exp(i(\omega_m - \omega_n)t)$$

$$+ \sum_{n=-\infty}^{+\infty} \sum_{m=-\infty}^{+\infty} \vec{\tilde{E}}_n^*(\vec{r}) \cdot i\omega_m \hat{\tilde{\varepsilon}}(\omega_m) \cdot \vec{\tilde{E}}_m(\vec{r}) \exp(i(\omega_m - \omega_n)t)$$

$$= - \sum_{n=-\infty}^{+\infty} \sum_{m=-\infty}^{+\infty} \vec{\tilde{E}}_n^*(\vec{r}) \cdot \vec{\tilde{J}}_m(\vec{r}) \exp(i(\omega_m - \omega_n)t). \tag{4.10}$$

The second and third terms are related to the stored magnetic and electric energy densities, respectively. In a similar approach to [51], one can divide the second term in (4.10) into two parts by changing the indices $m \to -n$ and $n \to -m$ to obtain

$$U_m = \frac{1}{2} \sum_{n=-\infty}^{+\infty} \sum_{m=-\infty}^{+\infty} \vec{\tilde{H}}_n^*(\vec{r}) \cdot i\omega_m \hat{\tilde{\mu}}(\omega_m) \cdot \vec{\tilde{H}}_m(\vec{r}) \exp(i(\omega_m - \omega_n)t)$$

$$+ \frac{1}{2} \sum_{n=-\infty}^{+\infty} \sum_{m=-\infty}^{+\infty} \vec{\tilde{H}}_m(\vec{r}) \cdot \left(-i\omega_n \hat{\tilde{\mu}}^*(\omega_n) \right) \cdot \vec{\tilde{H}}_n^*(\vec{r}) \exp(i(\omega_m - \omega_n)t)$$

$$= \frac{1}{2} \sum_{n=-\infty}^{+\infty} \sum_{m=-\infty}^{+\infty} \vec{\tilde{H}}_n^*(\vec{r}) \cdot \left(i\omega_m \hat{\tilde{\mu}}(\omega_m) - i\omega_n \hat{\tilde{\mu}}^\dagger(\omega_n) \right) \cdot \vec{\tilde{H}}_m(\vec{r}) \exp(i(\omega_m - \omega_n)t) \tag{4.11}$$

The term in brackets on the right-hand side of (4.11) implies the decomposition of the permeability tensor into two parts, Hermitian and anti-Hermitian, as $\tilde{\mu}(\omega_m) = \tilde{\mu}_H(\omega_m) + \tilde{\mu}_{AH}(\omega_m)$ [52] to obtain

$$U_m = \frac{1}{2} \sum_{n=-\infty}^{+\infty} \sum_{m=-\infty}^{+\infty} \vec{\tilde{H}}_n^*(\vec{r}) \cdot \left(i\omega_m \hat{\tilde{\mu}}_H(\omega_m) - i\omega_n \hat{\tilde{\mu}}_H(\omega_n) \right)$$

$$\cdot \vec{\tilde{H}}_m(\vec{r}) \exp(i(\omega_m - \omega_n)t)$$

$$+ \frac{1}{2} \sum_{n=-\infty}^{+\infty} \sum_{m=-\infty}^{+\infty} \vec{\tilde{H}}_n^*(\vec{r}) \cdot \left(i\omega_m \hat{\tilde{\mu}}_{AH}(\omega_m) + i\omega_n \hat{\tilde{\mu}}_{AH}(\omega_n) \right)$$

$$\cdot \vec{\tilde{H}}_m(\vec{r}) \exp(i(\omega_m - \omega_n)t) \tag{4.12}$$

The same can be done for the electric energy density (U_e). To account for the real valued-ness of these quantities, one can break the summations into four parts as

$\sum_{m=-\infty}^{+\infty} \rightarrow \sum_{m=-\infty}^{0} + \sum_{m=0}^{+\infty}$ and $\sum_{n=-\infty}^{+\infty} \rightarrow \sum_{n=-\infty}^{0} + \sum_{m=0}^{+\infty}$ and change the indices in the first parts by $m \rightarrow -m$ and $n \rightarrow -n$. By carrying out the same procedure for all the terms in (4.5a), one obtains the following term for the conservation of energy:

$$
\begin{aligned}
\text{Re}\Bigg\{ &\vec{E}_0(\vec{r}) \cdot \vec{J}_0(\vec{r}) + \nabla \cdot \left(\vec{E}_0(\vec{r}) \times \vec{H}_0(\vec{r}) \right) \\
&+ \sum_{n=1}^{+\infty} \sum_{l=-n}^{+\infty} \left(\vec{E}_n^*(\vec{r}) \cdot \vec{J}_{n+l}(\vec{r}) + \nabla \cdot \left(\vec{E}_n^*(\vec{r}) \times \vec{H}_{n+l}(\vec{r}) \right) \right. \\
&+ \frac{1}{2}\vec{H}_n^*(\vec{r}) \cdot \left(i\omega_{n+l}\hat{\bar{\mu}}_H(\omega_{n+l}) - i\omega_n\hat{\bar{\mu}}_H(\omega_n) \right) \cdot \vec{H}_{n+l}(\vec{r}) \\
&+ \frac{1}{2}\vec{H}_n^*(\vec{r}) \cdot \left(i\omega_{n+l}\hat{\bar{\mu}}_{AH}(\omega_{n+l}) + i\omega_n\hat{\bar{\mu}}_{AH}(\omega_n) \right) \cdot \vec{H}_{n+l}(\vec{r}) \\
&+ \frac{1}{2}\vec{E}_n^*(\vec{r}) \cdot \left(i\omega_{l+n}\hat{\bar{\varepsilon}}_H(\omega_{n+l}) - i\omega_n\hat{\bar{\varepsilon}}_H(\omega_n) \right) \cdot \vec{E}_{n+l}(\vec{r}) \\
&\left. + \frac{1}{2}\vec{E}_n^*(\vec{r}) \cdot \left(i\omega_{n+l}\hat{\bar{\varepsilon}}_{AH}(\omega_{n+l}) + i\omega_n\hat{\bar{\varepsilon}}_{AH}(\omega_n) \right) \cdot \vec{E}_{n+l}(\vec{r}) \right) \exp(i\omega_l t) \\
&+ \sum_{n=1}^{+\infty} \sum_{h=n}^{+\infty} \left(\vec{E}_n(\vec{r}) \cdot \vec{J}_{n-h}^*(\vec{r}) + \nabla \cdot \left(\vec{E}_n(\vec{r}) \times \vec{H}_{n-h}^*(\vec{r}) \right) \right. \\
&+ \frac{1}{2}\vec{H}_n(\vec{r}) \cdot \left(-i\omega_{n-h}\hat{\bar{\mu}}_H^*(\omega_{n-h}) + i\omega_n\hat{\bar{\mu}}_H^*(\omega_n) \right) \cdot \vec{H}_{n-h}^*(\vec{r}) \\
&+ \frac{1}{2}\vec{H}_n(\vec{r}) \cdot \left(-i\omega_{n-h}\hat{\bar{\mu}}_{AH}^*(\omega_{n-h}) - i\omega_n\hat{\bar{\mu}}_{AH}^*(\omega_n) \right) \cdot \vec{H}_{n-h}^*(\vec{r}) \\
&+ \frac{1}{2}\vec{E}_n(\vec{r}) \cdot \left(-i\omega_{n-h}\hat{\bar{\varepsilon}}_H^*(\omega_{n-h}) + i\omega_n\hat{\bar{\varepsilon}}_H^*(\omega_n) \right) \cdot \vec{E}_{n-h}^*(\vec{r}) \\
&\left. + \frac{1}{2}\vec{E}_n(\vec{r}) \cdot \left(-i\omega_{n-h}\hat{\bar{\varepsilon}}_{AH}^*(\omega_{n-h}) - i\omega_n\hat{\bar{\varepsilon}}_{AH}^*(\omega_n) \right) \cdot \vec{E}_{n-h}^*(\vec{r}) \right) \exp(i\omega_h t) \Bigg\} = 0
\end{aligned}
$$

$$(4.13)$$

This equation is the general equation of the conservation of the spontaneous power for the electromagnetic pulses, stating that the spontaneous power can be divided into two lower-frequency and higher-frequency parts, with harmonic frequencies denoted by $\omega_l = \omega_m - \omega_n$ and $\omega_h = \omega_m + \omega_n$, respectively. If the pulses are quasimonochromatic and the repetition rate is sufficiently slow that one can assume $T \rightarrow +\infty$, in a similar approach to [51], one can expand the terms $\omega_{n+l}\hat{\bar{\mu}}_H(\omega_{n+l})$, $\omega_{l+n}\hat{\bar{\varepsilon}}_H(\omega_{n+l})$, $\omega_{n-h}\hat{\bar{\mu}}_H^*(\omega_{n-h})$, and $\omega_{n-h}\hat{\bar{\varepsilon}}_H^*(\omega_{n-h})$ around ω_n with a Taylor series and attribute the corresponding terms to the derivatives in the material dispersion.

A well-defined power spectrum can be obtained by averaging the spontaneous power over the fundamental period as $\bar{p}_{av}(\omega_n) = \frac{1}{T}\int_{-T/2}^{T/2} P_{sp}(t, \omega_n)\,dt$. With such an assumption and using the identity $\delta_{n,m} = T\int_{-T/2}^{T/2} \exp(i(\omega_n - \omega_m)t)\,dt$ and

$\int_{-T/2}^{T/2} \exp(i(\omega_n + \omega_m)t)\, dt = 0$, while also neglecting the zero-order term, we obtain the following identity for the average power spectrum:

$$\sum_{n=1}^{+\infty} \omega_n \left\{ \frac{1}{\omega_n} \oiint_S \text{Re}\left\{ \vec{E}_n(\vec{r}) \times \vec{H}_n^*(\vec{r}) \right\}.ds - \iiint_v \vec{H}_n(\vec{r}) \cdot \text{Im}\left\{ \hat{\vec{\mu}}_{AH}(\omega_n) \right\} \cdot \vec{H}_n^*(\vec{r})\, dv \right.$$

$$\left. - \iiint_v \vec{E}_n(\vec{r}) \cdot \text{Im}\left\{ \hat{\vec{\varepsilon}}_{AH}(\omega_n) \right\} \cdot \vec{E}_n^*(\vec{r})\, dv + \frac{1}{\omega_n} \iiint_v \text{Re}\left\{ \vec{E}_n(\vec{r}) \cdot \vec{J}_n^*(\vec{r}) \right\} dv \right\} = 0 \quad (4.14)$$

Note that the integration over the volume of interaction has been performed explicitly. Since the summation should be zero for all the frequency components, an obvious solution to (4.14) is

$$\frac{-1}{\omega_n} \iiint_v \text{Re}\left\{ \vec{E}(\vec{r}, \omega_n) \cdot \vec{J}^*(\vec{r}, \omega_n) \right\} dv$$

$$= \frac{1}{\omega_n} \oiint_S \text{Re}\left\{ \vec{E}(\vec{r}, \omega_n) \times \vec{H}^*(\vec{r}, \omega_n) \right\} \cdot ds +$$

$$- \iiint_v \vec{E}(\vec{r}, \omega_n) \cdot \text{Im}\left\{ \hat{\vec{\varepsilon}}_{AH}(\omega_n) \right\} \cdot \vec{E}^*(\vec{r}, \omega_n) dv$$

$$- \iiint_v \vec{H}(\vec{r}, \omega_n) \cdot \text{Im}\left\{ \hat{\vec{\mu}}_{AH}(\omega_n) \right\} \cdot \vec{H}^*(\vec{r}, \omega_n) dv \quad (4.15)$$

It is interesting to note that the effective energy densities being stored at the volume of the interaction, in the forms of capacitive and inductive energies, do not contribute to the power spectrum averaged over the fundamental period, but only the dissipative terms appear in (4.15). However, it should be stated that the detected power is, in general, related to the impulse response of the detector, and averaging over the spontaneous power should be carried out considering the acquisition rate of the detectors.

It is easy to demonstrate that the first term in (4.15) is equivalent to the electron energy-loss spectra ($\Gamma^{\text{EELS}}(\omega)$) given in [4] within the undepleted pump approximation, only differing in the $\hbar$ coefficients. Considering the photon generation spectra given by

$$\Gamma^{PG}(\omega) = \frac{1}{\hbar \omega} \oiint_S \text{Re}\left\{ \vec{E}(\vec{r}, \omega) \times \vec{H}^*(\vec{r}, \omega) \right\}.ds \quad (4.16)$$

and EELS spectra given by

$$\Gamma^{\text{EELS}}(\omega) = \frac{-1}{\hbar \omega} \iiint_v \text{Re}\left\{ \vec{\tilde{E}}(\vec{r}, \omega) \cdot \vec{\tilde{J}}^*(\vec{r}, \omega) \right\} dv, \tag{4.17}$$

(4.15) is simplified into

$$\Gamma^{\text{EELS}}(\omega) = \Gamma^{PE}(\omega)$$

$$- \frac{1}{\hbar} \iiint_v \vec{\tilde{H}}(\vec{r}, \omega) . \text{Im}\left\{ \hat{\tilde{\mu}}_{AH}(\omega) \right\} . \vec{\tilde{H}}^*(\vec{r}, \omega) dv$$

$$- \frac{1}{\hbar} \iiint_v \vec{\tilde{E}}(\vec{r}, \omega) . \text{Im}\left\{ \hat{\tilde{\varepsilon}}_{AH}(\omega) \right\} . \vec{\tilde{E}}^*(\vec{r}, \omega) dv. \tag{4.18}$$

By assuming that the acquisition rate of the detector is equivalent to the emission rate of the source or locked to it, the PG and EEL spectra differ only in the dissipative losses occurring in the system. It should also be mentioned that the results derived here are in very good agreement with the results reported by Glasgow et al. [53]. However, to derive (4.13), the material properties are not restricted here to symmetric tensorial behaviours, and the Kramers-Kronig relation is not used here. The only necessary assumption to derive (4.18) is the real-valuedness of the field components and the tensors.

4.3 Summary

In this chapter, we exploited the mechanisms of the transverse and longitudinal classical recoils that are experienced by a moving electron interacting with the near-field zone of nanoobjects. We proposed a method to control the electron recoil by means of plasmonic nanoantennas positioned at two points along the electron trajectory and connected to a rib waveguide and a split-ring resonator. In this way, depending on the distance between the nanoantennas, either destructive or constructive interference will be observed in the photon generation probability and EEL spectra.

In the following chapters, we elaborate more on the EELS investigations of oligomer nanostructures supporting toroidal moments (Chap. 4) and the optical response of 3-dimensional gold tapers (Chap. 5).

References

1. N. Talebi, Electron-light interactions beyond the adiabatic approximation: recoil engineering and spectral interferometry AU - Talebi, Nahid. Adv. Phys.: X **3**(1), 1499438. (2018). https://doi.org/10.1080/23746149.2018.1499438
2. N. Talebi, A directional, ultrafast and integrated few-photon source utilizing the interaction of electron beams and plasmonic nanoantennas. New J. Phys. **16**(5), 053021 (2014). https://doi.org/10.1088/1367-2630/16/5/053021
3. R.H. Ritchie, A. Howie, Inelastic-scattering probabilities in scanning-transmission electron-microscopy (in English). Philos. Mag. A **58**(5), 753–767 (1988) (Online). Available: <Go to ISI>://WOS:A1988Q932100005
4. F.J.G. de Abajo, Optical excitations in electron microscopy (in English). Rev. Mod. Phys. **82**(1), 209–275 (2010). https://doi.org/10.1103/revmodphys.82.209
5. F.J.G. de Abajo, M. Kociak, Probing the photonic local density of states with electron energy loss spectroscopy," (in English). Phys. Rev. Lett. **100**(10) (2008). doi: Artn 106804 10.1103/Physrevlett.100.106804
6. P. Schattschneider, *Fundamentals of Inelastic Electron Scattering* (Springer, Wien, 1986)
7. A. Losquin, M. Kociak, Link between cathodoluminescence and electron energy loss spectroscopy and the radiative and full electromagnetic local density of states (in English). Acs Photonics **2**(11), 1619–1627 (2015). https://doi.org/10.1021/acsphotonics.5b00416
8. A. Losquin et al., Unveiling nanometer scale extinction and scattering phenomena through combined electron energy loss spectroscopy and cathodoluminescence measurements (in English). Nano Lett. **15**(2), 1229–1237 (2015). https://doi.org/10.1021/nl5043775
9. M. Kociak et al., Seeing and measuring in colours: electron microscopy and spectroscopies applied to nano-optics (in English). Cr Phys. **15**(2–3), 158–175 (2014). https://doi.org/10.1016/j.crhy.2013.10.003
10. T. Coenen, N.M. Haegel, Cathodoluminescence for the 21st century: Learning more from light (in English). Appl. Phys. Rev. **4**(3) (2017). doi: Artn 031103 10.1063/1.4985767
11. T. Coenen, S.V. den Hoedt, A. Polman, A new cathodoluminescence system for nanoscale optics, materials science, and geology. Microsc. Today **24**(3), 12–19 (2016). https://doi.org/10.1017/S1551929516000377
12. T. Coenen, E.J.R. Vesseur, A. Polman, Angle-resolved cathodoluminescence spectroscopy (in English). Appl. Phys. Lett. **99**(14) (2011). doi: Artn 143103 10.1063/1.3644985
13. F.J. García de Abajo, M. Kociak, Probing the photonic local density of states with electron energy loss spectroscopy. Phys. Rev. Lett. **100**(10), 106804 (2008). https://doi.org/10.1103/physrevlett.100.106804
14. J. Nelayah et al., Mapping surface plasmons on a single metallic nanoparticle (in English). Nat. Phys. **3**(5), 348–353 (2007). https://doi.org/10.1038/nphys575
15. N. Talebi, C. Ozsoy-Keskinbora, H. M. Benia, K. Kern, C.T. Koch, P.A. van Aken, Wedge Dyakonov waves and Dyakonov Plasmons in topological insulator Bi_2Se_3 probed by electron beams. ACS Nano **10**(7), 6988–6994 (2016). https://doi.org/10.1021/acsnano.6b02968
16. N. Talebi et al., Excitation of mesoscopic plasmonic tapers by relativistic electrons: phase matching versus eigenmode resonances (in English). ACS Nano **9**(7), 7641–7648 (2015). https://doi.org/10.1021/acsnano.5b03024
17. L. Gu et al., Resonant wedge-plasmon modes in single-crystalline gold nanoplatelets (in English). Phys. Rev. B **83**(19) (2011). doi: Artn 195433 10.1103/Physrevb.83.195433
18. B. Ögüt, N. Talebi, R. Vogelgesang, W. Sigle, P.A. van Aken, Toroidal plasmonic eigenmodes in oligomer nanocavities for the visible. Nano Lett. **12**(10), 5239–5244 (2012). https://doi.org/10.1021/nl302418n
19. E.P. Bellido, Y. Zhang, A. Manjavacas, P. Nordlander, G.A. Botton, Plasmonic coupling of multipolar edge modes and the formation of gap modes (in English). Acs Photonics **4**(6), 1558–1565 (2017). https://doi.org/10.1021/acsphotonics.7b00348
20. A. Campos et al., Plasmonic breathing and edge modes in aluminum nanotriangles (in English). Acs Photonics **4**(5), 1257–1263 (2017). https://doi.org/10.1021/acsphotonics.7b00204

21. Y. Fujiyoshi, T. Nemoto, H. Kurata, Studying substrate effects on localized surface plasmons in an individual silver nanoparticle using electron energy-loss spectroscopy (in English). Ultramicroscopy **175**, 116–120 (2017). https://doi.org/10.1016/j.ultramic.2017.01.006

22. D. Rossouw, G.A. Botton, Plasmonic response of bent silver nanowires for nanophotonic subwavelength waveguiding (in English). Phys. Rev. Lett. **110**(6) (2013). doi: Artn 066801 https://doi.org/10.1103/physrevlett.1https://doi.org/10.066801

23. P. Shekhar, M. Malac, V. Gaind, N. Dalili, A. Meldrum, Z. Jacob, Momentum-resolved electron energy loss spectroscopy for mapping the photonic density of states (in English). Acs Photonics **4**(4), 1009–1014 (2017). https://doi.org/10.1021/acsphotonics.7b00103

24. Y. Zhu, P.N.H. Nakashima, A.M. Funston, L. Bourgeois, J. Etheridge, Topologically enclosed aluminum voids as plasmonic nanostructures (in English). ACS Nano **11**(11), 11383–11392 (2017). https://doi.org/10.1021/acsnano.7b05944

25. G. Guzzinati, A. Beche, H. Lourenco-Martins, J. Martin, M. Kociak, J. Verbeeck, Probing the symmetry of the potential of localized surface plasmon resonances with phase-shaped electron beams (in English). Nat. Commun. **8** (2017). doi:Artn 14999 10.1038/Ncomms14999

26. L.F. Zagonel et al., Nanometer-scale monitoring of quantum-confined Stark effect and emission efficiency droop in multiple GaN/AlN quantum disks in nanowires. Phys. Rev. B **93**(20), 205410 (2016). https://doi.org/10.1103/physrevb.93.205410

27. N. Talebi, Schrödinger electrons interacting with optical gratings: quantum mechanical study of the inverse Smith–Purcell effect. New J. Phys. **18**(12), 123006 (2016). https://doi.org/10.1088/1367-2630/18/12/123006

28. O.L. Krivanek et al., Vibrational spectroscopy in the electron microscope (in English). Nature **514**(7521), 209–212 (2014). https://doi.org/10.1038/nature13870

29. A. Konečná et al.

30. P.R. Edwards, R.W. Martin, Cathodoluminescence nano-characterization of semiconductors. Semicond. Sci. Technol. **26**(6), 064005 (2011). https://doi.org/10.1088/0268-1242/26/6/064005

31. R.F. Egerton, Electron energy-loss spectroscopy in the TEM (in English). Rep. Prog. Phys. **72**(1) (2009). doi:Artn 016502 10.1088/0034-4885/72/1/016502

32. N. Talebi, W. Sigle, R. Vogelgesang, P. van Aken, Numerical simulations of interference effects in photon-assisted electron energy-loss spectroscopy. New J. Phys. **15**(5), 053013 (2013). https://doi.org/10.1088/1367-2630/15/5/053013

33. S. Guo et al., Reflection and phase matching in plasmonic gold tapers. Nano Lett. **16**(10), 6137–6144 (2016) https://doi.org/10.1021/acs.nanolett.6b02353

34. S.J. Smith, E.M. Purcell, Visible light from localized surface charges moving across a grating (in English). Phys. Rev. **92**(4), 1069–1069 (1953). doi:https://doi.org/10.1103/physrev.92.1069

35. A. Gover, P. Sprangle, A unified theory of magnetic Bremsstrahlung, electrostatic Bremsstrahlung, Compton-Raman scattering, and Cerenkov-Smith-Purcell free-electron lasers (in English). Ieee J. Quantum Elect. **17**(7), 1196–1215 (1981). https://doi.org/10.1109/Jqe.1981.1071257

36. N. Yamamoto, F.J.G. de Abajo, V. Myroshnychenko, Interference of surface plasmons and Smith-Purcell emission probed by angle-resolved cathodoluminescence spectroscopy (in English). Phys. Rev. B **91**(12) (2015). doi:Artn 125144 10.1103/Physrevb.91.125144

37. J.P. Verboncoeur, Particle simulation of plasmas: review and advances (in English). Plasma Phys. Contr. F **47**, A231–A260 (2005). https://doi.org/10.1088/0741-3335/47/5A/017

38. C.S. Meierbachtol, A.D. Greenwood, J.P. Verboncoeur, B. Shanker, Conformal electromagnetic particle in cell: a review (in English). IEEE T Plasma Sci. **43**(11), 3778–3793 (2015). https://doi.org/10.1109/Tps.2015.2487522

39. R. Lee, H.H. Klein, J.P. Boris, Computer-simulation of electromagnetic instabilities in a relativistic plasma (in English). B Am. Phys. Soc. **18**(4), 633–633 (1973) (Online). Available: <Go to ISI>://WOS:A1973P314600628

40. R. Lee, J.P. Boris, I. Haber, Electromagnetic simulation codes for relativistic plasmas (in English). B Am. Phys. Soc. **17**(11), 1048–1048 (1972) (Online). Available: <Go to ISI>://WOS:A1972N819700494

41. R.A. Fonseca et al., Exploiting multi-scale parallelism for large scale numerical modelling of laser wakefield accelerators (in English). Plasma Phys. Contr. F, **55**(12) (2013). doi:Artn 124011 10.1088/0741-3335/55/12/124011

42. C.G.R. Geddes et al., Laser guiding at relativistic intensities and wakefield particle acceleration in plasma channels (in English). Aip Conf. Proc. **737**, 521–527 (2004) (Online). Available: <Go to ISI>://WOS:000226549000055

43. A. Pukhov, Z.M. Sheng, J. Meyer-ter-Vehn, Particle acceleration in relativistic laser channels (in English). Phys. Plasmas **6**(7), 2847–2854 (1999). https://doi.org/10.1063/1.873242

44. N. Talebi et al., Merging transformation optics with electron-driven photon sources. Nat. Commun. (2019)

45. N. Talebi, Interaction of electron beams with optical nanostructures and metamaterials: from coherent photon sources towards shaping the wave function. J. Opt.-Uk **19**(10), 103001 (2017). https://doi.org/10.1088/2040-8986/aa8041

46. S.N. Lyle, *Self Force and Inertia, Old Light on New Ideas* (Springer, Heidelberg, 2010)

47. J.D. Jackson, *Classical Electrodynamics* (Wiley, USA, 1999)

48. A. Bek, R. Vogelgesang, K. Kern, Apertureless scanning near field optical microscope with sub-10 nm resolution. Rev. Sci. Instrum. **77**(4), 043703 (2006). https://doi.org/10.1063/1.2190211

49. T. Neuman, P. Alonso-González, A. Garcia-Etxarri, M. Schnell, R. Hillenbrand, J. Aizpurua, Mapping the near fields of plasmonic nanoantennas by scattering-type scanning near-field optical microscopy. Laser Photonics Rev. **9**(6), 637–649 (2015). https://doi.org/10.1002/lpor.201500031

50. B.J.M. Brenny, T. Coenen, A. Polman, Quantifying coherent and incoherent cathodoluminescence in semiconductors and metals. J. Appl. Phys. **115**(24), 244307 (2014). https://doi.org/10.1063/1.4885426

51. J.D. Jackson, *Classical Electrodynamics* (Wiley, USA, 1998)

52. M. Esslinger, R. Vogelgesang, Reciprocity theory of apertureless scanning near-field optical microscopy with point-dipole probes (in English). ACS Nano **6**(9), 8173–8182 (2012). https://doi.org/10.1021/Nn302864d

53. S. Glasgow, M. Ware, J. Peatross, Poynting's theorem and luminal total energy transport in passive dielectric media (in English). Phys. Rev. E **64**(4) (2001). doi: Artn 046610 https://doi.org/10.1103/physreve.64.046610

Chapter 5
Toroidal Moments Probed by Electron Beams

Abstract Dipole selection rules underpin much of our understanding of the characterization of matter and its interaction with external radiation. However, there are several examples where these selection rules simply break down and a more sophisticated knowledge of matter becomes necessary. An example, which is becoming increasingly more fascinating, is macroscopic toroidization (density of toroidal dipoles), which is a direct consequence of retardation. In fact, unlike the classical family of electric and magnetic multipoles that are outcomes of the Taylor expansion of the electromagnetic potentials and sources, toroidal dipoles are obtained by the decomposition of moment tensors. This chapter aims to discuss the fundamental and practical aspects of toroidal multipolar moments in electrodynamics, from their emergence in the expansion set, the electromagnetic field associated with them, and the unique characteristics of their interaction with external radiation and other moments to the recent attempts to realize pronounced toroidal resonances in smart configurations of meta-molecules (Talebi et al. in Nanophotonics 7:93, 2018, [1]). In particular, we outline our work in designing oligomeric meta-molecules that purely support toroidal moments within a frequency range using the Babinet's principle and duality of electromagnetics theory. We further discuss the radiation and coupling of toroidal moments in individual and merged oligomeric systems and experimentally probe those moments using electron beams.

5.1 Toroidal Moments: Their Emergence, Symmetry, and Charge-Current Configurations

In 1958, Zeldovich reported the possibility of elementary particles under the breakdown of spatial parity interacting with an electromagnetic wave in a peculiar form considering the interaction energy $H^{\text{int}} = -\vec{S} \cdot \vec{\nabla} \times \vec{H}$, where $\vec{S}$ is the spin and $\vec{H}$ is the magnetic field [5]. Considering this problem in classical electrodynam-

© Springer Nature Switzerland AG 2019

N. Talebi, *Near-Field-Mediated Photon–Electron Interactions*,
Springer Series in Optical Sciences 228,
https://doi.org/10.1007/978-3-030-33816-9_5

ics, in 1967, Dubovik noted the possibility of introducing a new class of moments, being excluded from the family of electric and magnetic moments due to different time-space symmetries but appearing in similar orders in the expansion set to the magnetic moments [6], the so-called toroidal moments. From those early days, discussions of the toroidal moment have created an impetus in both solid-state physics and electrodynamics.

What is so interesting about toroidal moments? This question is partly entangled with the human curiosity to find new states of matter and ordering and partly related to technological applications. Toroidal ordering in solid states can open up possibilities for a new kind of magnetoelectric (ME) phenomenon [6], with applications in data storage and sensing. Whereas toroidal moments have been reviewed in [6] with a focus on static spin-based toroidal moments and the ME effect in condensed-matter physics, we will focus this chapter on different expansion sets leading to the toroidal moment and on electrodynamic toroidal moments in metamaterials and free space. [7] has provided a concise review of this aspect, but we intend to provide a more complete answer to this question and invoke new directions in light-matter interactions involving toroidal moments. For this purpose, we start with the theoretical investigation of the multipole expansion sets leading to toroidal moments and the time-space symmetry characteristics of moments.

A direct Taylor expansion of the charge and current densities proposed by Dubovik in 1990 already demonstrated the existence of a toroidal moment, described as $\vec{T} = (1/10c) \int \left(\vec{r}\,\vec{r} \cdot \vec{J}(\vec{r}) - 2r^2 \vec{J}(\vec{r}) \right) d^3r$, which originates from the currents flowing along the meridians of an infinitesimal sphere as $\vec{J}_T = \vec{\nabla} \times \left(\vec{\nabla} \times \left(\vec{T}\delta(\vec{r}) \right) \right)$. $\vec{J}(\vec{r})$ then continues as the inner flow by extending from one pole to the other pole of the sphere [6, 8] (see Fig. 5.1a). Moreover, a toroidal moment was shown to be a multipole order originating from the transverse part of the current density $\vec{J}_\perp(\vec{r}, t)$, as the longitudinal current $\left(\vec{J}_\parallel(\vec{r}, t) \right)$ is not independent of the charge density because of the charge conservation criterion $\vec{\nabla} \cdot \vec{J}_\parallel(\vec{r}, t) = -\partial\rho(\vec{r}, t) / \partial t$ and hence is related to the multipole expansion of the scalar potential. Here, the longitudinal current is

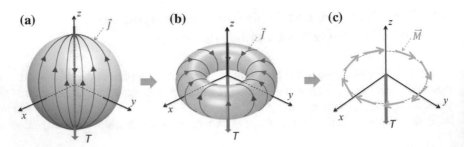

Fig. 5.1 Current density and magnetic moment distributions associated with the toroidal moment. Poloidal currents excited at the surface of **a** a sphere and **b** a toroid. **c** Equivalently, a ring of magnetic moments is also attributed to the excitation of a toroidal moment. Reprinted by permission from [1]

associated with the irrotational part of the current density vector field and points out from the charge density $\rho(\vec{r}, t)$ towards a direction normal to any surface enclosing the charge distribution, whereas $\vec{J}_\perp(\vec{r}, t)$ is parallel to such a surface. Obviously, a configuration of the current density distribution in the form of $\vec{J}_T$ is topologically similar to the poloidal currents flowing at the surface of a torus, hence leading to toroidal magnetic fields and the so-called polar toroidal moment. Interestingly, the poloidal current distribution leads to a diminishing of the magnetic quadrupole moment and hence an increase in the detectability of the toroidal moment [9]. As a simplification, a ring of either static [10, 11] or dynamic [12–15] magnetic moments is considered a configuration for the excitation of a polar toroidal moment or the so-called magnetic toroidal moment. A dual configuration to the polar toroidal moment is an axial toroidal moment (or electric toroidal moment), which is composed of a ring of electric dipolar configurations [16]. However, in contrast to a polar (magnetic) toroidal moment, the axial (electric) toroidal moment does not violate time and parity symmetry. In this paper, we mainly focus on the polar (magnetic) toroidal moment.

More often, the solutions to potentials rather than field components are considered. This is based on the fact that solutions of the Helmholtz equations, known as wave potentials, are well established in arbitrary coordinate systems. Moreover, it is possible to form a Helmholtz equation for potentials in complicated materials such as chiral and topological materials [17], which is not generally feasible for field components. In addition, these are the vector $\vec{A}(\vec{r}, t)$ and scalar $\varphi(\vec{r}, t)$ potentials that more often appear in the quantum mechanical Hamiltonian of the interaction of charged particles with light. However, the scalar and vector potentials are also not independent, and a gauge theory should be applied to pursue a physically relevant wave solution for electromagnetic fields [18]. In practice, considering each gauge, theory dictates a relation between the scalar and vector potentials, meaning that they are not independent. Hence, the multipoles obtained by expanding them are also not independent. Considering the Lorentz gauge as $\vec{\nabla} \cdot \vec{A}(\vec{r}, t) = -(1/c)\partial\varphi(\vec{r}, t)/\partial t$, the longitudinal part of the vector potential is also determined by the scalar potential. In other words, multipole expansion sets related to the scalar potential or the longitudinal part of the vector potential, along with the transverse part of the vector potential, form a complete basis of multipoles. Following Vrejoiu [19], the reduction of the multipole tensors in the form of symmetric traceless tensors demonstrates the existence of a toroidal moment, which has the same order as the magnetic quadrupole moment but complies to different symmetry rules. This is, however, in contrast to the static case, for which no toroidal moment is obtained even after a tedious reduction of the tensors according to the symmetry groups. Therefore, the family of toroidal moments is composed of hybrid moments, which only come into existence by expanding the transverse part of the vector potential. In this regard, the toroidal dipole and quadrupole tensors in Cartesian coordinates and in Gaussian units are obtained as:

$$T_i = \frac{1}{10c} \int \left[r_i \left(\vec{r} \cdot \vec{J}(r, t) \right) - 2r^2 J_i(r, t) \right] \mathrm{d}^3 r \tag{5.1}$$

$$T_{ik} = \frac{1}{42c} \int 4r_i r_k \left(\vec{r} \cdot \vec{J}(\vec{r}, t) \right) - 5r^2 (r_i J_k + r_k J_i) + 2r^2 \left(\vec{r} \cdot \vec{J}(\vec{r}, t) \delta_{ik} \right) d^3 r,$$

(5.2)

where $i, k \in (x, y, z)$ and δ_{ik} is the Kronecker delta function. It should be noted that an equivalent definition of the toroidal moment is often considered, namely, $\vec{T} = (1/6c) \int \vec{r} \times \left(\vec{r} \times \vec{J}(\vec{r}, t) \right) d^3 r$, which has an equivalent time-averaged quantity to the definition that was introduced in (5.1).

A direct expansion of fields may also be exploited, for which the outgoing-wave Green's function is expanded versus spherical harmonic multipoles [20]. By incorporating the Helmholtz decomposition of the current density function into the longitudinal and transverse toroidal and poloidal terms [21], the resulting electric field can be written with three components as $\vec{E}(\vec{r}, \omega) = -\vec{\nabla} e^L(\vec{r}, \omega) - \vec{r} \times \vec{\nabla} e^T(\vec{r}, \omega) - \vec{\nabla} \times \left(\vec{r} \times \vec{\nabla} e^P(\vec{r}, \omega) \right)$, where e^L, e^T, and e^P are the Debye scalar potentials associated with the longitudinal, toroidal, and poloidal currents. In this regard, toroidal currents produce toroidal electric and poloidal magnetic fields, and poloidal currents produce poloidal electric and toroidal magnetic fields.

In a different approach, Spaldin et al. [22] introduced the magnetization density $\vec{\mu}(\vec{r})$ in a solid-state system and its interaction energy with the external magnetic field as $H_{\text{int}} = -\int \vec{\mu}(\vec{r}) \cdot \vec{H}(\vec{r}) d^3 r$, where $\vec{\mu}(\vec{r})$ can have contributions from both the spin and orbital momenta. They further expanded the magnetic field in powers of the field gradient to obtain moment distributions in the form of (i) the net magnetic moment of the system $(\vec{m} = \int \vec{\mu}(\vec{r}) d^3 r)$, (ii) the pseudo-scalar quantity called the ME monopole $(a = (1/3) \int \vec{r} \cdot \vec{\mu}(\vec{r}) d^3 r)$, (iii) the toroidal moment vector dual to the antisymmetric part of the tensor $(\vec{t} = (1/2) \int \vec{r} \times \vec{\mu}(\vec{r}) d^3 r)$, and (iv) the quadruple magnetic moment of the system $(q_{ij} = (1/2) \int [r_i \mu_j + r_j \mu_i - (2/3) \delta_{ij} \vec{r} \cdot \vec{\mu}(\vec{r})] d^3 r)$, to name only the first few terms in the expansion set. The interaction energy can be further expanded as

$$H_{\text{int}} = -\vec{m} \cdot \vec{H}(\vec{r} = 0) - a \left(\vec{\nabla} \cdot \vec{H} \right)_{\vec{r}=0} - \vec{t} \cdot \left(\vec{\nabla} \times H \right)_{\vec{r}=0} - \sum_{ij} q_{ij} (\partial_i H_j + \partial_j H_i)_{\vec{r}=0},$$

where i, j are Cartesian directions and δ_{ij} is the Kronecker delta function.

In a Maxwell-Lorentz system of equations in which the only sources are charges and currents related to each other as $\vec{J} = \sum e_i \vec{v}_i \delta(\vec{r} - \vec{r}_i)$, the electric, magnetic, and toroidal moments form a complete system of moments, and there is no place for further generalizations. However, for the sake of completeness and symmetry, the consideration of magnetic charges is recommended as well [23] as $\vec{\nabla} \cdot \vec{B}(\vec{r}, t) = \rho_g(\vec{r}, t)$ and $\vec{\nabla} \times \vec{E}(\vec{r}, t) = -\left(\partial \vec{B}(\vec{r}, t) / \partial t \right) - \vec{J}_g(\vec{r}, t)$, where $\rho_g(\vec{r}, t)$ and $\vec{J}_g(\vec{r}, t)$ are the magnetic charge and magnetic current density distributions and $\vec{E}(\vec{r}, t)$ and $\vec{B}(\vec{r}, t)$ are the electric and magnetic flux components, respectively, at time t and a position shown by the displacement vector $\vec{r}$. Moreover, in contrast to the electric current distribution, which is a vector, the magnetic current distribution is an axial vector (pseudovector). Generalizing the multipole expansion schemes for including the magnetic sources, a new class of poloidal magnetic currents are unravelled that

Table 5.1 Transformation properties of electric (P), magnetic (M), polar toroidal (T), and axial toroidal (G) moments under the space inversion and time reversal operations

Multipole	Parity	
	Space	Time
P	−	+
M	+	−
T	−	−
G	+	+

give rise to the axial toroidal moments, which is an exact dual to the polar toroidal moment discussed previously. In analogy to the work carried out in [8], the axial toroidal moments are denoted here as G. Under space-time parity operations (space inversion and time reversal), the whole class of multipoles behaves as depicted in Table 5.1. The whole set of symmetry rules should be applicable to the vector fields acting on the multipoles, as the free energy of the system should be invariant upon space-time inversions. In this regard, the free energy of the interaction of an arbitrary system with the electromagnetic field is given by $\vec{H}_{int} = -\vec{P} \cdot \vec{E} - \vec{T} \cdot \left(\partial \vec{D} \big/ \partial t \right) - \vec{M} \cdot \vec{H} - \vec{G} \cdot \left(\partial \vec{B} \big/ \partial t \right)$ [24]. Moreover, Spaldin et al. argued that the cross-product quantity $\vec{P} \times \vec{M}$ sustains similar space-time symmetries to the toroidal moment, but it is not in fact a toroidal moment, as it interacts with the electromagnetic field according to the free-energy term $\left(\vec{P} \times \vec{M} \right) \cdot \left(\vec{E} \times \vec{H} \right)$. We only mention here that the local distribution of the vector $\vec{r} \times \vec{M}(\vec{r}) = \vec{r} \times \left(\vec{r} \times \vec{J}(\vec{r}) \right)$ might be described as the distribution of the toroidal moment when only the time-averaged quantity of this vector is anticipated and compared to $\left(\vec{r}\vec{r} \cdot \vec{J}(\vec{r}) - 2r^2 \vec{J}(\vec{r}) \right)$, as described in Sect. 5.3. Moreover, as will be discussed later in this review, a linear ME effect might occur in a system with a net toroidal moment, which means that the free-energy functional contains the contribution $\vec{T} \cdot \left(\vec{E} \times \vec{H} \right)$ [6].

Although parity considerations in space-time are considered to provide a complete class of multipoles, a more general picture beyond the quasi-static limits is to be considered when the retardation effects are taken into account [25], in which the Taylor expansion versus the wave number (k) is included as well. As this expansion is always ruled out versus $k^2 = (\omega/c)^2$ in free space, there is no room to discuss the momentum-frequency symmetries in classical electrodynamics. However, considering the induced currents in materials, anisotropic media, and material loss, it is well known that time-reversal symmetries might be violated in special cases [26]. In this regard, a complete picture including the k-symmetry is to be developed.

To conclude this section, we mention that despite several experimental realizations of toroidal moments in both solid-state systems and metamaterial systems, there are still debates about an individual class of toroidal moments to be excluded from the magnetic quadrupole terms. An interesting study in this regard is presented in [27], where the author has considered multipole expansions at both the potential and field levels but also in both Cartesian and spherical coordinates.

5.2   Dynamic Toroidal Moments in Artificial Metamolecules and Dielectric Nanostructures

In addition to the static toroidal moments in various condensed materials, engineered toroidal moments using the concepts in plasmonics and metamaterials [28] have attracted growing interest, as such moments can be excited by and interact with electromagnetic fields more efficiently than those in condensed matter, T_M [7]. This can help us understand the basis of T_M interacting with light and modifying the optical properties of materials.

However, the toroidal dipole response in electrodynamics is often masked by more dominant electric and magnetic multipoles at similar energies. Therefore, artificial toroidal metamaterials are initially designed to amplify T_M and suppress the competing electric and magnetic multipoles. This interesting field was established by mimicking toroidal coils [29, 30] at a microwave frequency to explore the toroidal dipole response in a great variety of structures and by entering the optical regime by scaling structures down to the nanoscale [31]. Figure 5.2 gives a brief but incomplete overview of the different investigated toroidal metamolecules in the catalogues of split-ring resonators and their variants, magnetic resonators, apertures, plasmonic cavities and structures and dielectric nanostructures. It should be mentioned that the classification of metamaterials is not restricted to a single catalogue; for example, metallic double disks can be considered as either magnetic resonators [32, 33] or plasmonic cavities [34]. There are also other novel designs, for example, a vertically assembled dumbbell-shaped aperture and split-ring resonator resulting in a horizontal toroidal response in the optical region [35, 36].

Very interestingly, the far-field radiation patterns of T_M are virtually identical to those of electric and magnetic moments of the same order, even though these moments are fundamentally different [7, 13] (see Sect. 5.4). In the past few years, the research focus has been gradually transitioning from suppressing electric and magnetic dipole excitation to making use of the destructive interference between the electric (magnetic) dipole and toroidal dipole to achieve non-radiation in the far-field region, so-called *anapole excitation* [53], *scattering transparency* [50] or, analogously, *electromagnetically induced transparency* [38]. Such an intriguing design has promising potential applications in designing low-loss, high-quality factor cavities for sensing, lasers, qubits, and non-scattering objects for cloaking behaviour.

In parallel to dealing with radiation losses in metamaterials, a dielectric metamaterial was proposed to overcome the dissipation loss of metals that is encountered in metal-based toroidal metamaterials [48]. It has been pointed out that the dissipation loss in metals originating from the ohmic resistance can hinder the excitation of toroidal multipoles, especially in the optical regime, resulting in their weak coupling to the external fields.

A fantastic characteristic of toroidal metamaterials as mentioned above is the feasibility of tuning toroidal responses via the size, shape, material [50] and spatial arrangement/symmetry [14] of the constitutive elements. To date, major toroidal metamolecules have been designed to achieve pronounced magnetic toroidal dipole

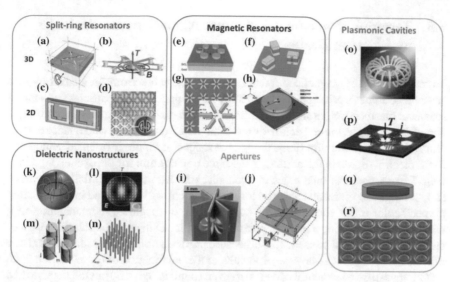

Fig. 5.2 Dynamic toroidal moments realized using metallic and dielectric meta-molecules. Split-ring resonators: (a) Schematic drawing of 3D SRR constituted by four rectangular metallic wire loops embedded in a dielectric slab [37], (b) A combined SRR sharing a central connecting bridge [38], (c) An asymmetric SRR-based planar toroidal metamaterial [39], (d) Planar conductive metamaterials formed by two symmetrical split rings [40]; Magnetic resonators: (e) an optical toroidal structure composed of a gold hexamer and metallic mirror separated by a dielectric layer [41], (f) Three magnetic resonators consisting of two metallic rods and a dielectric spacer [32], (g) An infrared toroid metamaterial composed of asymmetric double bars [42], (h) A THz flat-ring-dimer (metallic double disks) toroidal metamaterial [43]; Apertures: (i) Toroidal metamaterial arrays consisting of dumbbell-shaped apertures manifesting destructive interference between electric and toroidal dipole moments, leading to scattering transparency [44], (j) An electric toroidal dipolar response has been achieved by a metamaterial based on a sun-like aperture element at a microwave frequency [45]; Dielectric nanostructures: (k) Dielectric nanoparticles [46], (l) Dielectric nanodiscs with an illustration of the toroidal electric field distribution [47], (m) Dielectric cylinders [48], (n) Dielectric nanotubes [49]; Plasmonic cavities: (o) Core–shell nanoparticles support toroidal dipole excitation by a plane wave [50], (p) Plasmonic oligomer nanocavities with 7 nanoholes in metallic films sustain toroidal responses at visible wavelengths [4], (q) Toroidal modes are sustainable in the infrared and visible regime by a sidewall-coated plasmonic nanodisc antenna [51], (r) A circular V-groove array supports the plasmon toroidal mode at optical frequencies [52]; Figures reprinted with permission from the following: (a) [37], APS; (b) [38], WILEY-VCH Verlag GmbH & Co. KGaA, Weinheim; (c) [39], APS; (d) [40], APS; (e) [41], NPG; (f) [32], by courtesy of Jing Chen; (g) [42], AIP; (h) [43], Elsevier B.V.; (i) [44], NPG; (j) [45], AIP; (k) [46], OSA; (l) [47], NPG; (m) [48], APS; (n) [49], OSA; (o) [50], WILEY-VCH Verlag GmbH & Co. KGaA, Weinheim; (p) [4], AIP; (q) [51], ACS; (r) [52], OSA

responses due to their peculiar property of asymmetry in both spatial inversion and time reversal. In this case, introducing space-inversion asymmetry from either the geometry or excitation source is necessary for toroidal moment excitation, while breaking the time-reversal symmetry has already been intrinsically fulfilled by the magnetic dipole. These artificial toroidal structures open an avenue to study the interaction with electromagnetic radiation in both the far field and the near field.

5.3 Toroidal Moments in Oligomer Nano-Cavities Probed by Electron Beams

To realize a loop of magnetic moments, we use another electromagnetic theory—the so-called Babinet's principle. It is well known that metallic nanoparticles can support localized plasmon polaritons in the form of electric dipole resonances. According to Babinet's principle, one expects to observe magnetic moments in void nanostructures embedded in metallic thin films, though the magnetic dipole is oriented orthogonal to the electric dipole in the particle at the same polarization for the excited light [15] (Fig. 5.3). A chain of nanoholes in a metallic thin film thus potentially supports a ring of magnetic moments; hence, a toroidal moment will be excited in the direction perpendicular to the thin film. This configuration of magnetic moments, however, cannot be excited using a linearly polarized light; instead, radially polarized light or swift electrons should be used. This is because the orientation of the induced magnetic dipoles in the holes is always perpendicular to the polarization of the excitation field.

Our investigated structure consists of seven round holes of 60 nm diameter drilled in a free-standing 60 nm thick silver film [54]. We probe the optical modes of this nanostructure using EELS and EFTEM techniques. A fast-moving electron sustains a magnetic field with field lines circulating around the electron in the plane normal to the electron velocity. This magnetic field of the electron will induce magnetic moments in the holes, as shown in Fig. 5.4a. The heptamer nanocavity hence supports a number of optical excitations, attributed to the induced electric and magnetic dipoles and quadrupoles, and toroidal moments (Fig. 5.4b). At $E = 3.7$ eV, the toroidal moment is the dominant response. At this energy, the induced electric field is normal to the plane of the supporting film (Fig. 5.4c). The z-component of the induced magnetic field is concentrated within the holes in a spatial configuration showing the excitation of a ring of magnetic moments, as expected for the toroidal moment. In

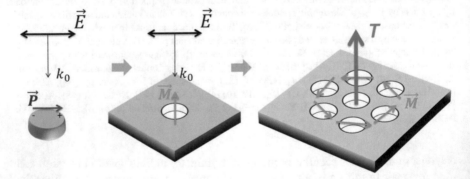

Fig. 5.3 A practical scheme for the realization of a toroidal moment using Babinet's principle. A metallic nanoparticle sustains an electric dipole moment when excited by a linearly polarized optical wave. A nanohole of the same size supports a magnetic moment at the same excitation condition. Thus, a chain of holes with cyclic symmetry supports a ring of magnetic moments and consequently a toroidal moment oriented perpendicular to the film

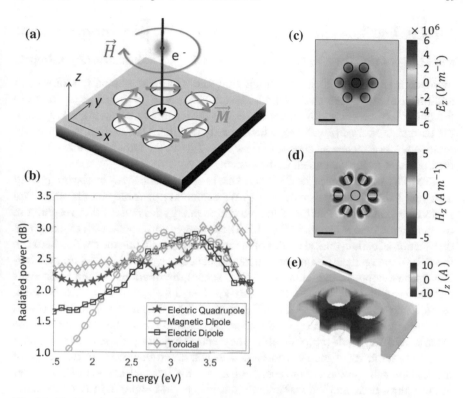

Fig. 5.4 Exciting toroidal moments in a heptamer nanocavity using swift electrons (Simulations).
a The magnetic field of a moving electron interacting with the holes generates a ring of induced
magnetic moments. **b** The radiated power from each individual class of moments, the z-components
of the induced **c** electric and **d** magnetic moments, and **e** the z-component of the induced current
density in the structure at an energy of $E = 4.7$ eV. The scale bars are 100 nm

this way, $\vec{T} = (1/6c) \int \vec{r} \times (\vec{r} \times \vec{J}(\vec{r}, t)) \, d^3r$ is also oriented normal to the plane,
along the z-axis. The free energy, including the interaction of the toroidal moment
with the evanescent field of the incoming electron, given by $H^{\text{int}} = -\vec{t} \cdot (\vec{\nabla} \times \vec{H})$,
thus exhibits a maximum, as does the power radiated to the far field. This radiation
power in total, including all the induced moments, is calculated as

$$I = \frac{2\omega^4}{3c^3}\left|\vec{P}\right|^2 + \frac{2\omega^4}{3c^3}\left|\vec{M}\right|^2 + \frac{2\omega^6}{3c^5}\left|\vec{T}\right|^2 + \frac{\omega^6}{5c^5}\sum_{\alpha,\,\beta \in x,\,y,\,z}\left|EQ_{\alpha\beta}\right|^2$$

$$+ \frac{\omega^6}{40c^5}\sum_{\alpha,\,\beta \in x,\,y,\,z}\left|MQ_{\alpha\beta}\right|^2 \tag{5.3}$$

where $\vec{P} = (i\omega)^{-1} \int d^3r \, \vec{J}$ is the electric dipole, $\vec{M} = (2c)^{-1} \int d^3r \, \vec{r} \times \vec{J}$ is the
magnetic dipole, $\vec{T} = (10c)^{-1} \int d^3r \left[(\vec{r} \times \vec{J})\vec{r} - 2r^2\vec{J}\right]$ is the toroidal moment,

$EQ_{\alpha\beta} = (2i\omega)^{-1} \int d^3r \left[r_\alpha J_\beta + r_\beta J_\alpha - (2/3)\left(\vec{r}\cdot\vec{J}\right)\delta_{\alpha\beta}\right]$ is the electric quadrupole

and $MQ_{\alpha\beta} = (3c)^{-1} \int d^3r \left[\left(\vec{r}\times\vec{J}\right)_\alpha r_\beta + \left(\vec{r}\times\vec{J}\right)_\beta r_\alpha\right]$ is the magnetic quadrupole
moment. It is obvious, then, that in addition to the toroidal moments, a number of
additional moments are also excited. Nevertheless, at $E = 3.7$ eV, very close to
the bulk plasmon energy for silver ($E = 3.8$ eV for a bulk plasmon), a longitudinal current is excited (Fig. 5.4e), which greatly facilitates the excitation of toroidal
moments. Nevertheless, at energies below $E = 2.4$ eV, the toroidal moment is again
the dominant induced moment in the system.

The EFTEM data-cube (see Chap. 1) of the heptamer nanocavity has been captured
using the Zeiss SESAM microscope [4]. We exploited a peak map finding algorithm
to find the spatial distribution of the maximum energy-loss intensities, resulting in
the grey-scale figures shown in Fig. 5.5. These figures correspond well to the spatial
distribution of certain optical modes of the system by eliminating the overlap between
the tails of energetically adjacent optical resonances, as probed by electron energy-
loss spectroscopy. In particular, 4 distinct spatial distributions for the peak maps
are observed at $E = 2.4$ eV, $E = 2.8$ eV, $E = 3.4$ eV, and $E = 3.6$ eV. Peak
maps at the energies of $E = 2.4$ eV and $E = 3.6$ eV have maximum intensities
at the centre of the heptamer. As the acquired electron energy loss is related to the
electric field amplitude projected along the electron trajectory, these two modes both
have z-components of the electric field with maximum intensities at the centre of
the heptamer. In contrast, at the energy of $E = 2.8$ eV, the EELS intensity vanishes
at the central hole, and the maximum intensities appear along the rim of the outer
holes, distributed symmetrically at two locations around the rim of each hole in the
azimuthal direction. At $E = 3.4$ eV, the maximum intensity again appears around
the rim of the outer holes; however, in contrast to the previous case, the peak intensity
is located at the maximum radial distance with respect to the centre of the structure.

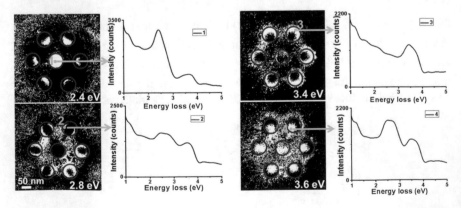

Fig. 5.5 Probing optical modes of a heptamer nanocavity using electron beams (Experiments).
Peak maps extracted from EFTEM series for the heptamer nanocavity at the depicted energies and
demonstrated energy-loss spectra at selected electron impacts. Figure reproduced with permission
from [4]

To investigate the role of symmetry in the excitation of the abovementioned EFTEM series, a modal decomposition technique is used to selectively excite all the possible optical modes of the system (Fig. 5.6). Optical excitations can easily be exploited for this purpose thanks to the polarization states of free-space optical waves. We use three kinds of polarization states within our numerical experiments, namely, dipoles and azimuthally polarized and radially polarized optical waves. An electric dipole positioned in the central hole can selectively excite toroidal moments due to the similar radiation pattern of the toroidal moments to the electric dipoles [7]. Using dipole excitations again, exactly two different toroidal moments are excited, which are spectrally located at $E = 2.5$ eV and $E = 3.7$ eV. The spatial distributions of the field components for both toroidal modes are very similar, with the only dissimilarity occurring in the spatial expansion of the induced current densities at the gaps between the holes and thereby the corresponding localization of the field distributions. When exciting the system with an azimuthally polarized light, a ring of electric dipoles is excited at the gaps between the outer holes, resulting in a dominant magnetic-dipolar excitation in the whole system. In contrast, when a radially polarized light is exploited, the electric dipoles are located transverse to the outer holes. In this way, the electric dipole response of the system becomes dominant at the energy of $E = 3.5$ eV (with an $E = 0.1$ eV energy shift compared to the experimental and simulated results when electron probes are used).

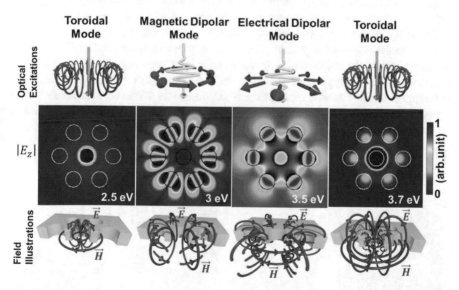

Fig. 5.6 Optical modes in a heptamer nanocavity [54]. The optical modes range from toroidal moments ($E = 2.5$ eV) to magnetic ($E = 3.0$ eV) and electric ($E = 3.5$ eV) dipoles to again a toroidal moment at $E = 3.7$ eV. Upper row: optical excitation used for decomposing the optical excitations with respect to their symmetries. Middle row: magnitude of the induced electric field calculated at the plane positioned 5 nm above the silver thin film. Lower row: 3D visualization of the vectorial field components for each excited mode

The central hole has been positioned for the purpose of facilitating a pure inelastic interaction of the electron with the optical modes by letting the electron pass through the hole and not the thin film. Nevertheless, the relativistic electrons have enough energy to pass through the 100 nm thick silver film and still allowing for a precise EELS measurement. For this purpose, it is illustrative to compare the heptamer structure with that of a hexamer, in which the central hole is not present. Figure 5.7 shows the peak maps obtained from the acquired EFTEM series for a hexamer structure composed of holes with a diameter of 100 nm, slightly larger than those in the previously investigated heptamer structure. For an electron traversing the structure at its centre, two prominent peaks are observed at energies of $E = 1.8\,\mathrm{eV}$ and $E = 3.8\,\mathrm{eV}$. The former is due to the excitation of the toroidal moment caused by the induced ring of the magnetic moments in the holes, similar to the heptamer structure.

However, the higher-order toroidal moment is completely masked by the excitation of the bulk plasmons, unlike the heptamer structure for which the bulk plasmon was not excited for the electrons passing through the hole. However, a higher-order mode can be excited at the energy of $E = 3.4\,\mathrm{eV}$ when the electron interacts with the structure at the inner rim of the outer holes (Fig. 5.7, lower left panel), and a second peak occurs at the energy $E = 2.8\,\mathrm{eV}$.

This peak is also observed for the electron impact position marked by #3. The spatial distribution of this mode, shown by the peak map image at the same energy (Fig. 5.7, lower right panel), demonstrates the excitation of the maximum energy-loss intensities at two points positioned along the rim, radially displaced with respect to the positions of the holes. Interestingly, at $E = 2.8\,\mathrm{eV}$, the maximum intensities of the electron energy-loss signal happen at the gap between the holes rather than at the inner rim of the holes, which is understood by comparing Fig. 5.7 to Fig. 5.5. Obviously, a highly symmetric structure such as a hexamer or heptamer nanocavity can support various numbers of optical excitations, which can be spectroscopically investigated by a technique such as EELS, hoping that the spectral overlaps between

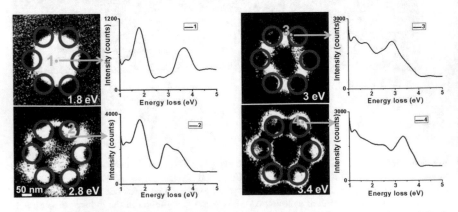

Fig. 5.7 Probing optical modes of a hexamer nanocavity using electron beams (Experiments). Peak maps extracted from the EFTEM series for the hexamer nanocavity at the depicted energies and demonstrated energy-loss spectra at selected electron impacts. Figure reproduced with permission from [4]; © Springer-Verlag Berlin Heidelberg 2014

the modes are not that critical and allowing the resolution of the full photonic states. Nevertheless, as EELS is blind to the phase of the optical excitations, the symmetry of the induced charges and moments is not easily captured—although it might be possible with some imagination. On the other hand, combining the EELS technique with inelastic electron holography [55] or spectral interferometry [56], allowing the spatial and spectroscopic resolution of the phase of the optical excitations, can be a tremendous step for advancing electron microscopy to gain better insight into the physics of observations. Another interesting approach might be to shape the electron beams according to the symmetry of the optical modes under investigation [57].

The symmetry groups of the optical excitations can be investigated, though numerically, by decomposing the optical modes using the polarization state of the light, as stated above. In Fig. 5.8, the z-components of the electric fields associated with the optical modes of both heptamer and hexamer structures are demonstrated and compared by sketching the distributions of the induced charges and electric dipoles. Here, the z-axis is parallel to the electron trajectory and perpendicular to the plane, so the EELS signal could be related to the z-component of the electric field if only each individual mode were excited (without spectral overlaps). Interestingly, for the hexamer structure, two different optical excitations, with the induced charges placed in the gap between the holes but with the negative and positive charges being placed close to the rim of the holes or in the centre of the gaps, are spectrally quite close

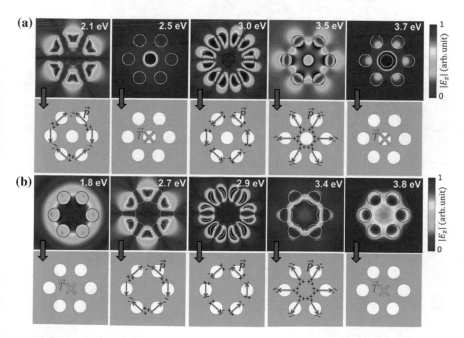

Fig. 5.8 Optical excitations in oligomeric nanocavities [4] . Demonstrated are the z-components of the electric field for the **a** heptamer and **b** hexamer nanostructures. The second row in each panel shows the distributions of induced charges and moments. Modified and reprinted by permission from [4]; © Springer-Verlag Berlin Heidelberg 2014

to each other, with their eigen-energies being equal to $E = 2.9\,\text{eV}$ and $E = 2.7\,\text{eV}$. Being spectrally different within only a 0.2 eV energy range, resolving these excitations using the EFTEM technique is critical and results in a picture formed by a superposition of the two states, as shown in Fig. 5.7 (lower left and upper right panels). The optical mode with the eigen-energy of $E = 2.9\,\text{eV}$ for the hexamer and $E = 3.0\,\text{eV}$ for the heptamer structure is formed by a ring of electric dipoles imposing a cyclic symmetry. However, the optical mode at the energy of $E = 2.7\,\text{eV}$ for the hexamer and $E = 2.1\,\text{eV}$ for the heptamer is composed of electric dipoles that are also situated along the ring supporting the outer holes but with opposite sign. This mode strongly depends on the size of the holes and the topology of the structure. If the holes were maintained with a similar size for both the heptamer and hexamer structures, there would be a negligible difference in the energy of this mode for both structures (see Fig. 5.9). This mode is extremely robust to changes in the topology as well, where removing one of the holes from the outer rim and making the structure completely asymmetrical will result in only a slight redshift of 0.2 eV in magnitude (Fig. 5.9c).

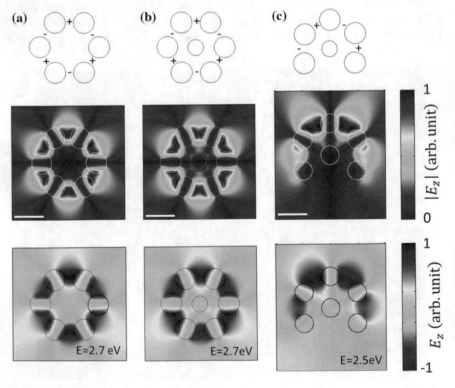

Fig. 5.9 The effect of symmetry and topology on the excitation of the radially polarized flipping optical mode. Demonstrated is the magnitude (second row) and amplitude (third row) of the z-component of the electric field for **a** cyclic symmetric hexamer, **b** cyclic symmetric heptamer, and **c** asymmetric hexamer structures

5.4   Radiation of Dynamic Toroidal Moments

Originally, plasmonic dark modes gained their name for their poor coupling to far-field plane waves under specific excitation or detection schemes [58]. Dark modes usually possess a vanishing or zero net-dipole moment [59], like the anti-symmetry mode in plasmonic dimers/trimers [60], the breathing mode in discs/triangles [61], and quadrupolar modes [62, 63]. However, dark modes can be switched to bright modes with a greater coupling to irradiation by tuning the excitation angle of the incident light [64] or using fast electrons [65]. Retardation can also cause the dark mode to radiate, while increasing the effective size of the sustained structures [66]. Therefore, the term *dark mode* cannot be quantitatively linked to nanoscale non-static charge configurations.

Toroidal moments have been addressed as dark modes or sub-radiant modes mainly by reference to highly radiative electric or magnetic dipolar modes in the context of constructing Fano resonances [67–70]. However, the name *dark mode* implies difficulties in coupling with light in both excitation and radiation processes. Certainly, for the excitation process, a normal optical planar wave cannot excite significant toroidal moments in most cases due to its complex and confined charge–current configuration. Toroidal moment s are then dark in this sense, especially in contrast to other optically excited electric or magnetic dipolar modes. On the other hand, dynamic toroidal dipoles are intrinsically radiative, and their scattering fields can be as strong as that of an electric dipole [47, 71]. In the literature, there exists confusion about the properties of dynamic toroidal dipoles, partially from a conception interchange with static toroidal dipoles and partially due to conflation with plasmonic dark modes. Clarification of this issue is urgently needed, as the research interest in dynamic toroidal moments is apparently increasing, even in the aspect of practical applications [72–74]. It is worth noting that the radiation properties of dynamic toroidal dipoles have been deduced from the results of a number of experiments [47, 75]. Here, we investigate the radiation characteristics of the toroidal moments excited in the heptamer structure discussed above.

As the first step of our twofold investigation approach, toroidal excitations in plasmonic heptamer nanocavities are experimentally inspected by electron energy-loss spectroscopy/scanning transmission electron microscopy (STEM-EELS) and energy-filtered transmission electron microscopy (EFTEM). Then, the corresponding far-field radiation behaviours of the toroidal excitations are investigated by means of cathodoluminescence (CL) spectroscopy. The incident electrons have an electromagnetic near field that polarizes plasmonic structures and excites plasmonic modes with conservation of energy and momentum. For EELS and EFTEM, the scattering fields of the excited modes in turn recoil upon the incident electrons along the trajectory, leading to energy losses [76]. For CL, the excited modes may couple to the far-field radiation and then be detected. EELS, EFTEM, and CL have been widely applied for characterizing plasmonic systems on the micro- and nanometric scales [77, 2]. EFTEM and EELS measurements are sensitive only to the associated electric fields of toroidal excitations projected along the electron trajectory. To confirm the toroidal excitations in our case, the characteristic vortex-like magnetic fields are revealed in parallel by finite-difference time-domain (FDTD) simulations. Interpretations of the experimental results are further supported by numerical calculations.

Our plasmonic heptamer cavity structure was patterned on a free-standing silver thin film with a focused ion beam system. As shown in Fig. 5.10a, the fabricated heptamer structure has a hole diameter of 80 ± 10 nm, and the thickness of the silver thin film is approximately 30 ± 15 nm, much thinner than the previously investigated heptamer. The incident electron beam direction is perpendicular to the structure, e.g., the thin film's surface, and parallel to the z axis. First, the single toroidal dipole moment is easily recognized in the energy-loss spectra by its characteristic excitation location. According to its field distribution, as described above, this mode has an electric field that is highly concentrated at the central hole and the nearby silver bridges [54], where exactly the corresponding energy-loss signal should be present (Fig. 5.10, vertical dashed red line, named T1 mode). The extracted experimental EEL spectra at the positions depicted by the coloured spots between the central and upper holes are displayed in Fig. 5.10c, left column. This finding shows that the T1 mode is excited at both the central hole and the neighbouring silver bridges. Furthermore, it has a broadband feature with a maximum centred at approximately 2.12 eV. On the other hand, there is another pronounced resonance at 2.58 eV, with the energy concentrated just at the silver bridges (Fig. 5.10, vertical dashed black line, named T2 mode). Contrary to the T1 mode, this mode is absent at the central hole but is excited at the silver bridges and the rims of the holes (Fig. 5.10c). Certainly, other cavity modes (radially and longitudinally polarized along the holes) can also be excited in this structure, as discussed in the previous chapter. However, here we only focus on the toroidal moments. The zig-zag curve above 2.6 eV is a camera artefact due to the afterglow of the zero-loss-peak (ZLP) but will not affect the data interpretation. The fact that the T2 mode could not be observed previously is attributed to the technique utilized (EFTEM instead of EELS) and the thickness of the silver film, as the 100 nm thickness makes the detection of optical excitations with their hot spots located inside the silver film a non-trivial task.

Figure 5.10b shows the corresponding simulated EEL spectra via the numerical FDTD calculations. The individual spectra extracted between the central and upper holes are shown in Fig. 5.10c, right column, with the same colour code used for the experimental spectra. Two pronounced resonances are observed at 2.1 and 2.5 eV (vertical dashed red and black lines). They show good agreement with the experimental T1 and T2 modes despite a slight energy shift of approximately 0.1 eV. The electromagnetic field distribution at 2.1 eV confirms the excitation of a single toroidal dipole moment with the clockwise rotation of the magnetic dipoles in the outer 6 nanoholes (Fig. 5.11a and the inset on the left) and a high electric field concentration at the central hole and the nearby silver bridges (Fig. 5.11b, left). Very interestingly, the resonance at 2.5 eV demonstrates an antiparallel pair of toroidal dipoles (Fig. 5.11a, right). The instant magnetic field distribution at the time of $P/8$, where P is the temporal duration of the optical cycle, shows a clockwise dipole loop in the upper 4 nanoholes (including the central hole) as well as a counter-clockwise dipole loop in the lower 4 nanoholes (including the central hole), indicating an antiparallel

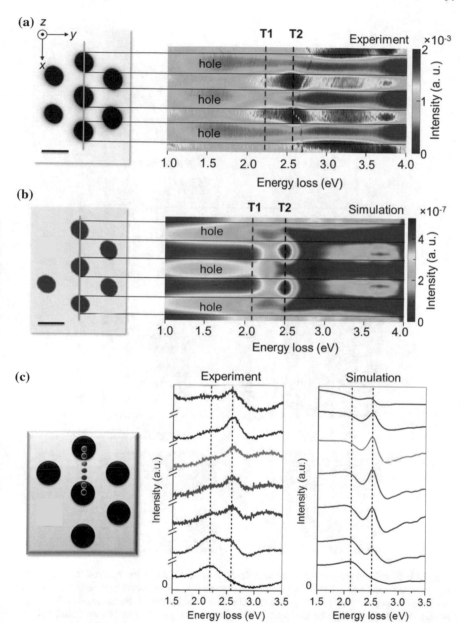

Fig. 5.10 EELS line scan for a single heptamer nanostructure containing 7 holes embedded in a 30-nm-thick silver film. **a** High-angle annular dark-field (HAADF) image and zero-loss-peak (ZLP)-normalized EEL spectra recorded along the green line of a fabricated heptamer nanocavity. **b** Image of the simulated heptamer cavity (left) and the corresponding EEL spectra along the cavity axis (green line). Scale bars are 100 nm. Vertical dashed red and black lines indicate the toroidal modes T1 and T2 at **a** 2.12 eV and 2.51 eV and **b** 2.2 eV and 2.58 eV, respectively. **c** Experimental (left) and simulated (right) EEL spectra of the investigated plasmonic heptamer cavity along the symmetric axis from the central to upper holes, as depicted by the coloured spots in the inset. Reprinted by permission from [78] under CC-BY license

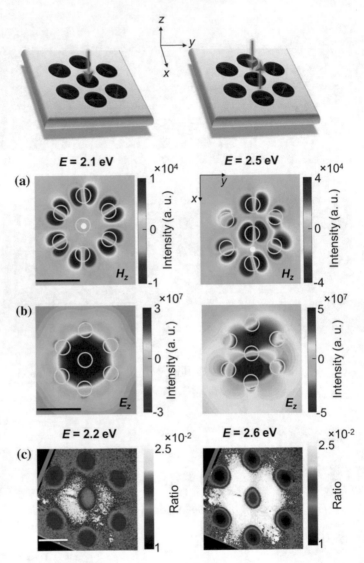

Fig. 5.11 Simulated magnetic and electric fields, H_z and E_z, along the electron trajectory of the heptamer structure **a** at 2.1 eV without a time delay and **b** at 2.5 eV with a time delay of $\pi/4$. Insets above are corresponding schematic illustrations. Grey circles denote the nanoholes. The impact locations of the electron probe are indicated by white dots. Red and green arrows denote magnetic and toroidal dipoles, respectively. **c** ZLP-normalized EFTEM images of the heptamer nanocavity at energy losses of 2.2 ± 0.1 and 1.7 ± 0.1 eV, respectively. The black triangular areas at the upper and lower left corners are beyond the acquisition area of the CCD camera. Scale bars are 200 nm. Reprinted by permission from [78] under CC-BY license

pair of toroidal dipoles perpendicular to the x–y plane (right inset above Fig. 5.11a). These two toroidal dipoles appear sequentially in the time domain with a time lag of $P/4$. As shown on the right side of Fig. 5.11b, the corresponding electric field distribution reveals that the energy of this mode is mainly concentrated between the holes (silver bridges) along the symmetry axis. The conclusion of the E_z field calculations at 2.1 and 2.5 eV is consistent with the experimental observations that at the centre of the heptamer structure, the T1 mode exclusively is excited, while the T2 mode is not. Their clear difference at the central hole is also unambiguously captured by the ZLP-normalized EFTEM images (Fig. 5.11c). Distinguishing features of the T2 mode include its higher excitation energy and the lack of an EELS signal at the central hole, as was expected from the previously demonstrated EELS line scan (Fig. 5.10a and b). This can serve as a fingerprint to distinguish radiation signals from these two modes later in CL spectra.

With the above knowledge of the toroidal moments in the investigated plasmonic heptamer nanocavity, we first calculated the corresponding CL spectra on the same structure along the symmetry axis in order to correlate the simulated EELS results in a controlled way (Fig. 5.12a). The qualitative correlation between the simulated CL and EEL spectra on the same structure can be applied later to interpret the experimental CL data. In the following, we consistently use vertical dashed red and black lines for the notation of the T1 and T2 modes at their free-space wavelengths, respectively.

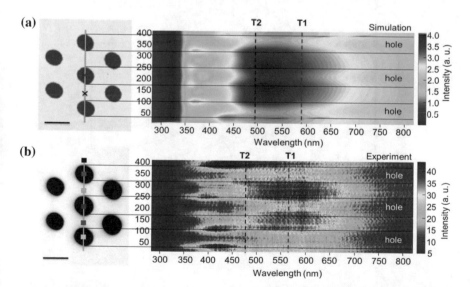

Fig. 5.12 CL response of the heptamer nanocavity. **a** Image of the simulated heptamer cavity (left) and the corresponding CL spectra along the cavity axis (green line). **b** HAADF image (left) and experimental CL spectra recorded along the green line of a fabricated heptamer nanocavity. Scale bars are 100 nm. Vertical dashed red and black lines mark the corresponding near-field resonances of the toroidal T1 and T2 modes at **a** 585 and 494 nm or **b** 595 and 480 nm, respectively. Reprinted by permission from [78] under CC-BY license

The main feature of the simulated CL spectra is a strong emission spanning from 500 to 650 nm present on the silver bridges, whereas this signal is weaker inside the central hole (Fig. 5.12a).

We compare the simulated CL and EEL spectra at the positions of the central hole and the neighbouring silver bridge, which are the representative excitation locations for the T1 and T2 modes, respectively (red and black crosses in Fig. 5.13). At the central hole, the T1 mode is exclusively excited according to the EEL spectrum (solid red curve in Fig. 5.13). The corresponding CL spectrum indicates a blueshift of the T1 mode to 510 nm with respect to its near-field resonance (blue arrow with red edge and the dotted red curve in Fig. 5.13). This is mainly ascribed to the different light collection geometries used in the experiment and simulations, rather than the electromagnetic dissipation of the excited modes [79]. In addition, the T1 mode displays a higher intensity in the silver bridges than in the central hole in both the CL (Fig. 5.12a and b) and EEL (Fig. 5.10a and b) spectra. This feature is likely attributed to the higher coupling efficiency of the induced polarized currents on the material rather than those in the void in order to form the toroidal moments.

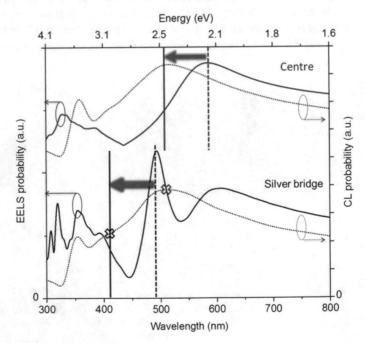

Fig. 5.13 Comparison between CL and EEL spectra. Simulated EELS and CL probabilities at the central hole and the centre of the silver bridge between the holes, as indicated by the red and black crosses in the left inset of Fig. 5.12a. The blue arrow with a red edge displays the blueshift of the T1 mode to 510 nm in the CL simulation (vertical solid red line). The blue arrow with a black edge indicates that the same degree of blueshift as experienced by the T1 mode is applied on the T2 mode to 419 nm in the CL simulation (vertical solid black line). Hollow red and black crosses show the calculated CL probabilities for the T1 and T2 modes, respectively. Reprinted by permission from [78] under CC-BY license

On the other hand, at the silver bridge, both the T1 and T2 modes can be excited (solid black curve in Fig. 5.13). However, only the peak of the T1 mode at 510 nm is observed (hollow red cross) in the simulated CL spectrum (dotted black curve). Assuming that the simulated CL signal of the T2 mode is subjected to the same amount of blueshift as the T1 mode, it is expected to see a peak at approximately 419 nm, as indicated by the vertical solid black line. However, no sharp peak but rather a very shallow rise is observed (dotted black curve). In fact, this shallow rise is also observed in the simulated CL spectrum at the centre of the central hole (dotted red curve), where the T2 mode is absent. Therefore, the shallow rise at 419 nm might be the signal of other cavity modes. Nevertheless, compared to the relative CL intensity of the T1 mode (hollow red cross), the CL probability at 419 nm is obviously lower (hollow black cross on the dotted black curve). This indicates a significant radiative behaviour of the T1 mode (a single dynamic toroidal dipole) but only a weakly radiating behaviour of the T2 mode (an antiparallel toroidal dipole pair).

The above observations obtained by combining the simulated EEL and CL spectra offer an important hint on how to interpret the corresponding experimental CL data.

Figure 5.12b shows the experimental CL spectra along the structural symmetry axis. Similar characteristics to those in the simulated CL spectra, such as a broad signal from 500 to 650 nm and a higher intensity at the silver bridges than in the central hole, appear. For further details, Fig. 5.14a and b shows the simulated CL

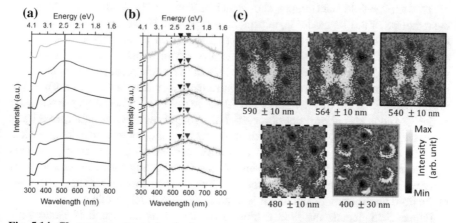

Fig. 5.14 CL spectra and map for the heptamer structure. **a** Simulated and **b** experimental CL spectra extracted from the 6 colour-marked locations in Fig. 5.12c. Each experimental spectrum is a sum over an area of 24 × 24 nm². Smoothed curves are superimposed on the raw data (light grey). The vertical solid red line marks the simulated radiation peak of the T1 mode at 510 nm. The vertical red and black dashed lines correspond to the T1 and T2 modes at 595 and 480 nm, respectively. The vertical orange and green solid lines label the wavelengths of the emission at 400 nm and the silver bulk plasmon at 330 nm. The inverse black and red triangles highlight the emission peaks at 540 and 590 nm. **d** CL chromatic maps showing the spatio-spectral dispersion of the emissions at 590 ± 10, 564 ± 10, 540 ± 10, 480 ± 10, and 400 ± 30 nm. Reprinted by permission from [78] under CC-BY license

spectra together with the experimental spectra extracted from similar locations along the structural symmetry axis (colour-coded locations in Fig. 5.14b). Despite their similarity, the measured emission of the single toroidal T1 mode has almost no spectral shift with respect to its EELS resonance at 2.2 eV (vertical red dashed line at 564 nm in Fig. 5.14b). This may imply a low damping rate of the single toroidal dipole mode in the investigated structure. However, there are two small peaks appearing at approximately 565 nm, highlighted by the inverse black and red triangles in Fig. 5.14b. To further investigate the possible difference between them, we examined the corresponding spectral-spatial distribution at these 3 resonances (590, 564 and 540 nm) by displaying their chromatic CL maps (first row in Fig. 5.14c). Interestingly, they all show the same distribution feature, resembling the electric field E_z of the T1 mode (Fig. 5.11a). This indicates the single excitation of the T1 mode and the broadband emission feature of the single dynamic toroidal dipole moment. As a speculation, these small peaks may be the intensity variation resulting from the far-field interference between multiple plasmonic modes. Regarding the T2 mode, at its free-space wavelength of 480 nm, the experimental CL spectra do not show a pronounced peak, especially at the characteristic excitation positions of the T2 mode (brown and green curves in Fig. 5.14b). The corresponding chromatic CL map (with the dashed black frame in Fig. 5.14c) also confirms no characteristic spatial excitation of the T2 mode in relation to its electric field E_z, as shown in Fig. 5.11b, right. This verifies the weakly radiating character of the T2 mode. Since these two toroidal dipoles excited at the T2 mode are not exactly out of phase (i.e., a phase shift of π), their far-field interference should not be completely destructive. Therefore, the radiation of the T2 modes is, in principle, expected. However, another important factor has to be taken into account in this case, which is retardation. Retardation causes radiation when the size of the structure becomes larger than the resonance wavelength. [66] In our case, the effective size of the single toroidal dipole at the T2 mode is approximately the radius of the entire heptamer structure (~175 nm), which is far smaller than the wavelength of the T2 mode (494 nm). Therefore, the radiation of the T2 mode was almost not observed, but stronger radiation is anticipated upon increasing the size of the heptamer structure. Furthermore, there is an evident peak at 400 nm (vertical orange line in Fig. 5.14b). The corresponding chromatic CL map is also present in Fig. 5.14c, with an orange frame. This clearly shows an excitation of a cavity mode that is neither the T1 nor the T2 mode. The corresponding simulations of the field distribution reveal that it is a radial electric dipole mode.

Figure 5.15 depicts the field evolution of the T2 mode over half a harmonic period of π, where the full harmonic period of the T2 mode takes 2π. In general, the magnetic field H_z (red curve) has a time shift of $\pi/2$ with respect to the electric field E_z (black curve) due to the dynamic nature of electromagnetism. At the beginning of the period (i.e., 0), a loop of magnetic dipoles is only observed in the lower 4 nanoholes (H_z field at the bottom in Fig. 5.15), and the corresponding electric field is mostly concentrated in the silver bridge between the lower 4 nanoholes. Hence, only one toroidal dipole is shown. After a time of $\pi/4$, an additional loop of magnetic dipoles appears in the upper 4 nanoholes, which assembles another toroidal dipole. Together with the toroidal dipole excited in the lower 4 holes, they now form an

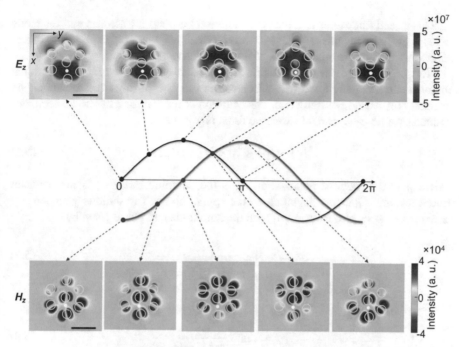

Fig. 5.15 Evolution of the simulated electric field E_z and magnetic field H_z of the T2 mode at 2.5 eV from 0 to π at π/4 intervals. Grey circles denote the nanoholes. The impact locations of the electron probe are indicated by the white dots. Red arrows show the directions of the magnetic dipoles. Scale bars are 200 nm. Reprinted by permission from [78] under CC-BY license

antiparallel pair of toroidal dipoles, as already presented in Fig. 5.11a at 2.5 eV. At time π/2, only the toroidal dipole in the upper 4 nanoholes is dominant (H_z field in Fig. 5.15). Therefore, these two constituent toroidal dipoles appear one after another in the time domain with a shift of π/2.

Indeed, electric and magnetic fields are synchronous in free space. This can be understood by simply applying the following Maxwell equation:

$$\vec{\nabla} \times \vec{H} = \frac{\partial D}{\partial t} \Rightarrow -i\vec{k} \times \vec{H} = i\omega\varepsilon_0 \vec{E} \Rightarrow \vec{E} = -\frac{1}{\omega\varepsilon_0}\vec{k} \times \vec{H} \tag{5.4}$$

where $\vec{k}$ is the wave vector and ε_0 is the free-space permittivity. Equation (5.1) tells us that the transverse components of the electric and magnetic fields are synchronous.

However, in this paper, we discuss the phase relation between electric and magnetic field components in matter. Assuming an electromagnetic field oscillating inside a bulk material, after some straight-forward algebra, we obtain the following relation between E_z and H_z:

$$H_z = -\frac{\omega\varepsilon_0\varepsilon_r k_x}{k_y k_z} E_z \tag{5.5}$$

This already indicates that the phase relation between the electric and magnetic fields is given by the phase of the permittivity as $\angle\varepsilon_r = \tan^{-1}\left(\varepsilon_r''/\varepsilon_r'\right)$, where ε_r' and ε_r'' are the real and imaginary parts of the complex-valued permittivity.

Next, we discuss the resonant conditions. The shape of the anti-symmetric toroidal moment can be nicely fitted with a Lorentzian function of the sort that is obtained normally for harmonic oscillators. The equation of motion for a harmonic oscillator excited by a time-dependent harmonic force is given by

$$\ddot{p} + \gamma\dot{p} + \omega_0^2 p = F_0 \cos(\omega t) \tag{5.6}$$

where p is the induced polarization, γ is the damping ratio, ω_0 is the resonant frequency, and F_0 is the magnitude of the applied force. The stationary response p is written as $p = p_0 \cos(\omega t + \varphi)$, with the amplitude and phase given by

$$p_0 = \frac{F_0}{\sqrt{\left(\omega_0^2 - \omega^2\right)^2 + \omega^2 \gamma^2}} \tag{5.7}$$

and

$$\varphi = \tan^{-1} \frac{\omega\gamma}{\omega_0^2 - \omega^2} \tag{5.8}$$

respectively. It is already seen that for $\omega = \omega_0$, we have $\varphi = \pi/2$. This means that the deriving phase at highly resonant conditions is given by $\varphi = \pi/2$, regardless of the damping of the oscillator.

To apply this analogy to our case, we briefly mention here that the toroidal moment is a hybrid solution and is not obtainable with a quasi-static analysis. This means that retardation plays an important role here. In this case, the force is applied by the magnetic-vortex circulating around the relativistic current distribution of the electron and this magnetic-field of the electron beam will excite the magnetic-dipole loop and hence the z-component of the toroidal moment excitation in our structure. In this regard, an exact shift of $\varphi = \pi/2$ will be obtained.

5.5   Coupling of the Dynamic Toroidal Moments

The coupling among toroidal moments is expected to form exotic new optical states, similar to the coupling of electric or magnetic dipoles tailoring the optical responses [80]. Here, we discuss the long-range coupling between toroidal moments excited in a merged system of heptamer nanocavities. This chapter is adapted from [3], with slight modifications.

Instead of creating two independent toroidal dipoles via two sets of heptamer cavities, we propose a merged structure where four nanoholes in the centre are mutually shared (Fig. 5.16, dashed circles). This entangled decamer cavity structure not only supports the spontaneous excitation of several toroidal moments but also creates a strong interaction between them, which serves the purpose of achieving strong coupling effects.

The fabricated decamer shown in Fig. 5.17a has a hole diameter d of 150 ± 10 nm and a thickness t of 67 ± 15 nm. The corresponding EEL spectra were acquired along the decamer symmetry axis in the x direction (green line) with the electron probe perpendicularly passing through the film along z direction. Along the symmetric axis, toroidal moments can be selectively excited, whereas other cavity modes such as azimuthally polarized modes can be effectively avoided. In Fig. 5.17a, the experimental zero-loss-peak (ZLP)-subtracted EEL spectra are presented versus the electron impact location. Because of elastic scattering in the silver film, both the zero-loss intensity and the plasmon intensity vary with the film thickness and, in particular, differ strongly between measurements in the hole and in the silver film. This effect can be eliminated by dividing the plasmonic signal by the ZLP intensity. We observe several resonances within the energy range from 0.6 to 3.8 eV. We first focus on the modes below 2.0 eV.

At energies below 2.0 eV, three significant maxima are marked by red inverse triangles and named as H1, H2 and H3, respectively (Fig. 5.17a). The individual EEL spectra obtained at the impact locations marked by coloured circles reveal the details of the distinguished resonances (Fig. 5.18 left column). H1, H2 and H3 resonances appear at energies of 1.22, 1.56, and 1.75 eV, respectively (dotted lines).

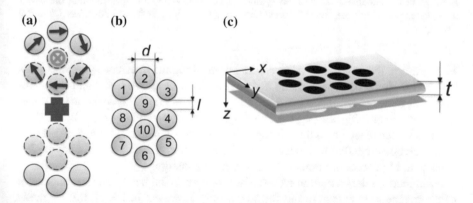

Fig. 5.16 Schematic illustration of a plasmonic decamer nanocavity consisting of 10 circular nanoholes originated from two heptamer nanocavities. **a** Red arrows and green cross represent magnetic dipoles and toroidal dipole, respectively. Circles with dashed frames in the heptamer cavities mark the mutually shared holes in the designed decamer cavity. **b** Each hole has a number index. **c** The symbols d, l and t denote the hole diameter, the distance between two holes from rim to rim and the thickness of the silver film, respectively. Reprinted by permission from [3]; Copyright © 2018 American Chemical Society

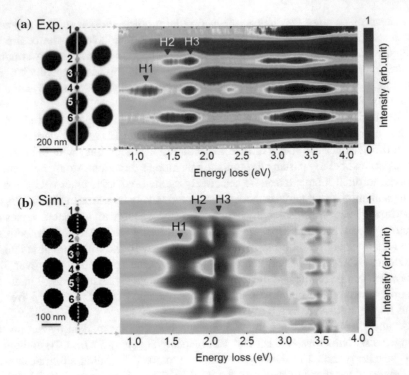

Fig. 5.17 EELS map of a decamer nanocavity formed by merging two heptamers. **a** High-angle annular dark-field image (left) and ZLP-normalized EEL spectra recorded along the green line. **b** Image of the simulated decamer cavity and the corresponding EEL spectra along the cavity axis (dotted green line). Reprinted by permission from [3]; Copyright © 2018 American Chemical Society

Furthermore, they appear at different locations along the decamer symmetry axis. The EELS signal of the H1 mode is found at the central two holes and the silver bridge in between (exp. spectra 3–5 in Fig. 5.18); mode H2 appears at the upper and lower silver bridges along the axis (exp. spectra 2 and 6 in Fig. 5.18); the third mode H3 is localized on all three bridges (exp. spectra 2, 4 and 6 in Fig. 5.18). These results demonstrate that fast electrons excite characteristic modes at displaying hot spots with high concentrations of electromagnetic energy.

Prominent to the excitation of toroidal moments is the head-to tail arrangement of the magnetic moments in the form of a ring. However, in EELS, the magnetic field is not probed, as this does not change the energy of the exciting electron. To explicitly understand the experimental findings, we replicated the experimental attempts by numerical FDTD calculations of the plasmonic decamer structure in the same excitation scenario (Fig. 5.17). For the simulated decamer structure, we used the following parameters: hole diameter $d = 100$ nm, rim-to-rim distance $l = 50$ nm and thickness $t = 50$ nm. In the simulated energy–distance EEL map, three

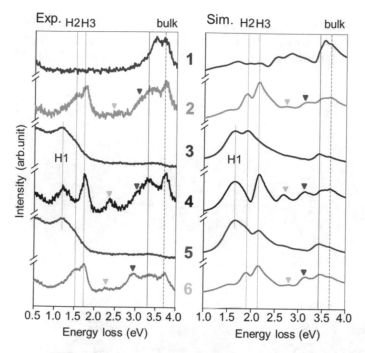

Fig. 5.18 EEL spectra extracted from a few electron impact positions. Experimental ZLP-subtracted and simulated EEL spectra extracted from the 6 marked locations in Fig. 5.17a and **b**, respectively. Dotted lines indicate the peak positions of the H1, H2 and H3 modes. Grey and magenta inverse triangles highlight the cavity modes. Localized surface plasmons confined between the holes are marked by a solid line just before the bulk mode (dashed line). Reprinted by permission from [3]; Copyright © 2018 American Chemical Society

maxima of the energy loss probability are well visible at energy losses of 1.7, 1.8 and 2.1 eV, respectively (red inverse triangles in Fig. 5.17). The spatial distribution of these modes is similar to the characteristics of the experimental H1, H2 and H3 modes but with a strong extension to the nearby nanoholes. The good correspondence between experiment and simulation is even more discernible in Fig. 5.18, where the simulated EEL spectra are presented alongside the experimental spectra extracted from similar locations in the corresponding structures (locations 1–6 in Fig. 5.17). Thus, we ascribe the three resonances at 1.7, 1.8, and 2.1 eV in the simulated EEL spectra to the excitations of the H1, H2 and H3 modes. The simulated data are blueshifted with respect to the experimental data because the simulated structure had a slightly smaller size.

Now, we show that the simulated magnetic field distributions of the H1, H2 and H3 modes help to unveil the physics behind our observations. First, we notice that each hole sustains a magnetic dipole moment in a Babinet analogue to nanodiscs that support electric dipoles. Second, the orientation of the excited magnetic dipoles at each mode can be understood from the corresponding simulated z-component of the magnetic field in the decamer structure (Fig. 5.19a), where magnetic dipoles are outlined from the negative to positive field amplitudes in each hole. In general, each hole holds a magnetic dipole for all the modes, except for holes 1, 3, 5 and 7 at 2.1 eV

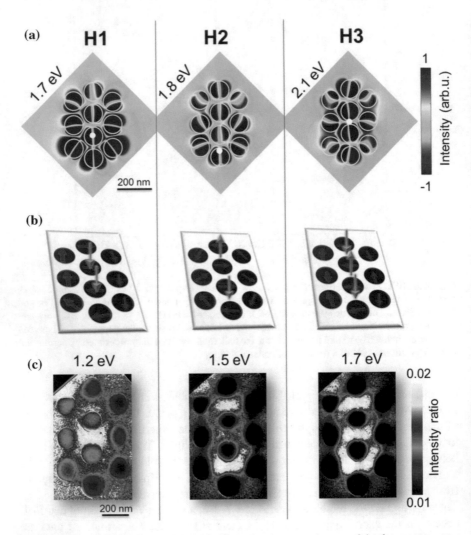

Fig. 5.19 **a** Simulated magnetic fields, H_z, along the electron trajectory of the decamer structure at the H1 (1.7 eV), H2 (1.8 eV), and H3 (2.1 eV) resonances. Grey circles denote nanoholes. The impact locations of the electron probe used for the calculation are indicated by the white dots in the corresponding figures. **b** Schematic illustration for the H1, H2 and H3 modes in the plasmonic decamer nanocavity. Red and green arrows represent magnetic and toroidal dipoles, respectively. **c** ZLP-normalized EFTEM images of the decamer nanocavity at energy losses of 1.2 ± 0.1, 1.5 ± 0.1, and 1.7 ± 0.1 eV, respectively. The black triangular areas at the upper left and right corners are beyond the acquisition area of the CCD camera. Reprinted by permission from [3]; Copyright © 2018, American Chemical Society

(for the hole index, see Fig. 5.16b), where the magnetic moments are slightly split into two dipoles in the case of the H3 mode. Several circles of magnetic dipoles can be distinguished for the different modes in Fig. 5.19b. For the H1 mode, head-to-tail circulating magnetic dipoles are displayed in the nanoholes 1-2-3-4-10-8 and 4-5-6-7-8-9 (red arrows), which indicates two parallel toroidal dipoles, each supported by 6 magnetic dipoles. The excited toroidal dipoles are located at the centres of holes 9 and 10 (green arrows) at a lateral distance of 252 nm. Accompanying that, the distorted magnetic loops along holes 8-9-4-10 and 1–8 indicate the coexistence of two toroidal moments, which are commonly located at the centre of the structure (between holes 9 and 10). All the excited toroidal moments in the H1 modes are degenerate and parallel to each other. For the H2 mode, two circles composed of 4 magnetic dipoles are found in nanoholes 1-2-3-9 and 10-7-6-5, indicating two antiparallel toroidal moments located at the silver bridges between holes 2-9 and 10-6, with a distance of 504 nm.

We note that, as a counterpart to mode H1, the antiparallel alignment of the two heptamer toroidal moments (along holes 1-2-3-4-10-8 and 9-8-7-6-5-4) could be expected. However, this would require that mutual nanoholes 8 and 4 spontaneously hold two differently oriented magnetic dipoles, which leads to a single moment, as shown in Fig. 5.19b. For the H3 mode, the three circles of magnetic dipoles displayed in nanoholes 1-2-3-9, 9-8-10-4 and 10-5-6-7 form three asymmetrically aligned toroidal moments located on the bridges between holes 2-9, 9-10 and 10-6. The lateral distance between the neighbouring toroidal moments for this case is approximately 252 nm. Very interestingly, the slight splitting of magnetic dipoles (in holes 1, 3, 5 and 7 at 2.1 eV) is observed only in this exotic coupling configuration, likely due to the compromise between assisting in the circular connection with neighbouring dipoles and keeping the electromagnetic energy of the system at a minimum.

To further confirm the excitation of toroidal moments, three-dimensional plots of both the magnetic and electric fields of the H1, H2 and H3 modes are provided in Fig. 5.20. It shows the essential character of the toroidal moments in all the cases; i.e., the magnetic fields have loops in the x-y plane, while the electric fields loop around the magnetic flow perpendicularly. Both the magnetic and electric fields are well confined in or near the structure region. The spatial distribution features of these three modes are well captured in the ZLP-normalized EFTEM measurements (Fig. 5.19c), again confirming the excitation of three different toroidal modes along the axis of the decamer cavity with high electromagnetic energy concentrations.

To date, we have understood that multiple toroidal dipoles consisting of 4, 6 or 8 magnetic dipoles were spontaneously excited in the designed decamer structure at 1.22, 1.56, and 1.7 eV. The excited toroidal dipoles are transversely aligned in either parallel (mode H1) or antiparallel (modes H2 and H3). The ascending energies of the observed modes H1, H2 and H3 (Fig. 5.21a) indicate a coupling effect between the excited toroidal moments, with a lower interaction energy for the parallel alignment than for the antiparallel alignment.

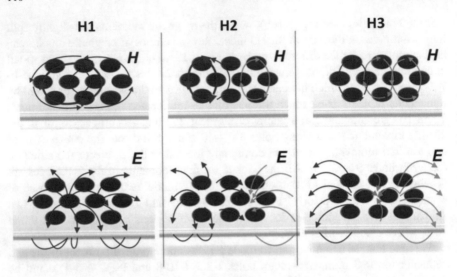

Fig. 5.20 Sketches of the three-dimensional field distributions. Magnetic (upper row) and electric fields (lower row) at the H1, H2 and H3 resonances according to the plots of the vectorial fields [3]. Different colours and thicknesses of the lines denote different groups and relative contributions. Reprinted by permission from [3]; Copyright © 2018 American Chemical Society

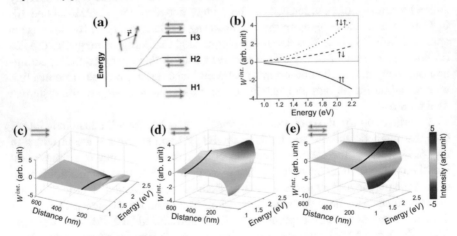

Fig. 5.21 Interaction between toroidal dipoles. **a** Energy level scheme of the experimental H1, H2 and H3 modes. Green arrows denote toroidal dipoles. The inset shows two coupled toroidal dipoles with a centre-to-centre distance r. **c–e** Interaction energies of two parallel aligned, two anti-parallel aligned, and three anti-parallel aligned toroidal dipoles as a function of the single toroidal moment energy and the dipole distance r. The same colour codes were applied to (**c–e**) to make them comparable. The black curves in (**c–e**) specify the interaction energy at dipole distances of 252 nm (**c, e**) and 504 nm (**d**). A comparison of the interaction energies specified by these black curves is presented in (**b**), where a horizontal red line denotes the interaction energy at zero

To understand and clarify the associated coupling phenomena, we outline a simple theoretical model. The interaction between two toroidal moments takes place through the external conduction and displacement currents, which are equivalent to a vortex of a magnetic field. In the case of two point-like toroidal dipoles in free space, the external conduction current is therefore zero, and the displacement current is related to the electric field by $\vec{D} = \varepsilon_0 \vec{E}$, i.e., the electric field associated with one toroidal moment is acting on the second toroidal moment. The electromagnetic field attributed to an oscillating toroidal moment has been derived elsewhere [81]. In free space, the interaction energy between two transversely coupled toroidal moments is given by

$$W^{\text{int}} = \left(\vec{T}_1 \cdot \vec{T}_2 \right) \left[\frac{\omega}{c} \cdot \frac{1}{r^3} - \left(\frac{\omega}{c} \right)^3 \cdot \frac{1}{r} \right], \tag{5.9}$$

where r is the centre-to-centre distance between two toroidal dipoles (inset in Fig. 5.21a). Equation (5.1) demonstrates the coexistence of two types of electromagnetic interaction, which in general are associated with the near-field and far-field contributions, with r^{-3} and r^{-1} dependences, respectively. Assuming that the two toroidal moments have identical energies, the calculated interaction energies for the parallel and antiparallel schemes are plotted as a function of distance r and the energy of a single toroidal moment in Fig. 5.21c and d, respectively.

Surprisingly, the toroidal dipoles can still interact at separations of several hundreds of nanometres. We note from (5.1) that the interaction energy flips its sign when the separation crosses a critical distance of $r_c = c\,\omega^{-1}$, which is a result of the competition between the near-field and far-field terms. This means that along a certain energy–distance line the near- and far-field terms cancel each other out. Within the energy range of 1–2.2 eV, the critical distance r_c is generally smaller than 200 nm. Above that range, in the long-distance regime ($r > r_c$), the interaction energy for parallel coupling is negative, in contrast to the positive interaction energy for the case of antiparallel coupling. We briefly note here that the transfer of the electromagnetic energy seems to be facilitated via the vacuum region; otherwise, the effective refractive index of the surface plasmons inside the silver would shorten the critical radius to only a few nanometres. When three identical toroidal dipole moments T_1, T_2, and T_3 couple with each other at a certain lateral distance r, the total interaction energy follows

$$W^{\text{int}}_{\uparrow\downarrow\uparrow} = W^{\text{int}}_{T_1,T_2}(r) + W^{\text{int}}_{T_2,T_3}(r) + W^{\text{int}}_{T_1,T_3}(2r) \tag{5.10}$$

The corresponding plot of the interaction energy is shown in Fig. 5.21e, which exhibits a similar behaviour as for the interaction energy of two antiparallel coupled toroidal moments, yet with a larger amplitude.

The above mentioned first principles describe the energy trends of modes H1, H2 and H3. For mode H1, the two initial toroidal moments have a distance of 252 nm (centre-to-centre distance between holes 9 and 10). The corresponding parallel coupling energy at this distance is highlighted in Fig. 5.21c by the black curve and replotted in Fig. 5.21b (also a black curve) for clarification. The two antiparallel toroidal moments of mode H2 have a separation of 504 nm (the distance between locations 2 and 6 in Fig. 5.17a). The corresponding interaction energy is highlighted (the black curve in Fig. 5.21d) and replotted as a dashed line in Fig. 5.21b. Finally, the excited toroidal moments of the mode H3 are 252 nm apart. The corresponding interaction energy is marked as a black curve in Fig. 5.21e and replotted as a dotted line in Fig. 5.21b. For the energy landscapes of the toroidal moments situated between 1 and 2.2 eV, the interaction energies calculated from the different coupling configurations clearly reveal that parallel coupling lowers the entire energy of the structure, whereas antiparallel coupling raises the entire energy. Moreover, the antiparallel coupling of the triple toroidal moment H3 has a larger interaction energy than that of the double antiparallel coupled toroidal moment H2. This difference becomes even more pronounced when increasing the single toroidal energy (Fig. 5.21d and e). These results fit very well with the experimentally observed energy tendencies of modes H1, H2 and H3.

Although theoretical calculations show that the long-range interaction between toroidal moments is relatively weak (Fig. 5.21c–e), the long-range coupling effect observed in experiments is rather pronounced. In theory, electromagnetic sources are considered as point sources. What we see from our experiments is that the toroidal modes are extended over hundreds of nanometres. Hence, they can couple to each other. In addition, the coupling is further enhanced by the retardation effect, which is described by the dynamic nature of toroidal moments.

Apart from these three coupled toroidal moments, we observe an additional resonance at 2.4 eV (grey inverse triangle in Fig. 5.18c). This feature was also found in the simulated electron energy-loss spectra at 2.7 eV (right column in 5.18c). By analysing the corresponding magnetic field (Fig. 5.22b), we find that at this resonance energy, both the magnetic dipoles and quadrupoles (holes 2 and 6) are simultaneously excited in the nanoholes. The configuration of the corresponding electric field is relatively simple. As shown in Fig. 5.22a, this finding indicates a flipping of the longitudinal polarization along the x-axis [4].

In addition, there is a resonance at a higher energy of 2.9 eV in the experimental EEL spectra, which corresponds to the double peak at approximately 3.1 eV in the simulated EEL spectrum (magenta inverse triangles in Fig. 5.18c). The corresponding electric field distribution (Fig. 5.22c) indicates radial polarization from the centre of the decamer structure, while the magnetic field (Fig. 5.22d) displays the coexistence of a magnetic quadrupole and a hexapole. This configuration is associated with an electric radial cavity mode. The imperfect structure in the fabricated decamer cavity results in a slight energy shift along the axis.

Close to the bulk plasmon energy at 3.8 eV, there is another spectral feature at approximately 3.4 eV resolved in both the simulated and experimental EEL spectra, as indicated by the black solid line in Fig. 5.18c. The energy range is typical for surface

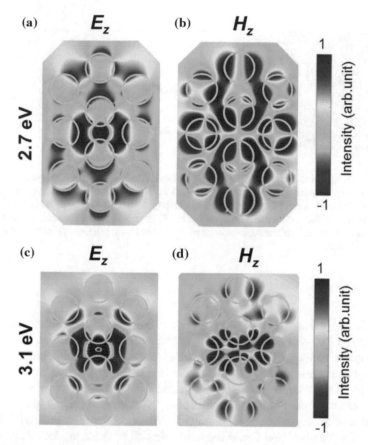

Fig. 5.22 Electric and magnetic field distributions projected along the z-direction in the x–y-plane for higher-order resonances. Field distributions for resonances at **a, b** 2.7 eV and **c, d** 3.1 eV. The diameter of the hole is 100 nm. Grey circles denote the locations of nanoholes. Reprinted by permission from [3]; Copyright © 2018 American Chemical Society

plasmon polariton excitation on infinite silver thin films (slab mode, see Fig. 5.23). However, the energy shifts when the film morphology changes. The observed peaks are probably due to the excitation of a localized surface mode, in which a small energy shift can be seen between the spectra extracted from location 1 (outside the structure) and the locations inside the structure. The thickness variations of the fabricated structure also have an effect on the energy and the width of the corresponding EELS signals (Fig. 5.23).

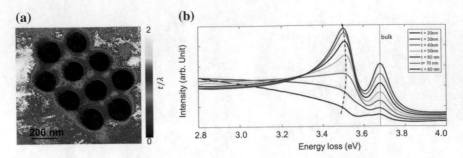

Fig. 5.23 Effect of the thickness of the silver thin film on the EEL spectra. **a** Thickness map of the plasmonic decamer cavity structure shown in Fig. 5.2a. **b** Simulated EEL spectra of infinite silver thin films with thicknesses from 20 to 80 nm at a step of 10 nm. The black dashed arrow indicates the energy shift of the surface plasmon polariton mode from 3.5 eV. The black solid line shows the location of the silver bulk plasmon. Reprinted by permission from [3]; Copyright © 2018 American Chemical Society

5.6   Summary

We explained in this chapter the emergence and theory of toroidal moments and realized them by means of oligomeric structures using Babinet's principle. The radiation and coupling of toroidal moments have also been studied using combined EELS and CL techniques. We also discussed the role of symmetry and topology in the excitation of the various modes of oligomeric nanostructures. These structures, in general, are very reach in the photonic density of states that they support. The optical modes of the oligomers cover radially polarized, azimuthally polarized, flipping azimuthally polarized and toroid moments, and all of these modes can be excited with electron beams. However, distinguishing between these modes is, in general, not straightforward with EELS and CL because of the lack of phase information in those techniques. In the coming chapter, we discuss the role of phase-matching in EELS when instead of a thin sample, micro-scale gold tapers are used.

References

1. N. Talebi, S. Guo, A. van Aken Peter, Theory and applications of toroidal moments in electrodynamics: their emergence, characteristics, and technological relevance. Nanophotonics **7**, p 93 (2018)
2. M. Kociak, L.F. Zagonel, Cathodoluminescence in the scanning transmission electron microscope. Ultramicroscopy **174**, 50–69 (2017). https://doi.org/10.1016/j.ultramic.2016.11.018
3. S. Guo, N. Talebi, P.A. van Aken, Long-range coupling of toroidal moments for the visible. ACS Photonics **5**(4), 1326–1333 (2018/04/18). https://doi.org/10.1021/acsphotonics.7b01313
4. N. Talebi, B. Ögüt, W. Sigle, R. Vogelgesang, P.A. van Aken, On the symmetry and topology of plasmonic eigenmodes in heptamer and hexamer nanocavities. Appl. Phys. A **116**(3), 947–954 (2014/09/01). https://doi.org/10.1007/s00339-014-8532-y

5. I.B. Zeldovich, Electromagnetic interaction with parity violation (in English). Sov. Phys. JETP-USSR **6**(6), 1184–1186 (1958). [Online]. Available: <Go to ISI>://WOS:A1958WT97900044

6. V.M. Dubovik, V.V. Tugushev, Toroid moments in electrodynamics and solid-state physics (in English). Phys. Rep. **187**(4), 145–202 (1990). https://doi.org/10.1016/0370-1573(90)90042-Z

7. N. Papasimakis, V.A. Fedotov, V. Savinov, T.A. Raybould, N.I. Zheludev, Electromagnetic toroidal excitations in matter and free space (in English). Nat. Mater. **15**(3), 263–271 (2016). https://doi.org/10.1038/NMAT4563

8. V.M. Dubovik, L.A. Tosunian, V.V. Tugushev, Axial toroidal moments in electrodynamics and solid-state physics (in Russian). Zh. Eksp. Teor. Fiz+. **90**(2), 590–605 (1986 Feb). [Online]. Available: <Go to ISI>://WOS:A1986A438000018

9. K. Marinov, A.D. Boardman, V.A. Fedotov, N. Zheludev, Toroidal metamaterial (in English). New J Phys. **9** (2007 Sep 14). doi:Artn 324, https://doi.org/10.1088/1367-2630/9/9/324

10. N.A. Spaldin, M. Fiebig, M. Mostovoy, The toroidal moment in condensed-matter physics and its relation to the magnetoelectric effect (in English). J. Phys. Condens. Mat. **20**(43) (2008 Oct 29). doi:Artn 434203, https://doi.org/10.1088/0953-8984/20/43/434203

11. A.S. Zimmermann, D. Meier, M. Fiebig, Ferroic nature of magnetic toroidal order (in English). Nat Commun **5** (2014 Sep). doi:Artn 4796, https://doi.org/10.1038/ncomms5796

12. V.A. Fedotov, A.V. Rogacheva, V. Savinov, D.P. Tsai, N.I. Zheludev, Resonant transparency and non-trivial non-radiating excitations in toroidal metamaterials (in English). Sci. Rep. **3** (2013 Oct 17). doi:Artn 2967, https://doi.org/10.1038/srep02967

13. V. Savinov, V.A. Fedotov, N.I. Zheludev, Toroidal dipolar excitation and macroscopic electromagnetic properties of metamaterials (in English). Phys. Rev. B **89**(20) (2014 May 14). doi:Artn 205112, https://doi.org/10.1103/physrevb.89.205112

14. T. Kaelberer, V.A. Fedotov, N. Papasimakis, D.P. Tsai, N.I. Zheludev, Toroidal dipolar response in a metamaterial (in English). Science **330**(6010), 1510–1512 (2010). https://doi.org/10.1126/science.1197172

15. Y.W. Huang et al., Design of plasmonic toroidal metamaterials at optical frequencies (in English). Opt. Express **20**(2), 1760–1768 (2012). https://doi.org/10.1364/Oe.20.001760

16. G. Thorner, J.M. Kiat, C. Bogicevic, I. Kornev, Axial hypertoroidal moment in a ferroelectric nanotorus: A way to switch local polarization (in English). Phys. Rev. B **89**(22) (2014 Jun 6). doi:Artn 220103, https://doi.org/10.1103/physrevb.89.220103

17. N. Talebi, Optical modes in slab waveguides with magnetoelectric effect (in English). J. Opt. UK **18**(5) (2016 May). doi:Artn 055607, https://doi.org/10.1088/2040-8978/18/5/055607

18. J.D. Jackson, From Lorenz to Coulomb and other explicit gauge transformations (in English). Am. J. Phys. **70**(9), 917–928 (2002). https://doi.org/10.1119/1.1491265

19. Vrejoiu, C.: Electromagnetic multipoles in Cartesian coordinates (in English). J. Phys. A Math. Gen. **35**(46), 9911–9922 (2002 Nov 22). https://doi.org/10.1088/0305-4470/35/46/313, Pii S0305-4470(02)39273-4

20. T. Gongora, E. Ley-Koo, Complete electromagnetic multipole expansion including toroidal moments. *Rev. Mex. Fis. E* **52**, 188–197 (2006)

21. C.G. Gray, Multipole expansions of electromagnetic-fields using debye potentials (in English). Am. J. Phys. **46**(2), 169–179 (1978). https://doi.org/10.1119/1.11364

22. N.A. Spaldin, M. Fechner, E. Bousquet, A. Balatsky, L. Nordstrom, Monopole-based formalism for the diagonal magnetoelectric response (in English). Phys. Rev. B **88**(9) (2013 Sep 23). doi:Artn 094429, https://doi.org/10.1103/physrevb.88.094429

23. J. Preskill, Magnetic monopoles (in English). Annu. Rev. Nucl. Part S **34**, 461–530 (1984). [Online]. Available: <Go to ISI>://WOS:A1984TU83600012

24. Y.A. Artamonov, A.A. Gorbatsevich, Symmetry and dynamics of systems with toroidal moments (in Russian). Zh. Eksp. Teor. Fiz+ **89**(3), 1078–1093 (1985). [Online]. Available: <Go to ISI>://WOS:A1985ASL8100033

25. E.E. Radescu, G. Vaman, Exact calculation of the angular momentum loss, recoil force, and radiation intensity for an arbitrary source in terms of electric, magnetic, and toroid multipoles. Phys. Rev. E Stat. Nonlinear Soft Matter Phys. **65**(4 Pt 2B), 046609 (2002 Apr). https://doi.org/10.1103/physreve.65.046609

26. A.S. Schwanecke, V.A. Fedotov, V.V. Khardikov, S.L. Prosvirnin, Y. Chen, N.I. Zheludev, Nanostructured metal film with asymmetric optical transmission (in English). Nano Lett. **8**(9), 2940–2943 (2008). https://doi.org/10.1021/nl801794d
27. S. Nanz, *Toroidal Multipole Moments in Classical Electrodynamics* (Springer Spektrum, Wiesbaden, 2016)
28. Y.M. Liu, X. Zhang, Metamaterials: a new frontier of science and technology (in English). Chem. Soc. Rev. **40**(5), 2494–2507 (2011). https://doi.org/10.1039/c0cs00184h
29. N. Papasimakis, V.A. Fedotov, K. Marinov, N.I. Zheludev, Gyrotropy of a metamolecule: wire on a torus. Phys. Rev. Lett. **103**(9), 093901 (2009 Aug 28). https://doi.org/10.1103/physrevlett.103.093901
30. K. Marinov, A.D. Boardman, V.A. Fedotov, N. Zheludev, Toroidal metamaterial. New J. Phys. **9**(9), 324 (2007). [Online]. Available: http://stacks.iop.org/1367-2630/9/i=9/a=324
31. Y.-W. Huang et al., Design of plasmonic toroidal metamaterials at optical frequencies. Opt. Express **20**(2), 1760–1768. (2012/01/16). https://doi.org/10.1364/oe.20.001760
32. C. Tang et al., Toroidal dipolar response in metamaterials composed of metal–dielectric–metal sandwich magnetic resonators. IEEE Photonics J. **8**(3), 1–9 (2016). https://doi.org/10.1109/JPHOT.2016.2574865
33. P.C. Wu et al., Plasmon coupling in vertical split-ring resonator metamolecules. Sci. Rep. **5**, 9726 (06/05/2015). https://doi.org/10.1038/srep09726
34. Z.-G. Dong, J. Zhu, X. Yin, J. Li, C. Lu, X. Zhang, All-optical Hall effect by the dynamic toroidal moment in a cavity-based metamaterial. Phys. Rev. B **87**(24), 245429 (06/24/2013). [Online]. Available: http://link.aps.org/doi/10.1103/PhysRevB.87.245429
35. P.R. Wu et al., Horizontal toroidal response in three-dimensional plasmonic (Conference Presentation), vol. 9921, pp. 992120–992120-1 (2016). [Online]. Available: http://dx.doi.org/10.1117/12.2236879
36. C.Y. Liao et al., Optical toroidal response in three-dimensional plasmonic metamaterial, vol. 9547, pp. 954724–954724-4 (2015). [Online]. Available: http://dx.doi.org/10.1117/12.2189052
37. T.A. Raybould et al., Toroidal circular dichroism. Phys. Rev. B **94**(3), 035119 (07/08/2016). [Online]. Available: http://link.aps.org/doi/10.1103/PhysRevB.94.035119
38. S. Han, L. Cong, F. Gao, R. Singh, H. Yang, Observation of Fano resonance and classical analog of electromagnetically induced transparency in toroidal metamaterials. Ann. Phys. **528**(5), 352–357 (2016). https://doi.org/10.1002/andp.201600016
39. Y. Fan, Z. Wei, H. Li, H. Chen, C.M. Soukoulis, Low-loss and high-Qplanar metamaterial with toroidal moment. Phys. Rev. B **87**(11) (2013). https://doi.org/10.1103/physrevb.87.115417
40. A.A. Basharin, V. Chuguevsky, N. Volsky, M. Kafesaki, E.N. Economou, Extremely high Q-factor metamaterials due to anapole excitation. Phys. Rev. B **95**(3), 035104 (01/03/2017). [Online]. Available: http://link.aps.org/doi/10.1103/PhysRevB.95.035104
41. Y. Bao, X. Zhu, Z. Fang, Plasmonic toroidal dipolar response under radially polarized excitation. Sci. Rep. **5**, 11793 (06/26/2015). https://doi.org/10.1038/srep11793, http://www.nature.com/articles/srep11793#supplementary-information
42. Z.-G. Dong et al., Optical toroidal dipolar response by an asymmetric double-bar metamaterial. Appl. Phys. Lett. **101**(14), 144105 (2012). https://doi.org/10.1063/1.4757613
43. J. Li et al., Optical responses of magnetic-vortex resonance in double-disk metamaterial variations. Phys. Lett. A **378**(26–27), 1871–1875 (5/16/2014). http://dx.doi.org/10.1016/j.physleta.2014.04.049
44. V.A. Fedotov, A.V. Rogacheva, V. Savinov, D.P. Tsai, N.I. Zheludev, Resonant transparency and non-trivial non-radiating excitations in toroidal metamaterials. Sci. Rep. **3**, 2967 (2013). https://doi.org/10.1038/srep02967
45. L.-Y. Guo, M.-H. Li, X.-J. Huang, H.-L. Yang, Electric toroidal metamaterial for resonant transparency and circular cross-polarization conversion. Appl. Phys. Lett. **105**(3), 033507 (2014). https://doi.org/10.1063/1.4891643
46. P.C. Wu et al., Vertical split-ring resonators for plasmon coupling, sensing and metasurface. **9544**, 954423–954423-4 (2015). [Online]. Available: http://dx.doi.org/10.1117/12.2189249

47. A.E. Miroshnichenko et al., Nonradiating anapole modes in dielectric nanoparticles. Nat. Commun. **6**, 8069 (08/27/2015). https://doi.org/10.1038/ncomms9069, https://www.nature.com/articles/ncomms9069#supplementary-information

48. A.A. Basharin et al., Dielectric metamaterials with toroidal dipolar response. Phys. Rev. X **5**(1), 011036 (03/27/2015). [Online]. Available: http://link.aps.org/doi/10.1103/PhysRevX.5.011036

49. J. Li, J. Shao, Y.-H. Wang, M.-J. Zhu, J.-Q. Li, Z.-G. Dong, Toroidal dipolar response by a dielectric microtube metamaterial in the terahertz regime. Opt. Express **23**(22), 29138–29144 (2015/11/02). https://doi.org/10.1364/oe.23.029138

50. W. Liu, J. Zhang, A.E. Miroshnichenko, Toroidal dipole-induced transparency in core–shell nanoparticles. Laser Photonics Rev. **9**(5), 564–570 (2015). https://doi.org/10.1002/lpor.201500102

51. Q. Zhang, J.J. Xiao, X.M. Zhang, D. Han, L. Gao, Core–shell-structured dielectric–metal circular nanodisk antenna: gap plasmon assisted magnetic toroid-like cavity modes. ACS Photonics **2**(1), 60–65 (2015/01/21). https://doi.org/10.1021/ph500229p

52. J. Li, Y. Zhang, R. Jin, Q. Wang, Q. Chen, Z. Dong, Excitation of plasmon toroidal mode at optical frequencies by angle-resolved reflection. Opt. Lett. **39**(23), 6683–6686 (2014/12/01). https://doi.org/10.1364/ol.39.006683

53. V.C. Alexey A. Basharin, N. Volsky, M. Kafesaki, E.N. Economou, Extremely high Q-factor metamaterials due to anapole excitation. [Online] Available: arXiv:1608.03233 [physics.class-ph]

54. B. Ögüt, N. Talebi, R. Vogelgesang, W. Sigle, P.A. van Aken, Toroidal plasmonic eigenmodes in oligomer nanocavities for the visible. Nano Lett. **12**(10), 5239–5244 (2012/10/10). https://doi.org/10.1021/nl302418n

55. F. Roder, H. Lichte, Inelastic electron holography—first results with surface plasmons (in English). Eur. Phys. J. Appl. Phys. **54**(3) (2011 June). doi:Artn 33504, https://doi.org/10.1051/epjap/2010100378

56. N. Talebi, Spectral interferometry with electron microscopes. Sci. Rep. **6**, 33874 (09/21/2016). https://doi.org/10.1038/srep33874, https://www.nature.com/articles/srep33874#supplementary-information

57. G. Guzzinati, A. Beche, H. Lourenco-Martins, J. Martin, M. Kociak, J. Verbeeck, Probing the symmetry of the potential of localized surface plasmon resonances with phase-shaped electron beams (in English). Nat. Commun. **8** (2017 Apr 12). doi:Artn 14999, https://doi.org/10.1038/ncomms14999

58. X. Zhang et al., Asymmetric excitation of surface plasmons by dark mode coupling. Sci. Adv. **2**(2) (2016). https://doi.org/10.1126/sciadv.1501142

59. S.J. Barrow, D. Rossouw, A.M. Funston, G.A. Botton, P. Mulvaney, Mapping bright and dark modes in gold nanoparticle chains using electron energy loss spectroscopy. Nano Lett. **14**(7), 3799–3808 (2014/07/09). https://doi.org/10.1021/nl5009053

60. A.L. Koh et al., Electron energy-loss spectroscopy (EELS) of surface plasmons in single silver nanoparticles and dimers: influence of beam damage and mapping of dark modes. ACS Nano **3**(10), 3015–3022 (2009/10/27). https://doi.org/10.1021/nn900922z

61. G. Fletcher, M.D. Arnold, T. Pedersen, V.J. Keast, M.B. Cortie, Multipolar and dark-mode plasmon resonances on drilled silver nano-triangles. Opt. Express **23**(14), 18002–18013 (2015/07/13). https://doi.org/10.1364/oe.23.018002

62. S.A. Maier, Plasmonics: the benefits of darkness. Nat. Mater. **8**(9), 699–700 (2009) https://doi.org/10.1038/nmat2522. [Online]. Available: http://dx.doi.org/10.1038/nmat2522

63. W.H. Yang, C. Zhang, S. Sun, J. Jing, Q. Song, S. Xiao, Dark plasmonic mode based perfect absorption and refractive index sensing. Nanoscale **9**(26), 8907–8912 (2017). https://doi.org/10.1039/c7nr02768k

64. W. Zhou, T.W. Odom, Tunable subradiant lattice plasmons by out-of-plane dipolar interactions. Nat. Nanotechnol. **6**, 423 (05/15/2011). https://doi.org/10.1038/nnano.2011.72, https://www.nature.com/articles/nnano.2011.72#supplementary-information

65. M.-W. Chu, V. Myroshnychenko, C.H. Chen, J.-P. Deng, C.-Y. Mou, F.J. García de Abajo, Probing bright and dark surface-plasmon modes in individual and coupled noble metal nanoparticles using an electron beam. Nano Lett. **9**(1), 399–404 (2009/01/14). https://doi.org/10.1021/nl803270x

66. F.-P. Schmidt, A. Losquin, F. Hofer, A. Hohenau, J.R. Krenn, M. Kociak, How dark are radial breathing modes in plasmonic nanodisks? ACS Photonics **5**(3), 861–866 (2018/03/21). https://doi.org/10.1021/acsphotonics.7b01060

67. M. Gupta et al., Sharp toroidal resonances in planar terahertz metasurfaces. Adv. Mater. **28**(37), 8206–8211 (2016). https://doi.org/10.1002/adma.201601611

68. B. Han, X. Li, C. Sui, J. Diao, X. Jing, Z. Hong, Analog of electromagnetically induced transparency in an E-shaped all-dielectric metasurface based on toroidal dipolar response. Opt. Mater. Express **8**(8), 2197–2207 (2018/08/01). https://doi.org/10.1364/ome.8.002197

69. D.W. Watson, S.D. Jenkins, J. Ruostekoski, V.A. Fedotov, N.I. Zheludev, Toroidal dipole excitations in metamolecules formed by interacting plasmonic nanorods. Phys. Rev. B **93**(12) (2016). https://doi.org/10.1103/physrevb.93.125420

70. L. Zhu et al., A low-loss electromagnetically induced transparency (EIT) metamaterial based on coupling between electric and toroidal dipoles. RSC Adv. **7**(88), 55897–55904 (2017). https://doi.org/10.1039/c7ra11175d

71. L. Wei, Z. Xi, N. Bhattacharya, H.P. Urbach, Excitation of the radiationless anapole mode. Optica **3**(8), 799–802 (2016). https://doi.org/10.1364/optica.3.000799

72. J.S. Totero Góngora, A.E. Miroshnichenko, Y.S. Kivshar, A. Fratalocchi, Anapole nanolasers for mode-locking and ultrafast pulse generation. Nat. Commun. **8**, 15535 (05/31/2017). https://doi.org/10.1038/ncomms15535, http://dharmasastra.live.cf.private.springer.com/articles/ncomms15535#supplementary-information

73. S.-J. Kim, S.-E. Mun, Y. Lee, H. Park, J. Hong, B. Lee, Nanofocusing of toroidal dipole for simultaneously enhanced electric and magnetic fields using plasmonic waveguide. J. Lightwave Technol. **36**(10), 1882–1889 (2018/05/15). [Online]. Available: http://jlt.osa.org/abstract.cfm?URI=jlt-36-10-1882

74. A. Ahmadivand et al., Rapid detection of infectious envelope proteins by magnetoplasmonic toroidal metasensors. ACS Sens. **2**(9), 1359–1368, 2017/09/22 2017, https://doi.org/10.1021/acssensors.7b00478

75. P.C. Wu et al., Optical anapole metamaterial. ACS Nano **12**(2), 1920–1927 (2018/02/27). https://doi.org/10.1021/acsnano.7b08828

76. F.J. García de Abajo, Optical excitations in electron microscopy. Rev. Modern Phys. **82**(1), 209–275 (02/03/2010). https://doi.org/10.1103/revmodphys.82.209

77. Y. Wu, G. Li, J.P. Camden, Probing nanoparticle plasmons with electron energy loss spectroscopy. Chem. Rev. (2017/12/07). https://doi.org/10.1021/acs.chemrev.7b00354

78. S. Guo, N. Talebi, A. Campos, M. Kociak, P.A. van Aken, Radiation of dynamic toroidal moments. ACS Photonics (2019/01/24). https://doi.org/10.1021/acsphotonics.8b01422

79. A. Losquin et al., Unveiling nanometer scale extinction and scattering phenomena through combined electron energy loss spectroscopy and cathodoluminescence measurements. Nano Letters **15**(2), 1229–1237 (2015/02/11). https://doi.org/10.1021/nl5043775

80. N. Liu, H. Giessen, Coupling effects in optical metamaterials (in English). Angew. Chem. Int. Edit. **49**(51), 9838–9852 (2010). https://doi.org/10.1002/anie.200906211

81. J.A. Heras, Electric and magnetic fields of a toroidal dipole in arbitrary motion (in English). Phys. Lett. A **249**(1–2), 1–9 (1998). https://doi.org/10.1016/S0375-9601(98)00712-9

Chapter 6
Optical Modes of Gold Tapers Probed by Electron Beams

Abstract The optical microscopy and spectroscopy of nano-objects and macro-molecules demands the efficient coupling of far-field light to the microscopic scale extended at only a few percent of the optical wavelength. Thus, a far-field-to-near-field coupler is a vital tool in optical metrology, providing the ability to combine well-advanced optical techniques such as pump-probe spectroscopy and spectral interferometry with the microscopic world. Three-dimensional optical tapers, particularly metallic tapers, have been proposed [1], intensively investigated, and employed in many groups worldwide for this purpose [2, 3]. The concept that allows an efficient coupling of far-field light onto single nano-objects is *adiabatic nanofocusing* [1]. In adiabatic nanofocusing, the optical modes of a metallic taper evolve, via propagation along the shaft towards the apex, in an extremely efficient mode-conversion procedure and result in localization of the electromagnetic energy only at the very apex, extended within a few atomic scales, provided that the apex is sharp enough.

To better understand how this mode-conversion process works practically, here, we theoretically and experimentally examine the optical modes of a gold taper using EELS and EFTEM. We first select a gold taper with an opening angle of 45° and then investigate in more detail tapers of different opening angles. We will show that two different mechanisms, namely, phase-matching and reflection from the apex, underpin the intensity maxima we observe in the EELS signal [4, 5].

6.1 Optical Modes of a Gold Taper with a 45° Opening Angle

Here, we selectively choose a gold taper with an extreme change in the local radius of the cross section, from the very low nanometre scale at the apex to the mesoscopic scale along the shaft. This dimensional aspect reveals an interesting selection rule

Portions of the text of this chapter have been re-published with permission from [4], Copyright © 2015, American Chemical Society; [5], Copyright © 2015, American Chemical Society.

© Springer Nature Switzerland AG 2019

N. Talebi, *Near-Field-Mediated Photon–Electron Interactions*,
Springer Series in Optical Sciences 228,
https://doi.org/10.1007/978-3-030-33816-9_6

for the electron-plasmon inelastic interaction. While at the apex, we can excite a rotationally symmetric optical mode with an extremely broad bandwidth, this mode is not efficiently excited along the shaft for a gold taper with a large opening angle. Rather, we detect the radiative modes of the gold taper along the shaft. Intriguingly, the excited modes dynamically interact with the swift electron in such a way that an interference pattern emerges in the EELS map displaying the energy *versus* impact parameter. We underpin our interpretation by analytical calculations of the EEL spectra, and we compare the data with the analytically calculated EELS of gold fibres. Considering all the observations, we propose a simple model for understanding the dispersion of the higher-order resonances we observe in EEL spectra by imposing a phase-matching criterion.

6.1.1   Experimental Results

We use single-crystalline gold tapers, whose particularly smooth surfaces eliminate SPP localization [6] and scattering losses along the taper shaft. Figure 6.1a shows a dark-field transmission electron microscope image of a taper with an opening angle of $\alpha = 45°$. Experimental and simulation results for a gold taper with a smaller opening angle of 19° are also shown in the supporting information.

EELS and energy-filtered transmission electron microscopy (EFTEM) experiments were conducted on a Zeiss SESAM microscope operated at an acceleration

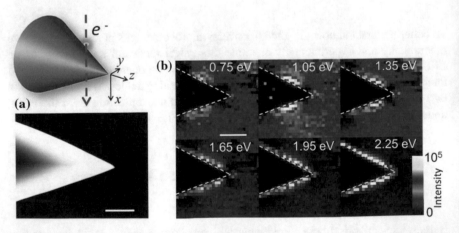

Fig. 6.1 Probing optical modes of a gold taper with EELS. **a** Dark-field transmission electron microscopy image of a taper with an opening angle of $\alpha = 45°$, the image at the top shows the 3-dimensional topology of the taper. **b** Images of the electron energy-loss intensity at 6 selected energy losses between 0.75 and 2.25 eV. These data are extracted from a series of 30×28 EEL spectra that were recorded at electron impact locations on a rectangular grid with a pixel size of 50 nm. The colour scales of the images are given in linear arbitrary units. Scale bar lengths are 500 nm. Reprinted with permission from [4]; Copyright © 2015 American Chemical Society

voltage of 200 kV. The EELS signal reflects the probability of the electron beam exciting surface plasmons during the interaction with the sample and thus senses the PLDOS (see Chap. 3 for more details). EELS allows measuring this PLDOS from a few tens of meV to hundreds of eV and thus is ideally suited for probing ultrabroad-band excitations. Figure 6.1b shows images at selected energy losses extracted from a series of EEL spectra with a spatial resolution of just 50 nm.

These overview maps already exhibit a spatially well-localized energy-loss sig-nal at the taper apex for energy losses below 2 eV. At higher energies, interband absorption in the gold sets in [7], resulting in a spatially homogeneous EELS signal in the vicinity of the metal surface. At lower energy losses ($E = 0.75$–1.35 eV), we find an additional pronounced EELS signature as bright lobes penetrating from the shaft surface into vacuum. These lobes are confined to a region of a few hundred nm along the shaft and are markedly displaced from the apex. As discussed below, this can be related to taper eigenmodes with higher-order angular momentum. The broadband ability of those tapers to localize light in exceedingly small volumes at the apex is seen in a series of EEL spectra recorded at various positions along the shaft (Fig. 6.2b) for a gold taper with a broadening angle of 49°. Near the apex, an intense, spatially localized and spectrally broadband EELS signal emerges, reflect-ing the confined localized surface plasmons (LSPs) at the taper apex. Interestingly, its intensity remains almost constant within the range from ~0.75 eV, a lower limit imposed by the difficulty of subtracting the zero-loss peak, to ~2.0 eV, the onset of interband absorption. Thus, the bandwidth of this resonance covers more than one octave, in distinct contrast to the spectrally narrower LSP resonances of metallic nanoparticles and to earlier EELS measurements on conical tapers of finite length [8].

When moving the electron excitation away from the apex, the EELS signal van-ishes entirely, until, at a finite distance from the apex, a second contribution sets in, leading to a sequence of distinct maxima in the energy-loss spectra (Fig. 6.2b) at energy losses below 2 eV. The spectral width of these resonances is in the range of a few hundred meV. As displayed in Fig. 6.2c, for each individual resonance in this sequence, the dispersion is well represented by a hyperbolic function of the form $E = E_0 + \kappa L^{-1}$, where $L = R csc(\alpha/2)$ is the distance from the apex and R is the local radius. Remarkably, the fitting parameters depend on the sequence number m in a linear fashion, resulting in the empirical relation

$$E = (0.11 \cdot m - 0.10)\text{eV} + \frac{(0.46 \cdot m - 0.23)}{R}\mu\text{m} \cdot \text{eV} \qquad (6.1)$$

As will be shown below, this resonance behaviour is a consequence of phase matching between the exciting electron and the local taper mode fields. The parameter m is directly related to the angular momentum of eigenmodes with higher-order azimuthal symmetry.

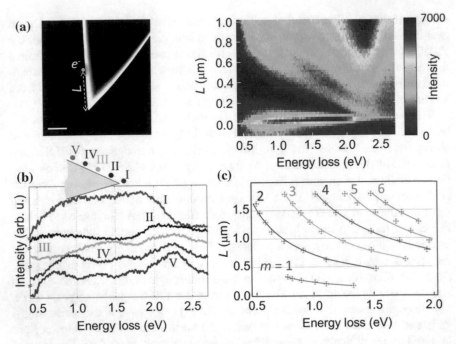

Fig. 6.2 Imaging and spectroscopy of a gold taper with an opening angle of $\alpha = 49°$. **a** Zero-loss-corrected electron energy loss intensity as a function of the impact location along the taper shaft. At certain distances L between electron impact and taper apex, pronounced resonance peaks are revealed. The length of the scale bar is 1 µm. **b** EEL spectra for electron impact at the apex (purple) and at distances of 276 nm (black), 506 nm (green), 736 nm (red), and 966 nm (blue) from the apex. The spectra are shifted vertically for clarity. The zero-loss peak contribution was subtracted from the individual spectra by using a power-law fit. **c** Dispersions of the maxima in the EEL spectra *versus* the distance from the taper. These maxima are denoted by their corresponding mode numbers $(m = 1, 2, \ldots)$. Curves are least-square fits to a hyperbolic function $E = E_0 + \kappa/L$, where κ and E_0 are constants. Reprinted by permission from [4]; Copyright © 2015 American Chemical Society

6.1.2 Numerical Results

To understand the physics behind our experimental observations, we have to consider the whole set of possible eigenmodes of the system. Scattering by conical structures has been studied in numerous works over the past decades [9, 10]. Explicit solutions are still the subject of ongoing research, involving rather advanced applications of integral transforms [11] or novel approximation methods based on the quasi-separation of variables [12]. Here, we adopt the more intuitive stance of viewing the metallic cone as a 3-dimensional taper formed from metallic fibre segments of infinitesimal height. In particular, the local near-field behaviour of the taper eigenmodes can be described effectively by a superposition of the eigenmodes of fibres with continuously varying radii. The fibre eigenmodes are calculated by solving the Helmholtz equation in a cylindrical coordinate system [13, 14]. They are characterized by the complex wavenumber k_z along the fibre axis and the azimuthal angular momentum number $m = 0, \pm1, \pm2, \ldots$. The mode field outside the fibre can thus be derived from a potential of the form (see also supporting information)

$$\Phi_{m,k_z,\omega}(x, y, z, t) \propto H_m^{(1)}\left(k_\rho \rho\right)e^{ik_z z}e^{im\varphi}e^{-i\omega t} + c.c., \tag{6.2}$$

where H_m^1 is a Hankel function of the first kind, $(\rho \cos\phi, \rho \sin\phi) = (x, y)$, $k_\rho^2 = k_0^2 - k_z^2$, and $\omega = c_0 k_0$ is the angular frequency of the mode. It is well known that only the fundamental $m = 0$ mode of a metallic taper is evanescently bound to the surface, regardless of the local radius [15]. It propagates to the apex with high efficiency and with a concomitant increase in the effective refractive index and hence the localization of light [16, 17]. This process requires a smooth transition of the local radius of the taper. In addition to the propagating $m = 0$ mode, there are also higher-order modes, characterized by $|m| > 1$, that may propagate with a low loss towards the apex as bound modes, at least where the local radius is larger than a mode-specific critical value. At the critical radii, however, they couple to the radiative (unbound) continuum of photonic modes. For instance, on a gold taper excited at a vacuum wavelength of 800 nm, the critical radii of the $|m| = 1$ and $|m| = 2$ modes are approximately $R \approx 100$ nm and $R \approx 600$ nm, respectively [18]. It should be noted that, in addition to the modes mentioned above, for any given parameter pair $\{m, \omega\}$, there exists a continuum of discrete values in the complex k_z plane that satisfy the eigenmode equation. They are characterized by rapidly increasing imaginary parts, and the corresponding modes may contribute to an acute localization along the taper shaft but decay quickly upon propagation. That is because they sustain an oscillatory amplitude inside the core along the radial direction, though they are evanescent in the air region. To understand if and how these features of a local fibre description of a plasmonic taper may be used to explain the experimentally observed EELS results, we discuss in the following the first numerical simulations of the electron energy-loss process near three-dimensional taper structures. These and the experimental results are then compared to an analytical model for EELS near fibres, which allows a detailed discussion of the individual contributions of different k_z and m components.

We perform numerical finite-difference time-domain (FDTD) calculations of three-dimensional tapers with an embedded relativistic electron source [19]. The electron is assumed to pass the taper on a straight line 1 nm from its surface. A speed of $V \approx 0.695c_0$ is chosen (unless specified otherwise) in correspondence with an acceleration voltage of 200 kV used in the experiments. Figure 6.3a displays an instantaneous distribution of the $E_z(\vec{r}, t)$ field component on the taper surface at a moment of approximately 1 fs before the electron approaches closest to the taper, approximately 900 nm from the apex. Further time frames from the same simulation are shown in Fig. 6.3c for cross sections normal to the taper axis. An analysis of the spatial and temporal evolution of the fields in such simulations reveals two main features. First, a wave packet of radiation is emerging from the interaction volume as an almost spherical, ultrashort light pulse, propagating away freely from the electron at approximately the vacuum speed of light. Second, a comparatively minor part of the transferred energy is converted into evanescent surface modes that continue

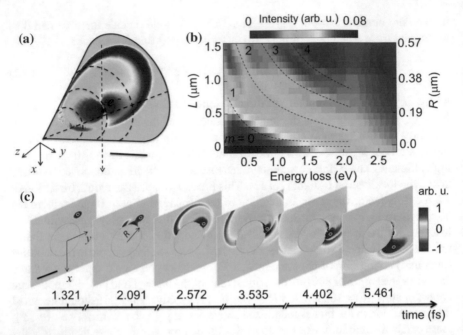

Fig. 6.3 Dynamical simulation of the energy loss of relativistic electrons upon passing a gold taper with an opening angle of $\alpha = 45°$. The taper axis is denoted as z, and the electron trajectory is assumed to be parallel to the x-axis. **a** Snapshot of the instantaneous component E_z of the total electric field at the surface of the taper for an electron impacting just outside the surface at a distance of $L = 900$ nm from the apex, corresponding to a local radius of $R = 344$ nm. **b** Computed EEL spectra as a function of the energy loss and impact location along the taper shaft. The dashed lines are guides for the eyes. **c** Series of instantaneous total electric field distributions measured at several time points during the passage of the electrons by the taper at $L = 900$ nm. The panels show $E_z(x, y, t)$ in an xy-plane located $\delta z = 5$ nm behind the electron trajectory. Reprinted by permission from [4]; Copyright © 2015 American Chemical Society

to propagate along the z-axis. Evidently, the electron impulsively launches a wave packet composed of modes in a very broad energy range, with a wide distribution of azimuthal orders m and wavenumbers k_z along the taper axis. For the local radius of $R = 344$ nm at the electron impact, only fibre modes with $m = 0$ and $|m| = 1$ exhibit a (locally) bound character and may propagate along the taper shaft. Other modes are either unbound (contributing to the free light pulse) or are strongly evanescent along the taper shaft as well as normal to the surface. The full distribution of the evanescent and radiative modes at each local radius of the taper is discussed below.

In Fig. 6.3b, the calculated EEL spectra are shown for different distances L from the taper apex. Clear signatures of discrete resonances emerge, in full accord with the experimental findings displayed in Fig. 6.2a. At this point, however, the time domain calculations by themselves offer little further insight into the characteristic hyperbolic dispersions of EELS probability maxima as functions of the local taper radius.

Fig. 6.4 A material fibre
with radius α interacting
with a relativistic electron at
the impact of y_0

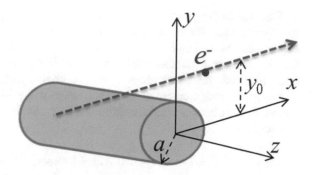

6.1.3 EELS of Gold Fibres

We have therefore developed an analytical description of the electron energy loss
mechanism for an electron passing an infinitely long fibre in a scenario as depicted in
Fig. 6.4. To this end, we adopt the standard symmetry decomposition for representing
the electromagnetic potentials associated with the incident electromagnetic fields of
the electron and those scattered by the fibre. A solution in cylindrical coordinates
can be written as

$$A(x, y, z, t) = \frac{1}{2\pi} \int_{-\infty}^{+\infty} d\omega \, e^{-i\omega t} \frac{1}{2\pi} \int_{-\infty}^{+\infty} dk_z \, e^{ik_z z} \sum_{m=-\infty}^{+\infty} e^{im\varphi} A_m(k_z, \omega) \cdot B(k_\rho \rho),$$

(6.3)

where $(x, y) = \rho(\cos(\varphi), \sin(\varphi))$, $k_\rho^2 = k_0^2 - k_z^2$, and $B_m(k_\rho \rho)$ is a Bessel or Hankel
function of order m, which is chosen such that singularities are avoided and the
Sommerfeld radiation condition is fulfilled [20]. $A_m(k_z, \omega)$ is the amplitude function
to be determined.

The steps necessary to evaluate the energy loss are the following: First, we decompose the incident electron fields in this symmetry, so they can be evaluated at the
boundary of the fibre, where they experience reflection and refraction. Second, we
formulate the reflected field (outside the fibre) as well as the refracted field (inside
the fibre) using the four tangential boundary conditions for each symmetry component individually. Third, we use the scattered field outside the fibre to evaluate the
electric field along the electron trajectory and integrate it over the entire electron's
trajectory to obtain the total energy loss. Last, we decompose the total energy loss
into its symmetry components.

$$\vec{J}(x, y, z, \omega) = \hat{x} q \delta(y - y_0) \delta(z - z_0) e^{i \frac{\omega}{v_{el}} x}$$

$$= \hat{x} \frac{q}{2\pi} \delta(y - y_0) e^{i \frac{\omega}{v_{el}} x} \int_{-\infty}^{+\infty} e^{ik_z z} dk_z.$$

(6.4)

We consider the geometry of the system shown in Fig. 6.4. The incident field is due to the current density associated with a single electron of charge $q = -e_0$ moving in *empty* space parallel to the $\hat{x}$-axis at the speed of v_{el}. It is given by $(y, z) = (y_0, 0)$ is the electron impact parameter in the yz-plane. To represent the incident and scattered electromagnetic fields, we use a vector potential approach, considering the electric and magnetic vector potentials $\vec{F}$ and $\vec{A}$, respectively, in a Lorentz gauge [21]. The generic relation between the electric and magnetic fields and the electric and magnetic potentials is

$$
\begin{aligned}
\vec{E}(\vec{r}, \omega) &= -\vec{\nabla} \times \vec{F}(\vec{r}, \omega) - (i\omega\varepsilon_0\varepsilon_r)^{-1}\vec{\nabla} \times \vec{\nabla} \times \vec{A}(\vec{r}, \omega) \\
\vec{H}(\vec{r}, \omega) &= \vec{\nabla} \times \vec{A}(\vec{r}, \omega) - (i\omega\mu_0)^{-1}\vec{\nabla} \times \vec{\nabla} \times \vec{F}(\vec{r}, \omega).
\end{aligned}
\tag{6.5}
$$

For the incident electron fields, however, it is sufficient to evaluate only the magnetic potential of the current distribution and to set $\vec{F}^{inc} = 0$. We have to solve

$$
\nabla^2 \vec{A}^{inc}(\vec{r}, \omega) + k_0^2 \vec{A}^{inc}(\vec{r}, \omega) = -\vec{J}(\vec{r}, \omega).
\tag{6.6}
$$

The translational invariance along $\hat{x}$ implies that the incident potentials have the same phase dependence as the current density; hence, $A_x^{inc}(x = 0, y, k_z, \omega) \exp(i\omega x / v_{el})$ is the only nontrivial component of the vector potential. It is determined from (6.6) as [22]

$$
A_x^{inc}(x = 0, y, k_z; \omega) = \frac{-q}{2ik_y} e^{ik_y|y-y_0|}.
\tag{6.7}
$$

In the given geometry, we need to evaluate the incident fields for $y \leq y_0$ at the surface of an infinitely long cylinder. Therefore, we transform this expression to cylindrical coordinates. As illustrated in Fig. 6.5, for the wavevector describing the field of the electron in reciprocal space, we define the perpendicular vector component by its magnitude $k_\rho^2 = k_x^2 + k_y^2 = k_0^2 - k_z^2$ and the direction angle by $\tan(\varphi_e) = -k_y/k_x = -k_y/(\omega/v_{el})$. For the location (x, y), we write $\rho^2 = x^2 + y^2$ and $\tan(\varphi) = y/x$. This enables the following algebraic substitution:

$$
e^{-ik_y y} e^{i\frac{\omega}{v_{el}}x} = e^{ik_\rho(\cos\varphi_e x + \sin\varphi_e y)}
$$

$$
= e^{ik_\rho\rho\cos(\varphi-\varphi_e)} = \sum_{m=-\infty}^{+\infty} i^m J_m(k_\rho\rho) e^{im(\varphi-\varphi_e)},
\tag{6.8}
$$

where $(\varphi - \varphi_e)$ is the angle between (x, y) and $(-k_x, k_y)$. With these preparations, the magnetic vector potential is decomposed into the form of (6.3) as

$$
A_x^{inc}(\vec{r}, \omega) = \frac{-q}{4\pi i} \int_{-\infty}^{+\infty} dk_z \frac{e^{ik_y y_0} e^{ik_z z}}{k_y} \sum_{m=-\infty}^{+\infty} i^m J_m(k_\rho\rho) e^{im(\varphi-\varphi_e)}
\tag{6.9}
$$

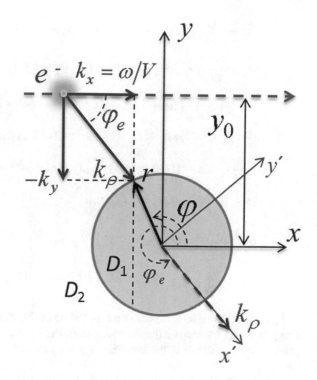

From this magnetic vector potential, the incident electron field components are obtained with the help of (6.5). Specifically, the tangential $\hat{z}$- and $\hat{\varphi}$-components take the form

$$H_\varphi^{\text{inc}}(\vec{r}, \omega) = \frac{-q}{4\pi i} \int\limits_{-\infty}^{+\infty} dk_z \sum_{m=-\infty}^{+\infty} i^m \tilde{h}_\varphi e^{im(\varphi-\varphi_e)} e^{ik_z z} e^{ik_y y_0}$$

$$H_z^{\text{inc}}(\vec{r}, \omega) = \frac{-q}{4\pi i} \int\limits_{-\infty}^{+\infty} dk_z \sum_{m=-\infty}^{+\infty} i^m \tilde{h}_z e^{im(\varphi-\varphi_e)} e^{ik_z z} e^{ik_y y_0}$$

$$E_\varphi^{\text{inc}}(\vec{r}, \omega) = \frac{1}{i\omega\varepsilon_0} \frac{q}{4\pi i} \int\limits_{-\infty}^{+\infty} dk_z \sum_{m=-\infty}^{+\infty} i^m \tilde{e}_\varphi e^{im(\varphi-\varphi_e)} e^{ik_z z} e^{ik_y y_0}$$

$$E_z^{\text{inc}}(\vec{r}, \omega) = \frac{1}{i\omega\varepsilon_0} \frac{q}{4\pi i} \int\limits_{-\infty}^{+\infty} dk_z \sum_{m=-\infty}^{+\infty} i^m \tilde{e}_z e^{im(\varphi-\varphi_e)} e^{ik_z z} e^{ik_y y_0} \tag{6.10}$$

where the amplitude functions are given by:

$$\tilde{h}_\varphi = \frac{k_z}{2k_y}\left[J_{m-1}(k_\rho\rho)e^{i\varphi_e} - J_{m+1}(k_\rho\rho)e^{-i\varphi_e}\right]$$

$$\tilde{h}_z = \frac{k_\rho}{2k_y}\left[J'_{m+1}(k_\rho\rho)e^{-i\varphi_e} + J'_{m-1}(k_\rho\rho)e^{i\varphi_e}\right]$$

$$+ \frac{1}{2\rho k_y}\left[(m+1)J_{m+1}(k_\rho\rho)e^{-i\varphi_e} - (m-1)J_{m-1}(k_\rho\rho)e^{i\varphi_e}\right]$$

$$\tilde{e}_\varphi = \frac{1}{2k_y}\left[\left(k_z^2 J_{m-1}(k_\rho\rho) - k_\rho^2 J''_{m-1}(k_\rho\rho)\right)e^{i\varphi_e} + \left(k_z^2 J_{m+1}(k_\rho\rho) - k_\rho^2 J''_{m+1}(k_\rho\rho)\right)e^{-i\varphi_e}\right]$$

$$+ \frac{1}{2\rho k_y}\left[-(m-1)\left(\frac{1}{\rho}J_{m-1}(k_\rho\rho) - k_\rho J'_{m-1}(k_\rho\rho)\right)e^{i\varphi_e} + \cdots\right.$$

$$\left. +(m+1)\left(\frac{1}{\rho}J_{m+1}(k_\rho\rho) - k_\rho J'_{m+1}(k_\rho\rho)\right)e^{-i\varphi_e}\right]$$

$$\tilde{e}_z = \frac{k_z k_\rho}{2k_y}\left[J'_{m-1}(k_\rho\rho)e^{i\varphi_e} - J'_{m+1}(k_\rho\rho)e^{-i\varphi_e}\right]$$

$$- \frac{k_z}{2\rho k_y}\left[(m-1)J_{m-1}(k_\rho\rho)e^{i\varphi_e} + (m+1)J_{m+1}(k_\rho\rho)e^{-i\varphi_e}\right]. \qquad (6.11)$$

Having obtained an analytical representation of the dynamic electromagnetic field dragged along by the relativistic electron in the cylindrical coordinate system, we proceed to evaluate it at the boundary ($\rho = a$) in order to determine the scattered fields.

Suitable choices for the magnetic and electric potentials of the fields scattered by the infinitely long cylinder are $\vec{A} = (0, 0, A_z)$ and $\vec{F} = (0, 0, F_z)$, respectively. We assign one pair of potentials for the inside domain (D_1), which in a space-frequency description reads

$$A_z^{D_1}(\vec{r}, \omega) = \int\limits_{-\infty}^{+\infty} dk_z \sum_{m=-\infty}^{+\infty} i^m C_{A,m}^{D_1}(k_z, \omega) J_m(k_\rho^{D_1}\rho)e^{im(\varphi-\varphi_e)}e^{ik_z z}e^{ik_y y_0},$$

$$F_z^{D_1}(\vec{r}, \omega) = \int\limits_{-\infty}^{+\infty} dk_z \sum_{m=-\infty}^{+\infty} i^m C_{F,m}^{D_1}(k_z, \omega) J_m(k_\rho^{D_1}\rho)e^{im(\varphi-\varphi_e)}e^{ik_z z}e^{ik_y y_0}. \qquad (6.12)$$

Note that the radial wavenumber here assumes the form $k_\rho^{D_1} = \sqrt{\varepsilon_r k_0^2 - k_z^2}$, where ε_r is the relative permittivity of the fibre material. The relative permeability is assumed to be equal to unity. Outside the fibre (see Fig. 6.5), an independent second potential pair is introduced according to

$$A_z^{D_2}(\hat{r}, \omega) = \int\limits_{-\infty}^{+\infty} dk_z \sum_{m=-\infty}^{+\infty} i^m C_{A,m}^{D_2}(k_z, \omega) H_m^{(1)}(k_\rho\rho)e^{im(\varphi-\varphi_e)}e^{ik_z z}e^{ik_y y_0},$$

$$F_z^{D_2}(\vec{r}, \omega) = \int_{-\infty}^{+\infty} dk_z \sum_{m=-\infty}^{+\infty} i^m C_{F,m}^{D_2}(k_z, \omega) H_m^{(1)}(k_\rho \rho) e^{im(\varphi-\varphi_e)} e^{ik_z z} e^{ik_y y_0}. \quad (6.13)$$

$C_{A,m}^{D_1}(k_z, \omega)$, $C_{A,m}^{D_2}(k_z, \omega)$, $C_{F,m}^{D_1}(k_z, \omega)$, and $C_{F,m}^{D_2}(k_z, \omega)$ are thus far unspecified coefficient functions. The tangential field components are obtained using (S3) and then used for satisfying the boundary conditions for each symmetry component $\{m, k_z, \omega\}$. For the outside domain, the coefficients read

$$C_{F,m}^{D_2} = \frac{1}{\Delta} \frac{q}{4\pi\varepsilon_0\omega} \left\{ -\left(\left(\tilde{e}_\varphi + \frac{m}{a} \frac{k_z}{\left(k_\rho^{D_1}\right)^2} \tilde{e}_z \right) J_m + \frac{k_0^2}{k_\rho^{D_1}} J_m' \tilde{h}_z \right) \frac{k_\rho}{k_\rho^{D_1}} \left(\varepsilon_r k_\rho H_m^{(1)} J_m' - k_\rho^{D_1} H_m^{(1)'} J_m \right) \right.$$

$$+ \cdots + \left. \left(\left(\tilde{h}_\varphi + \frac{m}{a} \frac{k_z}{\left(k_\rho^{D_1}\right)^2} \tilde{h}_z \right) J_m + \frac{\varepsilon_r}{k_\rho^{D_1}} J_m' \tilde{e}_z \right) \left(\left(\frac{k_\rho}{k_\rho^{D_1}} \right)^2 - 1 \right) \frac{m}{a} k_z H_m^{(1)} J_m \right\},$$

$$C_{A,m}^{D_2} = \frac{1}{\Delta} \frac{q}{4\pi i} \left\{ -\left(\left(\tilde{e}_\varphi + \frac{m}{a} \frac{k_z}{\left(k_\rho^{D_1}\right)^2} \tilde{e}_z \right) J_m + \frac{k_0^2}{k_\rho^{D_1}} J_m' \tilde{h}_z \right) \left(\left(\frac{k_\rho}{k_\rho^{D_1}} \right)^2 - 1 \right) \frac{m}{a} \frac{k_z}{k_0^2} H_m^{(1)} J_m \right.$$

$$+ \cdots + \left. \left(\left(\tilde{h}_\varphi + \frac{m}{a} \frac{k_z}{\left(k_\rho^{D_1}\right)^2} \tilde{h}_z \right) J_m + \frac{\varepsilon_r}{k_\rho^{D_1}} J_m' \tilde{e}_z \right) \frac{k_\rho}{k_\rho^{D_1}} \left(k_\rho H_m^{(1)} J_m' - k_\rho^{D_1} H_m^{(1)'} J_m \right) \right\}$$

$$(6.14)$$

Here,

$$\Delta = \left(\frac{k_\rho}{k_\rho^{D_1}} \right)^2 \left(k_\rho H_m^{(1)} J_m' - k_\rho^{D_1} H_m^{(1)'} J_m \right) \left(\varepsilon_r k_\rho H_m^{(1)} J_m' - k_\rho^{D_1} H_m^{(1)'} J_m \right)$$

$$- \left[\left(\left(\frac{k_\rho}{k_\rho^{D_1}} \right)^2 - 1 \right) \frac{m}{a} \frac{k_z}{k_0} H_m^{(1)} J_m \right]^2 \quad (6.15)$$

is the determinant that describes the eigenmodes of the fibre, which are found whenever it vanishes [23, 24].

Finally, the EEL spectral density, i.e., the probability of recording an energy loss between ω and $\omega + d\omega$, is given by the overlap integral (see Chap. 3) [25]

$$\Gamma^{\text{EELS}}(\omega) = \Re \frac{-1}{\pi\hbar\omega} \iiint d^3r \, \vec{J}(\vec{r}, \omega)^* \cdot \vec{E}^{\text{sca}}(\vec{r}, \omega). \quad (6.16)$$

To the lowest order of approximation, we assume no multiple interactions between the electron and the structure, and the total effect of the energy loss is negligible. That is, the action on the scattering structures is that of an electron moving at constant velocity on a straight line, as described by (6.4). Considering also the other separable symmetries, we disperse the EELS probability further according to the azimuthal order m and the wavenumber k_z along the fibre axis,

$$\Gamma^{\mathrm{EELS}}(\omega) = \int\limits_{-\infty}^{+\infty} dk_z \sum_m \Gamma_m^{\mathrm{EELS}}(k_z, \omega), \qquad (6.17)$$

which leads to the central result of this derivation. For a given set of symmetry parameters $\{m, k_z, \omega\}$, the EELS probability reads

$$\Gamma_m^{\mathrm{EELS}}(k_z, \omega) = \left(\frac{-q}{\pi\hbar\omega}\right) \Re \int\limits_{-\infty}^{+\infty} dx e^{-i\frac{\omega}{V}x} \times i^m e^{im(\varphi - \varphi_e)} e^{ik_y y_0} \times$$

$$\left[\left\{-\frac{k_\rho k_z}{\omega\varepsilon_0} H_m^{(1)\prime}(k_\rho\rho) C_{A,m}^{D_2} - \frac{im}{\rho} H_m^{(1)}(k_\rho\rho) C_{F,m}^{D_2}\right\} \cos\varphi \right.$$

$$\left. + \left\{\frac{im}{\rho}\frac{k_z}{\omega\varepsilon_0} H_m^{(1)}(k_\rho\rho) C_{A,m}^{D_2} - k_\rho H_m^{(1)\prime}(k_\rho\rho) C_{F,m}^{D_2}\right\} \sin\varphi\right] \qquad (6.18)$$

Using the recurrence formulae for Bessel functions, it can be further simplified to:

$$\Gamma_m^{\mathrm{EELS}}(k_z, \omega) = \left(\frac{-q}{\pi\hbar\omega}\right) \Re i^m \frac{k_\rho}{2} e^{-im\varphi_e} e^{ik_y y_0}$$

$$\times \left\{\left(-\frac{k_z}{\omega\varepsilon_0} C_{A,m}^{D_2}(k_z, \omega) - i C_{F,m}^{D_2}(k_z, \omega)\right) \int\limits_{-\infty}^{+\infty} dx e^{-i\frac{\omega}{V}x} H_{m-1}^{(1)}(k_\rho\rho) e^{i(m-1)\varphi}\right.$$

$$\left. + \left(\frac{k_z}{\omega\varepsilon_0} C_{A,m}^{D_2}(k_z, \omega) - i C_{F,m}^{D_2}(k_z, \omega)\right) \int\limits_{-\infty}^{+\infty} dx e^{-i\frac{\omega}{V}x} H_{m+1}^{(1)}(k_\rho\rho) e^{i(m+1)\varphi}\right\}$$

$$(6.19)$$

Equation (6.19) provides an instructive distinction between two very different mechanisms of obtaining resonant EELS signals. The first factors are the scattering coefficients. Notably, all four coefficient functions share a common resonance denominator Δ given by (6.15). The vanishing of this term is precisely the condition for the eigenmodes of the fibres. That is, whenever the excitation due to the swift electron is such that the field scattered by the fibre closely resembles an eigenmode, the divergence of the scattering coefficients implies a strong EELS signal.

A second curious feature of the fully expanded EELS probability $\Gamma_m^{\mathrm{EELS}}(k_z, \omega)$ is the integrals that are evaluated along the electron trajectory, $\vec{r}(x) = (x, y_0, 0) = (\rho\cos(\phi), \rho\sin(\phi), 0)$. They have the form of an overlap integral,

$$\int_{-\infty}^{+\infty} dx \left[e^{-i\frac{\omega}{v_{el}}x} \right] \cdot \left[H^{(1)}_{m\pm1}(k_\rho\rho)e^{i(m\pm1)\varphi} \right] \tag{6.20}$$

with factors resembling the spatial part $\exp(-i\omega v_{el}^{-1}x)$ of the Fourier component of the electron field at frequency ω and that of the fibre eigenmodes [cf. (6.3)]. Thus, a resonant transfer of energy from the electron to the plasmonic structure can be expected whenever these two factors exhibit closely matching spatial phase patterns. This is indeed what we observe in the FDTD calculations, as Fig. 6.6 shows. For a fixed frequency $\omega = 1.6\,\mathrm{eV}\,\hbar^{-1}$, the electron field oscillates along the trajectory with a constant wavelength of approximately 539 nm (Fig. 6.6b, right panel). Fourier transforms of the FDTD results to the frequency domain, evaluated for the same frequency ω, show that the scattered fields are particularly strong only when the electron passes the taper at specific local radii (Fig. 6.6). The field distributions, in turn,

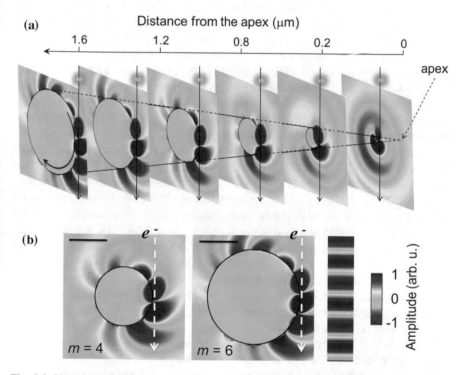

Fig. 6.6 Phase-matching condition imposed by the EELS overlap integral. Maps of the scattered electric field component, E_z, at an energy loss of 1.6 eV for **a** various electron impacts and **b** more specifically, electron impacts at distances of $L = 1\,\mu\mathrm{m}$ (left, $R = 383$ nm) and $L = 1.6\,\mu\mathrm{m}$ (middle, $R = 612$ nm) from the taper apex. The right panel shows the plane wave component $k_x = \omega v_{el}^{-1} = 11.7\,\mathrm{rad}\,\mu\mathrm{m}^{-1}$ ($2\pi/k_x = 539$ nm) of the electric field associated with the relativistic electron, which corresponds to an energy of 1.6 eV. Scale bars are 500 nm. Panel **b** is reproduced by permission from [4]; Copyright © 2015 American Chemical Society

resemble the mode field patterns of individual fibre modes (here, $m = 4$ and $m = 6$) but not exactly, since the interaction of the electron with the taper involves a multitude of eigenmodes.

An approximate selection rule for observing the phase matching resonance in EELS of fibres follows from finding a condition for which the overlap integral (6.20) attains its maximal value. A crude estimate is obtained if the integral is evaluated only over a certain "interaction length", where the integrand attains its largest magnitude, i.e., close to the fibre. This truncated interval becomes maximal when the phases of the electron field and the fibre mode field match over the interaction length. The (unrealistic) assumption that the interaction length extends only over small angles $\phi = \arctan(x/y_0) \approx x/R$ leads to the condition

$$\hbar\omega \approx (m \pm 1)\frac{\hbar v_{el}}{R} \qquad (6.21)$$

In comparing this to the empirical result from Fig. 6.2c, (6.1), we find that this expression already captures the qualitative features quite well. The hyperbolic functional relation between the resonance energy and the local radius is found to have a scaling factor of $\hbar v_{el} = 0.137\,\mu m \cdot eV$, which is on the same order of magnitude as the empirical value around $0.104\,\mu m \cdot eV$. As in the experiment, we find equidistant energy spacing between successive resonances at a given local radius. Nevertheless, quantitative agreement with the experimental data is not achieved by this simplistic phase-matching condition, indicating the need for a more extensive analytical treatment of (6.19) for much longer interaction.

Whereas in EELS experiments, the loss probability is dispersed only into its energy spectrum, (6.19) affords us with the opportunity to expand it further into its angular momentum about the fibre axis and the linear momentum along the fibre axis. Figure 6.7 shows an MREELS map, calculated for a gold fibre with a small radius of $R = 50$ nm and summed over all the m orders. Figure 6.7c shows individual MREELS maps for angular mode orders of $m = 0$ and $m = 1$. Evidently, the rotationally symmetric $m = 0$ mode makes the dominant contribution to the sum. In fact, the $m = 1$ mode contributes strongly only at energies above 2.0 eV, where the volume plasmon in gold is excited. The EELS signal of this comparatively thin fibre closely resembles the $m = 0$ eigenmode dispersion with a narrow spectral width of less than 100 meV. This is due mainly to the resonant denominator of (6.15). The phase-matching resonances according to (6.17) are not relevant.

However, this situation changes drastically when the gold fibre has a larger radius, which increases the effective "interaction length". In Fig. 6.8, the calculated MREELS map is shown for a fibre of radius $R = 400$ nm. Somewhat surprisingly, it is the radiative modes—characterized by $k_z < k_0$, where $k_0 = E/\hbar c$ is the vacuum wavenumber of light—that dominate the total EELS signal. Significant contributions due to the bulk plasmon are observed again for energies above 2.0 eV, and the evanescent contributions (for $k_z > k_0$) are negligible in comparison, with only faint hints of fibre eigenmode resonances for lower values of $|m|$. The EEL spectra for individual angular momentum orders, as shown in Fig. 6.8, elucidate the physics behind the

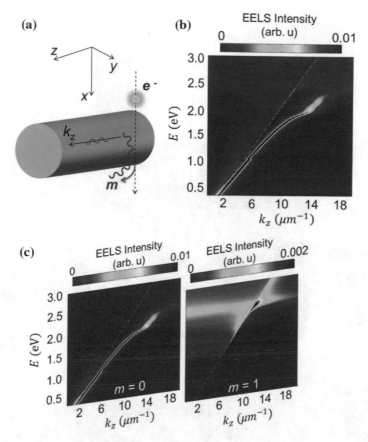

Fig. 6.7 Interaction of a swift electron with a 50 nm infinitely long gold fibre. **a** A swift electron interacting with a metallic fibre can excite both evanescent SPP modes that propagate along the fibre as well as radiative modes of the fibres with different angular momentum values. The colour scale was normalized individually for each panel. (Calculated MREELS maps versus the linear momentum along the axis of a gold fibre. **b** Summed over all the angular momentum contributions. **c** For the individual excited angular momentum orders $m = 0$ and $m = 1$. The straight dashed lines indicate the light line $(E = \hbar c_0 k_z)$

experimentally observed resonances in the EEL spectra. In fact, the radiative part of the EEL spectra at each momentum contribution exhibits a clear peak versus energy, the position of which depends on m. The spectral width of these resonances of a few hundred meV is well in accord with that of the experimentally observed resonances. Consequently, for such larger radii, the EELS signal is mostly influenced by the spatial phase matching along the electron trajectory and much less by resonances due to excitation of fiber eigenmodes.

To further underpin our interpretation, we have calculated the EELS spectra for increasing fiber radius, from $R = 40$ nm to $R = 800$ nm, thus representing the taper in the local radius approximation. Figure 6.9 shows the total computed EELS

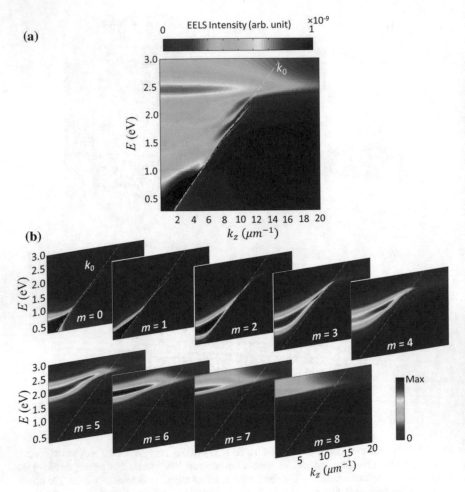

Fig. 6.8 Interaction of a swift electron with a 400 nm infinitely long gold fibre. **a** Calculated MREELS maps versus the linear momentum along the axis of a gold fibre. **b** Summed over all the angular momentum contributions. **c** For the individual excited angular momentum orders $m = 0$ to $m = 8$. The straight dashed lines indicate the light line ($E = \hbar c_0 k_z$). The colour scale was normalized individually for each panel

intensity, summed over m and integrated over k_z. At low energy loss we find that bound eigenmodes of the fiber are very dominant. Experimentally, though, we do not observe sharp fibermode-like resonance dispersions because the taper with its opening angle of 49° deviates too strongly from the geometry of fibres. Tapers with a much shallower opening angle (below 10°) would be required for that. Indeed, a taper with an opening angle of 49° cannot support the fibre eigenmodes, since such modes will couple strongly to the radiation continuum. The resulting map of the energy loss probability *versus* energy and local radius is displayed in Fig. 6.9b. Its resemblance with the experimental data from Fig. 6.2 is striking. For a one-to-one comparison, the

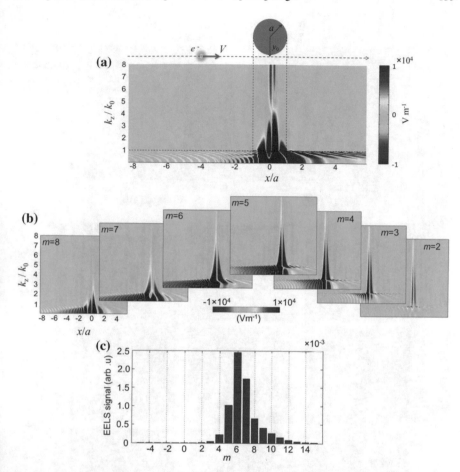

Fig. 6.9 Calculated electron-induced field projected along the electron trajectory versus the normalized linear-momentum k_z/k_0. Only in the immediate neighbourhood of the gold fibre within the range of $-0.5a < x < 0.5a$ are the evanescent waves excited. In the region outside this limit, radiating modes will be excited. **a** The total induced field and **b** the induced field for each azimuthal order. The lower panel shows the contribution of each azimuthal order to the total EELS signal

circles mark selected experimentally observed resonance maxima in Fig. 6.9b. This is a strong indication that the nearly hyperbolically dispersed resonance maxima are due to phase matching over extended parts of the electron trajectory, as suggested by the overlap integral (6.20). A final proof of this conclusion is presented in Fig. 6.9c, where individual panels for azimuthal orders $m = 0 \ldots 8$ are shown. Clearly, each of the observed resonance dispersions is associated with a different order.

At each specific electron energy-loss and for a given fibre radius, a certain mode with a given azimuthal order is hence the dominant contribution to the EELS signal, though a range of various m-orders is always excited (Figs. 6.9 and 6.10). This is the reason that the electron-induced fields shown in Fig. 6.6 do not sustain the

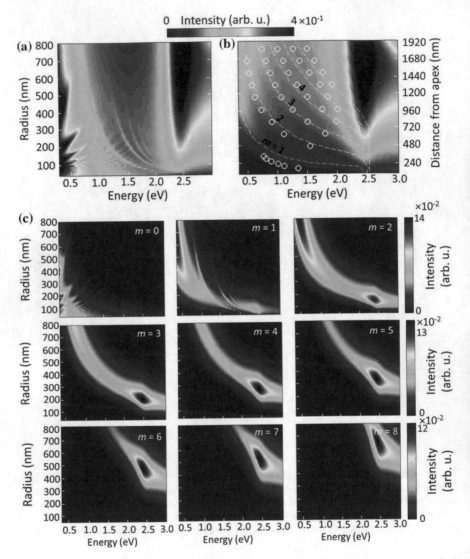

Fig. 6.10 Computed electron energy-loss probability as a function of the energy loss and fibre radius, **a** including and **b** excluding the contribution of the *evanescent* modes with $k_z > k_0$. The electron is travelling at a distance of 1 nm from the surface. Experimental data are depicted as solid circles. The dashed lines track the maximum EELS signal according to (6.21) for a given radius and mode order. **c** Computed EEL spectra from each angular momentum order. Reprinted by permission from [4]; Copyright © 2015 American Chemical Society

complete symmetry of a certain m-order. Moreover, the interaction length of the electron with the taper is rather extended much beyond the few nanometres in the very neighbourhood of the taper and is increasingly scaled by the radius of the fibre. In particular, higher-order modes contribute most significantly to the radiation and present a larger interaction length compared to the lower-order modes.

In summary, EELS maps were recorded along the shaft of conical gold tapers. At the apex, these confirm the ultrabroadband ability of such tapers to concentrate light into nanometric volumes. At remote locations along the shaft, unexpected resonance dispersions are found. All these observations are confirmed by FDTD calculations. For deeper insight into their nature, we adopt the framework of viewing the taper as being composed of infinitesimal fibre segments and develop an analytical expression for the EELS signal from the fibres. In addition to the possibility of exciting eigenmodes, a phase-matching condition is identified, which also results in the emergence of resonance maxima. In the application of this analysis, we identify the scattering of radiative modes as being responsible for the observed resonance dispersions. Our results thus verify the excitation of modes with certain angular momentum orders, placing an emphasis on the crucial role that phase matching plays in the interpretation of EELS data. It should be noted that reflection from the apex can also modulate the EELS signal due to the formation of a standing wave along the taper shaft. However, as is apparent from the movies provided as supporting information, the contribution of the reflection from the apex is much weaker than the contribution of the radiative modes for a tape with an opening angle of 45°, and thus, it has minor importance under our experimental conditions.

Our findings highlight an important caveat to the well-known intimate relation between the EELS probabilities and the PLDOS near a specimen under study. Strictly, these two quantities are proportional only for structures that are translationally invariant (or periodic) along the electron trajectory. Also, in thin planar sample geometries, the EELS probability closely mimics the PLDOS of the structure. Generally, though, for finite sample structures with dimensions along the electron trajectory that are larger than the inverse wavenumber $k_x^{-1} = v_{el}\omega^{-1}$ of the electromagnetic fields of the electron along this direction, phase matching must be taken into consideration, and the results do depend on the electron speed. Moreover, specific electron velocities are required to couple with radiative modes with $m > 1$ in mesoscopic metallic fibres, tapers, and spheres; specifically, electrons at energies lower than 100 keV cannot detect the kind of phase matching related resonances observed in the present study. To further verify the soundness of this statement, we have calculated the EELS for fibres with different radii and for various electron energies, as shown in Fig. 6.11. To calculate the EELS maps, the MREELS contributions were all calculated, including $m = 0$ to $m = 20$, and the evanescent mode contributions were discarded by limiting the integration to the range of $k_z < k_0$. Additionally, the whole EELS signal was calculated by summing over all the MREELS orders. Figure 6.11 demonstrates the important effect of the electron kinetic energy and hence its velocity on the calculated EELS signal, which is due to the dependence of the momentum selection rule and thus the phase-matching condition on the electron velocity.

However, all the discussions above are valid only for tapers with large opening angles, where the propagation of the electromagnetic energy along the shaft is largely suppressed thanks to the non-adiabatic behaviour of the taper. In the following, we switch to tapers of various opening angles and describe the effect of the opening angle on the PLDOS that these tapers can sustain and, correspondingly, the EELS signal.

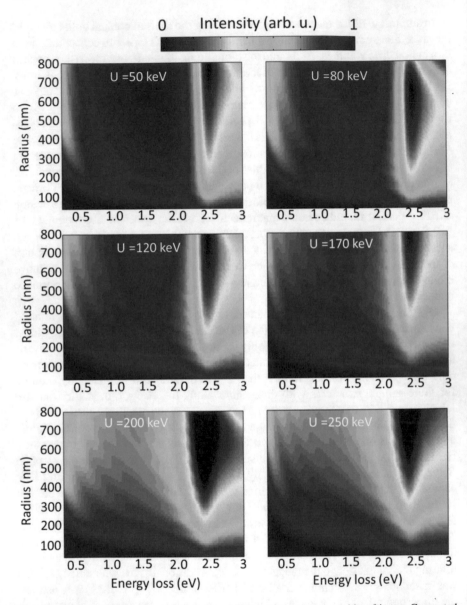

Fig. 6.11 Impact of electron energy on the visibility of the phase-matching fringes. Computed electron energy-loss probability as a function of the energy loss and fibre radius for various electron energies as depicted on each frame, summed over all m orders from $m = 0$ to $m = 20$. Reprinted by permission from [4]; Copyright © 2015 American Chemical Society

6.2   Reflection from the Apex in Gold Tapers with Small Opening Angles

A comprehensive understanding of the surface plasmon polaritons (SPPs) sustained by conical gold tapers, especially the nonlocal behaviour of the excited modes, such as their corresponding phase constant and attenuation, is requested to enable better manipulation and application. As is already known, a radially symmetric surface plasmon polariton mode (the fundamental TM mode with a vanishing angular momentum, $m = 0$) is supported by tapers to achieve adiabatic nanofocusing [26]. On the other hand, higher-order modes with larger angular momenta are strongly coupled to the radiation continuum at critical radii and thus cannot be preserved at the apex [4, 27]. The behaviours of all these modes can be significantly tailored by the geometry of the gold tapers, such as the opening angle and the radius of curvature at the apex [28].

Recently, two papers [4, 29] reported on the investigation of surface plasmons on three-dimensional, single-crystalline gold tapers with different opening angles using EELS. The signatures of multiple plasmonic modes were observed as resonances in the EEL spectra in both studies. Although the experimental results in these two studies showed important similarities, the suggested mechanisms behind the observed resonances were fundamentally different. One interpretation, based on studies of tapers with small opening angles, suggested that the EELS signatures of the investigated tapers were due to the interference between the forwardly propagating fundamental azimuthal surface plasmon polariton mode ($m = 0$ with radial symmetry) and its **reflection** from the apex (Fig. 6.12) [29]. Another interpretation, based on studies of tapers with larger opening angles, suggested that the EELS resonances were consequences of the **phase matching** between the exciting electron wave packet and the electric field re-emitted by higher m-order modes of the taper (Fig. 6.12) [4]. In other words, the swift electron dynamically interacts with the plasmonic modes of the taper and experiences a force induced by the electromagnetic field re-emitted by those modes. Hence, a maximum, or scattering resonance, appears in the EELS signal whenever the electron self-interferes constructively with the electromagnetic field re-emitted by the excited plasmon modes. Both interpretations match well with numerical simulations of Maxwell's equations for the corresponding taper geometries in the time domain. Very recently, an analytical model for the EEL spectra of conical metal tapers has been introduced [30] that confirmed the occurrence of these two different mechanisms in cylindrical nanostructures.

In the present article, we aim to clarify the origin of the EELS resonances observed for conical gold tapers by systematically investigating tapers of varying opening angles using both electron energy-loss spectroscopy and numerical finite-difference time-domain (FDTD) simulations. We show that for all tapers, both phase matching and reflection contribute to the EELS signal. For small taper angles, the apex reflection of the SPP wave packet launched by the swift electrons dominates. The resulting interferogram is highly sensitive to the temporal chirp of the time-delayed SPP wave

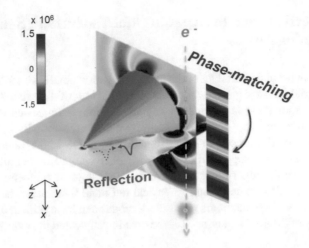

Fig. 6.12 Schematic drawing of the dominant optical modes of a conical gold taper excited by a relativistic electron. The passing electron excites both evanescent SPP modes and radiative, higher-order angular momentum modes. Only the lowest-order, $M = 0$, SPP mode of the taper is guided towards and reflected at the apex. It gives rise to a time-delayed evanescent field, acting back on the electron and inducing reflection resonances in the EEL spectra. In contrast, the excitation of radiative, higher-order angular momentum modes of the taper induces the emission of electromagnetic fields. Their back-action on the passing electron leads to phase-matching resonances in the EEL spectra. The corresponding electric field distributions in both the xy- and yz-planes are presented, as well as the x component of the Fourier-transformed oscillating electric field associated with the exciting electron at the corresponding energy. Reprinted by permission from [5]; Copyright © 2016 American Chemical Society

packet. For large taper angles, however, the apex reflections contribute less, and the excitation of radiative higher-order angular momentum modes of the taper gives rise to resonances in the EEL spectra.

Recent experimental and theoretical results [4, 27, 29, 30] indicate that the opening angle of conical gold tapers plays a key role in tailoring the properties of the excited surface plasmons sustained by this structure. To analyse the effect of the opening angle on the EEL spectra and to elucidate the interplay between the reflection and phase-matching resonances, we perform numerical FDTD calculations on two tapers with extremely small and large full opening angles (5° and 30°). In the following, we use the notations "n" and "m" to indicate the order of multiple resonances dominated by reflection and phase matching, respectively.

For this, we first simulate the electric fields generated by a swift relativistic electron, moving perpendicular to the taper axis. The electron, with a voltage kinetic energy of 200 keV and moving at $V = 0.67c$ (c: speed of light in vacuum), is modelled as a Gaussian wave packet with a transverse width of 8 nm and a time duration of less than 100 as, as in [19]. For each taper, the component of the scattered electric field projected along the electron trajectory, $|E_x|$, at a distance $L = 1460$ nm from the apex is plotted on a logarithmic scale as a function of the electron trajectory and time. This field component is chosen because it is the only component that contributes to the EELS signal. The large travelling distance of 1460 nm is intentionally set to separate the signals of the reflected wave and the phase-matched wave along the time axis.

We first analyse the results for a 5° taper (Fig. 6.13a). Three different contributions to the re-emitted fields can be distinguished. First, the excitation of plasmon oscillations gives rise to the emission of propagating electromagnetic fields. These propagate away from the taper axis at the speed of light c to the free space, and retardation results in field maxima that are inclined with respect to the time axis at an angle indicated by the dashed yellow line. In the time domain, the excitation of different radiative, higher-order angular momentum SPP modes of the taper gives rise to a complex interference pattern dying out within a few fs after the arrival of the electron beam. Bound, evanescent SPP modes of the taper are also excited by the electron beam. For a 5° taper, only the lowest-order mode, $m = 0$, is a bound mode and can propagate away from the excitation spot towards the apex. Due to this

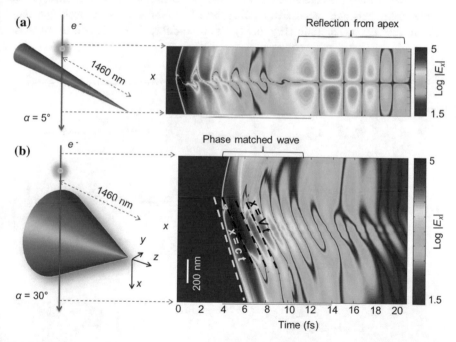

Fig. 6.13 Different mechanisms contributing to the EEL resonances: phase-matching versus reflection. FDTD simulations of the total scattered electric field along the electron trajectory $|E_x|$, plotted on a logarithmic scale versus the electron trajectory (nm) and time (fs) when a relativistic electron is passing at a constant distance of $L = 1460$ nm for gold tapers with full opening angles α of **a** 5° and **b** 30°. L, the distance between the taper apex and the electron impact location along the taper surface, is sufficiently large to separate the phase-matched and reflected waves along the time axis. The slope of the black dashed line denotes the velocity of the exciting electron, which is approximately 2/3 of the speed of light in vacuum [shown as yellow dashed lines in (**a**) and (**b**)]. Reprinted by permission from [5]; Copyright © 2016 American Chemical Society

propagation, the contribution of the bound mode to the re-emitted field lasts only for a few fs. Importantly, the $m = 0$ mode remains a bound mode even for vanishingly small local taper radii and hence will generate a spatially highly confined electromagnetic field near the apex [3, 31, 32]. For finite taper radii, the incident mode is partially reflected [29, 33]. This reflection is clearly seen as a pronounced, time-delayed evanescent field burst arriving at $\Delta t \sim 11$ fs. Since only the radial (E_r) and on-axis (E_z) electric field components of the $m = 0$ mode are nonzero [26], the projection along the electron trajectory imposes a null intensity at the nearest distance of the electron to the taper. The time delay of the reflected SPP field provides a measure of the path-integrated local effective refractive index $n_{eff}(r)$ of the $m = 0$ mode, $\Delta t = 2 \int_{z_e}^{z_a} \frac{n_{eff}(z)}{c} dz$, between the impact point of the electron at z_e and the apex at z_a. From this, we estimate an average effective refractive index of $n_{eff} \approx 1.15$, reasonably close to what is expected based on the tabulated values of bulk gold [34]. The possible time delays [35] due to the local increase in n_{eff} near the apex [32] seem too small to be reliably extracted. Instead, a pronounced chirp of the reflected SPP wave packet is visible in the FDTD simulation as a clear decrease in time delay between successive oscillations of the reflected evanescent plasmon field. In the spectral domain (Fig. 6.14), the time-delay SPP oscillations give rise

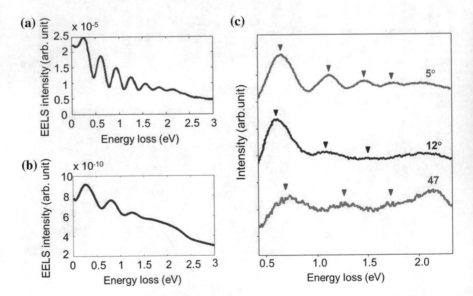

Fig. 6.14 Calculated EEL spectra for tapers with different opening angles. **a** Calculated EEL spectrum from the system depicted in Fig. 6.12a, which is dominated by the reflection mechanism. **b** Calculated EEL spectrum from the system of Fig. 6.12b, which is dominated by the phase-matching mechanism. **c** Experimental zero-loss-corrected EEL spectra at an equivalent distance from the apex of $L = 800$ nm for gold tapers with opening angles of 47°, 12° and 5° in the energy-loss range from 0.5 to 2 eV. The zero-loss peak contribution was subtracted from the individual spectra by using a power-law fit, and the maxima of the spectra are normalized and vertically shifted for comparison. Reprinted by permission from [5]; Copyright © 2016 American Chemical Society

to pronounced oscillations in the EEL spectra, [29, 30] with a fringe spacing that decreases with the increasing loss energy. The loss energy dependence of the EELS fringes hence provides a quantitative measure of the chirp of the bound $m = 0$ SPP wave packet upon its propagation from the injection point to the apex and back.

A fundamentally different time structure of the re-emitted field is observed for the excitation of a taper with a large opening angle of $30°$ (Fig. 6.13b). Here, again, the radiative higher-order angular momentum modes of the taper are impulsively excited. Their re-emitted fields give rise to a strong burst of light that propagates away from the taper surface at the speed of light in vacuum (dashed yellow line). For this taper, not only the $m = 0$ mode but also the $m = 1$ and $m = 2$ modes are bound, evanescent SPP modes, and their excitation launches a coherent SPP wave packet that propagates towards the apex. The characteristic of this wave packet is the different inclination angle (black dashed line), defined by the velocity v_{el} of the electron. The large radius of this taper leads to a longer interaction length along the electron trajectory and hence a finite propagation distance of the SPP wave packet during the interaction. The SPP propagation along the taper surface then results in the observed decrease in this component within the first few femtoseconds. Both the $m = 1$ and $m = 2$ modes are coupled into the far field at finite distances from the apex. The $m = 0$ mode again propagates to the very apex of the taper where it is partially converted into far-field radiation and partially back-reflected. For an opening angle of $30°$, however, the amplitude of the back-reflected field is fairly weak. Hence, the modulation contrast in the EEL spectra (Fig. 6.14b), which is much smaller than that for a $5°$ taper (Fig. 6.14a), arises from the action of the higher-order angular momentum modes, $m > 0$, which are all coupled into far-field radiation before reaching the apex region. Therefore, phase matching clearly makes the dominant contribution to the modulation contrast for such tapers with large opening angles.

To verify these theoretical findings, we now compare experimental EEL spectra of gold tapers with various full opening angles from $5°$ to $47°$ at equivalent distances of $L = 800$ nm from their apex. The full opening angles for the gold tapers investigated in this paper are $5° \pm 1°$, $12° \pm 2°$, $19° \pm 1°$ and $47° \pm 4°$. Obviously, there are a larger number of resonances for the gold taper with the $5°$ opening angle with a fringe spacing and modulation depth that approximately match those in the simulations. Evidently, these fringes are reflection-dominated, as concluded in [30].

The modulation depth is much more pronounced than that of the resonances observed for the gold taper with a $47°$ opening angle, which is evidently dominated by phase matching. Interestingly, the EELS resonances of the gold taper with a $12°$ opening angle show a much reduced fringe contrast compared to those of the two gold tapers with $5°$ and $47°$ opening angles. This is an unexpected result that cannot easily be understood by considering only either the phase matching or the reflection-type contribution to the EELS fringes. Instead, we conclude that for the $12°$ taper, both phase matching and reflection are active. Their overlap leads to a smearing of the EELS signal because both resonances appear within the same energy range. In addition to showing an individual EEL spectrum at a fixed location L, we also compute and measure the EEL spectra as a function of the distance (distance–energy

EELS maps) to analyse the spectral features in more detail. This allows us to directly compare the simulated and experimental distance–energy EELS maps.

As shown in Fig. 6.15, there is good agreement between the experimental and computed EELS maps of gold tapers with different opening angles. Slight differences are attributed to a non-perfect apex shape, surface roughness or non-constant opening angle. For all the tapers, there is an extremely broadband resonance near the apex, extending from 0.5 eV, a lower limit set by the finite energetic width of the zero-loss peak, to 2.0 eV, an upper limit set by the onset of the interband absorption of gold. This broad resonance shows that the taper can act as a broadband waveguide that can localize surface plasmon polariton fields, i.e., the $M = 0$ mode, at its very apex. In fact, the broadband signal at the apex reflects the capturing of the full local

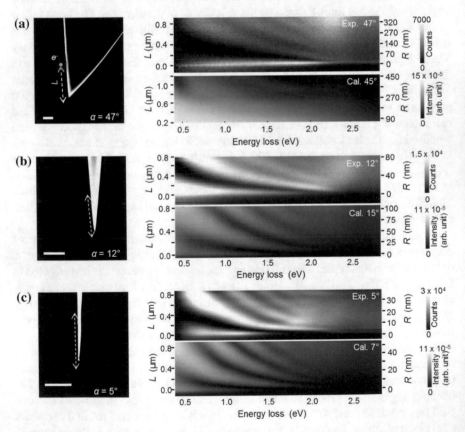

Fig. 6.15 Experimental and numerical energy-distance EELS maps. Dark-field imaging and spectroscopy of gold tapers with full opening angles of 47° **a**, 12° **b** and 5° **c**. The position $L = 0$ μm corresponds to the taper apex, and the scanning direction is upward. Both the experimental and computed electron energy-loss spectra are plotted as functions of the impact location along the taper shaft and the corresponding local radius. In the experimental EEL spectra, the contribution of the elastic zero-loss peak was subtracted by a power-law fit. Scale bar lengths are 500 nm. Reprinted by permission from [5]; Copyright © 2016 American Chemical Society

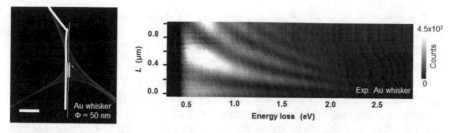

Fig. 6.16 STEM annular dark-field image and zero-loss-corrected distance–energy EELS map of a gold whisker with a diameter of ~50 nm. The location $L = 0\,\mu$m corresponds to the left end of the whisker, and the scanning direction is towards the top. The scale bar length is 200 nm. Reprinted by permission from [5]; Copyright © 2016 American Chemical Society

density of states by the taper and the perfect quality of our tapers. It also shows that such tapers, when excited at the apex, can capture the electron-induced radiation with high collection efficiency, as investigated elsewhere [37]. Discrete resonances along the shaft are observed in both the experimental and computed EELS maps. When decreasing the opening angle, more resonances in each sequence are observed within a certain distance from the apex. Furthermore, the reflection-dominated EELS map displays a more distinct contrast among the multiple resonances (Fig. 6.15c) than the phase-matching-dominated one in Fig. 6.15a. Indeed, by further decreasing the opening angle, the taper asymptotically evolves into a fibre with a small radius, for which only the reflection from the ends will be responsible for the observed resonances in the EELS signal.

In essence, the experimental distance–energy EELS maps (Fig. 6.15) show a gradual transition from reflection to phase-matching with the opening angle. To further support this observation, Fig. 6.17a compares the total scattered electric field $\log |E_x|$ (superposition of all the excited plasmons) of a half-infinite gold whisker (see Fig. 6.16) with a 50 nm diameter with that of gold tapers with opening angles of 10°, 20° and 50° at a constant distance of $L = 1460$ nm. For this gold whisker, only the fundamental $m = 0$ mode is expected to be sustained by this structure according to previous theoretical calculations [4]. Note that phase matching can also become relevant for whiskers at a sufficiently large radius. Therefore, this gold whisker represents the case of pure reflection. As the opening angle of the gold taper increases to 10° and 20°, the contribution of the phase-matched waves becomes apparent. When reaching approximately 50°, only phase-matched waves are visible in the corresponding $\log |E_x|$ plot.

Although the energy dispersions of the reflection-dominated and the phase-matching-dominated EELS resonances appear very similar, the physics for forming this energy dispersion is intrinsically different. We therefore apply curve fitting using hyperbolic functions to the experimental EELS resonances of a half-infinite gold whisker [38] with a 50 nm diameter as a reference of pure reflection and to those of a gold taper with a 47° opening angle as a representative of phase matching.

The reflection-induced resonances have energy dispersions proportional to the inverse distance between the impact location and the end of the whisker with a virtual length increase $E = \kappa_L/(L + L_0)$ as shown in Fig. 6.17b, where L_0 is almost a constant of (0.202 ± 0.024) μm for all the resonances $N = 1, 2, 3$ and 4. The other fitting parameter κ_L depends on the sequence number N in a linear fashion as $\kappa_L = (0.379N + 0.1553)$ μm eV. Figure 6.18a illustrates that the observed EELS resonances of the gold whisker can be understood as a reflection-type resonance [39, 40]. Therefore, the fitting parameter L_0 is interpreted in terms of a length increase, as shown in Fig. 6.18a, which is linked to a phase shift $\Delta\varphi$ at the whisker end in the energy range between 0.5 and 2.0 eV.

On the other hand, the dispersion of the EELS resonances of a gold taper with a 47° opening angle can be well captured by a hyperbolic function of the form $E = \kappa/(L + L_0)$. With $L = R/\sin(\alpha/2)$ being the distance from the apex and

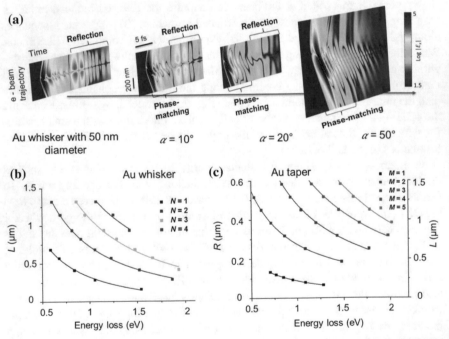

Fig. 6.17 Reflection and phase-matched waves in the scattered electric field. **a** Transition from reflection to phase matching: numerically calculated total scattered electric field along the electron trajectory, $\log|E_x|$, versus electron trajectory (nm) and time (fs) for a gold whisker and tapers with different opening angles of 10°, 20° and 50° when a relativistic electron is traversing the taper surface at a constant distance of $L = 1460$ nm from the apex. **b** Relations of maxima in EEL spectra versus the distance from the end of a gold whisker with a 50 nm diameter (dots). Curves are fits to a hyperbolic function $E = \kappa_L/(L + L_0)$, where κ_L and L_0 are constants. **c** Dispersions of maxima in EEL spectra versus the distance from the end of the gold taper with a 47° opening angle (dots). Curves are fits to a hyperbolic function $E = \kappa_R/(R + R_0)$, where κ_R and R_0 are constants. Reprinted by permission from [5]; Copyright © 2016 American Chemical Society

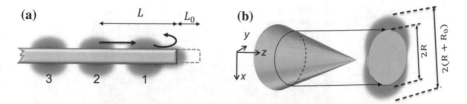

Fig. 6.18 Schematic of different phenomena involved in the observation of the resonances. **a** Reflection-type (standing wave anti-node) mechanism for the SPP resonances on the gold whisker. The physical meaning of the fitting parameter L_0 is related to the phase shift at the end of the gold whisker. **b** Phase-matching principle for large-opening-angle tapers along the electron trajectory. The physical meaning of the fitting parameter R_0 is relevant to the effective interaction length along the electron trajectory. Reprinted by permission from [5]; Copyright © 2016 American Chemical Society

R and α being the local radius and the full opening angle, respectively, the fitting function can be transformed into $E = \kappa_R/(R + R_0)$. Contrary to the gold whisker, the fitting parameters κ_R and R_0 for the taper with a 47° opening angle have different meanings since the phase-matching principle is dominant in this system. Figure 6.18b illustrates that the fitting parameter R_0 can be viewed as an extension of the geometric interaction length between the SPP field and the fast electron along the trajectory beyond $2R$.

The interaction length extension R_0 may be because the interaction already starts once the fast electron enters the near-field region of the taper structure [4]. In addition, R_0 also includes the phase lag between the SPP and the fast electron. κ_R is therefore associated with the phase matching, linearly depending on the azimuthal mode order $\kappa_R = (Am + B)$ μm eV, with the linear fitting parameters $A = 0.254$ and $B = 0.201$. Considering this, the swift electron acquires a phase of $\omega(2R + 2R_0)/v_{el}$ by moving through the near-field region of the taper, whereas the excited plasmons with an azimuthal order m acquire a phase of $m\Delta\varphi$. In the phase-matching picture, a resonance is observed in the distance–energy map whenever these two phases are equivalent, at the energy of $E = m\Delta\varphi\hbar v_{el}/(2R + 2R_0)$, where $E = \hbar\omega$ is the energy loss due to the excited plasmon and $\Delta\varphi$ is the angular rotation of the plasmon along the circumference. This simple model matches perfectly with the hyperbolic relation stated above. The value of $\Delta\varphi$ can be thus deduced as a function of the angular momentum order $\Delta\varphi = (0.363m + 1.308)$ radian. The significance of this model is its ability to determine the parameters $\Delta\varphi$ and R_0, by comparison to experimental results. In particular, the parameter R_0 demonstrates the actual increase in the EELS cross section compared to the real size of the structure. Therefore, it seems to be worthwhile to define an electron-induced cross section, in analogy to the so-called scattering, extinction, and absorption cross sections in far-field optical experiments.

In addition, the hyperbolic fitting function $E = \kappa_{R,L}/(R(L) + R_0(L_0))$ introduced here can help us to distinguish between the phase-matching and reflection mechanisms. To better demonstrate this, we have plotted the values of the slope of $\kappa_{R,L}$ (as it is linearly dependent on the EELS resonance order) for gold tapers with

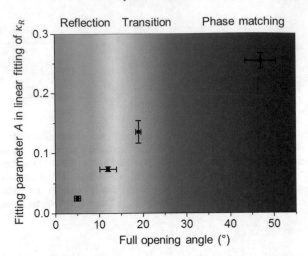

Fig. 6.19 Differentiation of experimental EEL resonances of gold tapers with different opening angles of 5°, 12°, 19° and 47° by empirical hyperbolic fitting $E = \kappa_{R,L}/(R(L) + R_0(L_0))$. The fit parameter $\kappa_{R,L}$ for each taper linearly depends on its resonance order as $\kappa_R = (Am + B)$ or $\kappa_L = (An + B)$. The linear fitting parameter A (slope of $\kappa_{R,L}$) exhibits a monotonic increase as a function of the full opening angle. The uncertainties of the measured opening angles and of the fit parameters are attached horizontally and vertically to each data point. Reprinted by permission from [5]; Copyright © 2016 American Chemical Society

full opening angles of 5°, 12°, 19° and 47°, as shown in Fig. 6.19. The slope of the fit parameter $\kappa_{R,L}$ increases monotonically with the opening angle. In other words, the value of the phase-matching-induced EELS resonance is almost 10 times larger than that of the reflection-induced EELS resonances. For intermediate opening angles (transition between reflection and phase matching), the linear fitting parameter A also exhibits intermediate values. Another way to distinguish phase matching from reflection would be to vary the kinetic energy of the incoming electrons, which only affects phase matching.

In summary, we have clarified the link between the phase matching and the reflection mechanisms contributing to EELS resonances in electron-induced surface plasmons of three-dimensional gold tapers. This was achieved by studying a large range of opening angles both experimentally and by simulation. Phase matching and reflection coexist in the system of gold tapers and can both contribute to the EELS signal. The EELS signals due to phase matching and reflection show close similarity. However, the contribution of either reflection or phase matching varies with both the local radius of the gold taper at the impact location and the distance to the taper apex along the shaft. Near the apex (<2–3 μm), the opening angle of the taper can be considered a crucial parameter. At small opening angles, the reflection of surface plasmon polaritons from the taper apex is dominant since small radii limit the excitation of higher-order azimuthal plasmons. Moreover, the long-range propagation of the plasmons along the taper apex is better supported by such tapers. In contrast, phase matching manifests itself in the maxima of the interference pattern recorded in the energy–distance EELS map at larger opening angles, mostly because of the longer interaction length between the electron and the taper along the electron trajectory. In between, a gradual transition from the reflection-based mechanism towards the phase-matching principle is observed by increasing the opening angle.

Additionally, the energy loss due to the reflection-induced EELS resonances is proportional to the inverse distance between the impact location and the reflection end with a virtual length increase, where the virtual length increase represents the phase shift at the reflection end. On the other hand, the energy dispersion of phase-matching-induced EELS resonances is proportional to the inverse local radius with an extension of the interaction length along the electron trajectory. This stems from the fact that the initial interaction starts when the electron enters the near-field region of the taper.

6.3 Summary

In this chapter, we studied the optical density of states in three-dimensional gold tapers with various opening angles, numerically and experimentally using EELS. We demonstrated that for tapers with large opening angles (larger than 30°), phase-matching between the evanescent field of a swift electron and the excited plasmons with angular momentum orders $|m| > 0$ results in peaks in the observed EEL spectra. However, for tapers with small opening angles, reflection from the apex is the main mechanism. Our observations demonstrate that the suitability of EELS for mapping the PLDOS of samples applies only to geometrically thin structures.

References

1. M.I. Stockman, Nanofocusing of optical energy in tapered plasmonic waveguides (in English). Phys. Rev. Lett. **93**(13) (2004). doi: Artn 137404 https://doi.org/10.1103/physrevlett.93.137404
2. S. Schmidt et al., Adiabatic nanofocusing on ultrasmooth single-crystalline gold tapers creates a 10-nm-sized light source with few-cycle time resolution. ACS Nano **6**(7), 6040–6048 (2012). https://doi.org/10.1021/nn301121h
3. C. Ropers, C.C. Neacsu, T. Elsaesser, M. Albrecht, M.B. Raschke, C. Lienau, Grating-coupling of surface plasmons onto metallic tips: a nanoconfined light source. Nano Lett. **7**(9), 2784–2788 (2007). https://doi.org/10.1021/nl071340m
4. N. Talebi et al., Excitation of mesoscopic plasmonic tapers by relativistic electrons: phase matching versus eigenmode resonances (in English). ACS Nano **9**(7), 7641–7648 (2015). https://doi.org/10.1021/acsnano.5b03024
5. S. Guo et al., Reflection and phase matching in plasmonic gold tapers. Nano Lett. **16**(10), 6137–6144 (2016). https://doi.org/10.1021/acs.nanolett.6b02353
6. S. Grésillon et al., Experimental observation of localized optical excitations in random metal-dielectric films. Phys. Rev. Lett. **82**(22), 4520–4523 (1999). https://doi.org/10.1103/physrevlett.82.4520
7. P.G. Etchegoin, E.C.L. Ru, M. Meyer, An analytic model for the optical properties of gold. J. Chem. Phys. **125**(16), 164705 (2006). https://doi.org/10.1063/1.2360270

8. F. Huth et al., Resonant antenna probes for tip-enhanced infrared near-field microscopy. Nano Lett. **13**(3), 1065–1072 (2013). https://doi.org/10.1021/nl304289g
9. L.B. Felsen, N. Marcuvitz, *Radiation and Scattering of Waves* (Prentice Hall: Englewood Cliff, New Jersey, 1973)
10. J.J. Bowman, T.B.A. Senior, P.L.E. Uslenghi, *Electromagnetic and Acoustic Scattering by Simple Shapes* (Hemisphere, New York, 1987)
11. M.A. Lyalinov, Electromagnetic scattering by a circular impedance cone: diffraction coefficients and surface waves. IMA J. Appl. Math. **79**(3), 393–430 (2014). https://doi.org/10.1093/imamat/hxs072
12. K. Kurihara, A. Otomo, A. Syouji, J. Takahara, K. Suzuki, S. Yokoyama, Superfocusing modes of surface plasmon polaritons in conical geometry based on the quasi-separation of variables approach. J. Phys. A: Math. Theor. **40**(41), 12479–12503 (2007). https://doi.org/10.1088/1751-8113/40/41/015
13. J.C. Ashley, L.C. Emerson, Dispersion relations for non-radiative surface plasmons on cylinders. Surf. Sci. **41**(2), 615–618 (1974). doi:https://doi.org/10.1016/0039-6028(74)90080-6
14. C.A. Pfeiffer, E.N. Economou, K.L. Ngai, Surface polaritons in a circularly cylindrical interface: surface plasmons. Phys. Rev. B **10**(8), 3038–3051. (1974). https://doi.org/10.1103/physrevb.10.3038
15. L. Novotny, C. Hafner, Light propagation in a cylindrical waveguide with a complex, metallic, dielectric function. Phys. Rev. E **50**(5), 4094–4106. (1994). https://doi.org/10.1103/physreve.50.4094
16. M.I. Stockman, Nanofocusing of optical energy in tapered plasmonic waveguides (vol 93, 137404, 2004) (in English). Phys. Rev. Lett. **106**(1) (2011). doi: Artn 019901 https://doi.org/10.1103/physrevlett.106.019901
17. N. Issa, R. Guckenberger, Optical nanofocusing on tapered metallic waveguides. Plasmonics **2**, 31–37 (2007)
18. M. Esmann et al., K-space imaging of the eigenmodes on a sharp gold taper for near-field scanning optical microscopy. Beilstein J. Nanotechnol. **4**, 603–610 (2013)
19. N. Talebi, W. Sigle, R. Vogelgesang, P. van Aken, Numerical simulations of interference effects in photon-assisted electron energy-loss spectroscopy. New J. Phys. **15**(5), 053013 (2013). https://doi.org/10.1088/1367-2630/15/5/053013
20. J.A. Stratton, *Electromagnetic Theory* (Wiley, New York, 2007)
21. R.F. Harrington, *Time-Harmonic Electromagnetic Fields* (McGraw-Hill Book Company, New York, 1961)
22. R. Garciamolina, A. Grasmarti, A. Howie, R.H. Ritchie, Retardation effects in the interaction of charged-particle beams with bounded condensed media (in English). J. Phys. C Solid State **18**(27), 5335–5345 (1985). doi:https://doi.org/10.1088/0022-3719/18/27/019
23. J.C. Ashley, L.C. Emerson, Dispersion-relations for non-radiative surface plasmons on cylinders. Surf. Sci. **41**(2), 615–618 (1974). https://doi.org/10.1016/0039-6028(74)90080-6
24. C.A. Pfeiffer, E.N. Economou, K.L. Ngai, Surface polaritons in a circularly cylindrical interface—surface plasmons. Phys. Rev. B **10**(8), 3038–3051 (1974). https://doi.org/10.1103/physrevb.10.3038
25. F.J.G. de Abajo, M. Kociak, Probing the photonic local density of states with electron energy loss spectroscopy (in English) Phys. Rev. Lett. Article **100**(10), 4, Art no. 106804. https://doi.org/10.1103/physrevlett.100.106804
26. C.C. Neacsu, S. Berweger, R.L. Olmon, L.V. Saraf, C. Ropers, M.B. Raschke, Near-field localization in plasmonic superfocusing: a nanoemitter on a tip. Nano Lett. **10**(2), 592–596 (2010). https://doi.org/10.1021/nl903574a
27. M. Esmann et al., k-space imaging of the eigenmodes of sharp gold tapers for scanning near-field optical microscopy, (in English). Beilstein J. Nanotech **4**, 603–610 (2013). doi:https://doi.org/10.3762/bjnano.4.67
28. S. Thomas, G. Wachter, C. Lemell, J. Burgdörfer, P. Hommelhoff, Large optical field enhancement for nanotips with large opening angles. New J. Phys. **17**(6), 063010 (2015). (Online). Available: http://stacks.iop.org/1367-2630/17/i=6/a=063010

29. B. Schröder et al., Real-space imaging of nanotip plasmons using electron energy loss spectroscopy. Phys. Rev. B **92**(8), 085411 (2015). https://doi.org/10.1103/PhysRevB.92.085411

30. S.V. Yalunin, B. Schröder, C. Ropers, Theory of electron energy loss near plasmonic wires, nanorods, and cones. Phys. Rev. B **93**(11), 115408 (2016) (Online). Available: http://link.aps.org/doi/10.1103/PhysRevB.93.115408

31. A.J. Babadjanyan, N.L. Margaryan, K.V. Nerkararyan, Superfocusing of surface polaritons in the conical structure. J. Appl. Phys. **87**(8), 3785–3788 (2000). https://doi.org/10.1063/1.372414

32. M. Stockman, Nanofocusing of optical energy in tapered plasmonic waveguides. Phys. Rev. Lett. **93**(13) (2004). https://doi.org/10.1103/physrevlett.93.137404

33. M.S. Jang, H. Atwater, Plasmonic rainbow trapping structures for light localization and spectrum splitting. Phys. Rev. Lett. **107**(20), 207401 (2011) (Online). Available: http://link.aps.org/doi/10.1103/PhysRevLett.107.207401

34. P.B. Johnson, R.W. Christy, Optical constants of the noble metals. Phys. Rev. B **6**(12), 4370–4379 (1972) (Online). Available: http://link.aps.org/doi/10.1103/PhysRevB.6.4370

35. V. Kravtsov, J.M. Atkin, M.B. Raschke, Group delay and dispersion in adiabatic plasmonic nanofocusing. Opt. Lett. **38**(8), 1322–1324 (2013). https://doi.org/10.1364/ol.38.001322

36. S.R. Guo et al., Reflection and phase matching in plasmonic gold tapers (in English). Nano Lett. **16**(10), 6137–6144 (2016). https://doi.org/10.1021/acs.nanolett.6b02353

37. P. Groß, M. Esmann, S.F. Becker, J. Vogelsang, N. Talebi, C. Lienau, Plasmonic nanofocusing—grey holes for light. Adv. Phys. X, 1–34 (2016). https://doi.org/10.1080/23746149.2016.1177469

38. G. Richter, K. Hillerich, D.S. Gianola, R. Mönig, O. Kraft, C.A. Volkert, Ultrahigh strength single crystalline nanowhiskers grown by physical vapor deposition. Nano Lett. **9**(8), 3048–3052 (2009). https://doi.org/10.1021/nl9015107

39. J. Dorfmüller et al., Fabry-Pérot resonances in one-dimensional plasmonic nanostructures. Nano Lett. **9**(6), 2372–2377 (2009). https://doi.org/10.1021/nl900900r

40. X. Zhou, A. Hörl, A. Trügler, U. Hohenester, T.B. Norris, A.A. Herzing, Effect of multipole excitations in electron energy-loss spectroscopy of surface plasmon modes in silver nanowires. J. Appl. Phys. **116**(22), 223101 (2014). https://doi.org/10.1063/1.4903535

Chapter 7
Photon–Induced and Photon—Assisted Domains

Abstract Combining laser and electron guns in electron microscopes has created a plethora of opportunities for characterizing the chemical reactions and near-field distributions of nanostructures (Park et al. in New J Phys 12:123028, 2010 [1]; Siwick et al. in Science 302:1382–1385, 2003 [2]; Barwick et al. in Nature 462:902–906, 2009 [3]; Zewail and Thomas in 4D electron microscopy, imaging in space and time. Imperial College Press, Singapore, 2010 [4]; Zewail in Science 328:187–193, 2010 [5]; Sciaini and Miller in Rep Prog Phys 74:096101, 2011 [6]; Miller in Annu Rev Phys Chem 65:583–604, 2014 [7]; Vanacore et al. in Nano Today 11:228–249, 2016 [8]). This method, which is called ultrafast electron diffraction or photon-induced near-field electron microscopy (PINEM), where for the former, diffraction patterns are acquired, and for the latter, spectra are acquired, has recently been further developed by several groups around the world (Lee et al. in Struct Dynam-Us 4:044023, 2017 [9]; Feist et al. in Nature 521:200–203, 2015 [10]; Piazza et al. in Nat Commun 6:6407, 2015 [11]; Bücker et al. in European Microscopy Congress 2016: Proceedings. Wiley-VCH Verlag GmbH & Co. KGaA, 2016 [12]; Cao et al. in Sci Rep 5:8404, 2015 [13]) into a time-resolved pump-probe characterization methodology. Photoemission electron guns are currently controlled to create sub-picosecond electron pulses with great spatial coherence, almost at the same level as that of field-driven electron guns (Ehberger et al. in Phys Rev Lett 114:227601, 2015 [14]). Moreover, by controlling the laser-electron jitter by means of microwave or THz cavities, the longitudinal broadening of the electron pulses has been considerably reduced (Walbran et al. in Phys Rev Appl 4:044013, 2015 [15]; Verhoeven et al. in Ultramicroscopy 188:85–89, 2018 [16]; Verhoeven et al. in Struct Dynam-Us 3:054303, 2016 [17]; Maxson et al. in Phys Rev Lett 118:154802, 2017 [18]; Yang et al. in Nat Commun 7:11232, 2016 [19]; Weathersby et al. in Rev Sci Instrum 86:073702, 2015 [20]). Additionally, it was recently demonstrated that the interaction of electron pulses with the laser-induced near field of nanostructures will cause the

Portions of the text of this chapter have been re-published with permission from [22], re-printed under the CC BY license; [23], re-printed under the CC BY license; [24], re-printed under the CC BY license; [25], re-printed under the CC BY license;

© Springer Nature Switzerland AG 2019
N. Talebi, *Near-Field-Mediated Photon–Electron Interactions*,
Springer Series in Optical Sciences 228,
https://doi.org/10.1007/978-3-030-33816-9_7

153

attosecond bunching of single-electron pulses in space-time (Priebe et al. in Nat Photonics 11:793–797, 2017 [21]). Here, we briefly discuss recent achievements in PINEM, with a focus on the electron-light interactions in the near field of nanoobjects and their combination with time-resolved spectroscopy. After that, we introduce new possibilities by exploring the photon-assisted domain and propose novel methods for revealing the spectral phase and investigating the mutual correlations among photonic systems.

7.1 Electron-Energy Gain Spectroscopy

In 2008, Garcia de Abajo and Kociak introduced EEGS as a tool to improve the energy resolution of electron microscopes [26]. In an EELS setup, electrons undergo spontaneous emission upon interaction with nanostructures and lose energy. Hence, in the presence of external continuous-wave (CW) laser excitations, stimulated photon emission and photon absorption processes will also be possible. In other words, at sufficiently high laser energies, electron energy-gain and energy-loss processes are both possible. Based on Fermi's golden rule, they obtained an energy gain expression for single photon-electron interactions as

$$\Gamma^{\text{EEGS}}(\omega) = \left(\frac{e}{\hbar\omega}\right)^2 \left|\tilde{E}_z\big(x_0, y_0, k_z = \omega/v_{\text{el}}; \omega\big)\right|^2 \tag{7.1}$$

which is related to the intensity of the electric field, in contrast with the EELS formalism [see (7.2)]. Obviously, using the classical approach, as used to obtain the EELS probability in Chap. 2, an incorrect relation for the gain probability is derived, by which Γ^{EEGS} is linearly related to the electric field amplitude [26]. Moreover, negative probabilities can be obtained if (7.1) is incorrectly used for calculating the energy gain/loss spectra in the presence of external laser excitation.

7.2 Photon-Induced Near-Field Electron Microscopy

In PINEM, a pulsed laser excitation is synchronized by the electrons emitted from a photoemission electron gun within a photon-pump and electron-probe time-resolved spectroscopy apparatus (see Fig. 7.1a). Observations of the energy gain and loss processes in a series of spectacular experiments carried out in the group of Zewail and co-workers demonstrated a series of loss and gain peaks in the spectra of carbon nanotubes (see Fig. 7.1b) [1, 3, 27]. Indeed, those experiments established the possibility of multiple-photon gain and loss processes, in addition to the single-photon processes. Similar multiphoton processes were observed in the interaction of electrons with a protein vesicle (see Fig. 7.1c). To describe the effects theoretically, a scattering theory based on semiclassical propagators for the Schrödinger equation

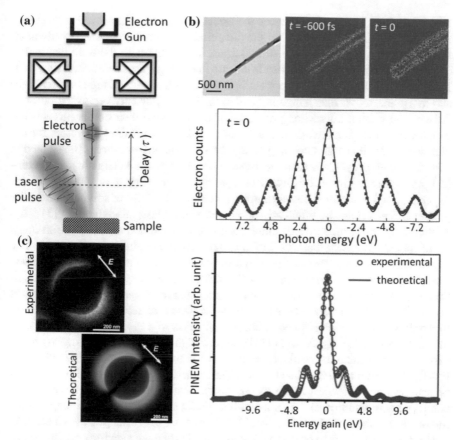

Fig. 7.1 Photon-induced near-field electron microscopy. **a** The setup is a pump-probe spectroscopy apparatus where the pump is a laser pulse and the probe is an electron pulse. The delay between these two pulses is precisely controlled using an optical delay line. **b** top: PINEM images on a single carbon nanotube, shown at two different delays, bottom: PINEM intensity versus energy at a delay time $t = 0$ [3]. **c** Left: Normalized intensity of the first PINEM peak for a truncated copper thin film and a silver triangular nanoparticle versus the distance from the edge. For the copper sample, two regions at distances smaller and larger than 100 nm can be distinguished, for which dynamical and kinematical processes are respectively dominant. Right: Energy exchanges between electrons and photons in dynamical and kinematical processes [1]. Re-printed by permission from [22] under the CC BY license

has been developed by Park and coworkers and has been demonstrated to be in good agreement with experimental results [1]. Followed by the pioneering experiments mentioned above, many other groups have already advanced the PINEM technique towards realizing better temporal and spatial resolutions. We outline here only works based on spectroscopy; for ultrafast electron diffraction, the readers are referred to the review by Zewail [28] and more recent works [29]. Feist et al. recently demonstrated that the strong resonant interaction between the laser beam and electrons in the vicinity of a nanostructure can be used to induce multilevel Rabi oscillations, with equal Rabi frequencies between the levels, into an electron wave function (see Fig. 7.2a) [10]. The probability for the transition between the photonic levels in an equal-Rabi case is given by Bessel functions, in contrast with the Rabi oscillation in a harmonic oscillator system that is describable using a Poisson distribution [30]. This behaviour results in a strong oscillation in the intensity of the observed gain and loss peaks rather than a monotonic decrease in the intensity versus the order of the electron-photon interaction processes. They described their observations using a quantum optical framework on annihilation and creation ladder operators for the plasmon modes that commute (neglecting the spontaneous emission). Based on this approach, they proposed a more accurate description of the kinematic and dynamical processes mentioned by Zewail in a previous work (Fig. 7.1c). This approach in general, which is called boson sampling, holds the promise of solving intractable problems that cannot be efficiently simulated using classical computers [31, 32]. Additionally, they proposed a quantum coherent manipulation of free-electron states using optical metrology that was further experimentally confirmed and expanded by demonstrating a Ramsey-type phase control of free-electron motions [21, 33]. Interestingly, such near-field interactions with electron wave packets leads to the generation of a train of attosecond electron pulses propagating in space-time, which might be further utilized for ultrafast diffraction [34]. Piazza et al. recently demonstrated up to 9 orders of photon emission and absorption peaks in a PINEM time-energy map of silver nanowires (Fig. 7.2b) [11]. They also highlighted an interesting aspect of PINEM for recording the spatial dependence of individual loss and gain peaks [35]. Finally, Ryabov and Baum have shown that transverse electron recoil can indeed be acquired and utilized to map the spatiotemporal distribution of vectorial field components in addition to the gain and loss peaks accumulated in the longitudinal momentum of the free-electron waves [36]. They used THz laser pulses and electron pulses as short as 80 fs to respectively initiate and probe the electromagnetic excitations in a split-ring resonator (Fig. 7.2c). Their unique observations in this field are intriguing to the fields of ultrafast science and electron microscopy in principle, as they clearly demonstrate the advantage of recording the transverse momentum of the electron in addition to its longitudinal momentum in a PINEM setup.

7.3   Ultrafast Point Projection Electron Microscopy

Complementary to the developments of ultrafast techniques in TEMs, ultrafast point projection electron microscopy (PPM) has recently appeared as a probe of ultrafast electron dynamics in samples. PPM holds the advantage of a less complicated setup without the need for the massive magnetic lenses used in TEMs, though the spatial resolution has yet to be improved enough to compete with ultrafast TEMs. The key element behind the development of PPM is the recent progress in controlling the ultrafast dynamics of photoelectrons emitted from sharp tips, either by employing adiabatic nanofocusing [37] for transferring the grating-coupled light from the shaft at several micrometres away from the apex to the apex itself (see Fig. 7.3a and b) or by

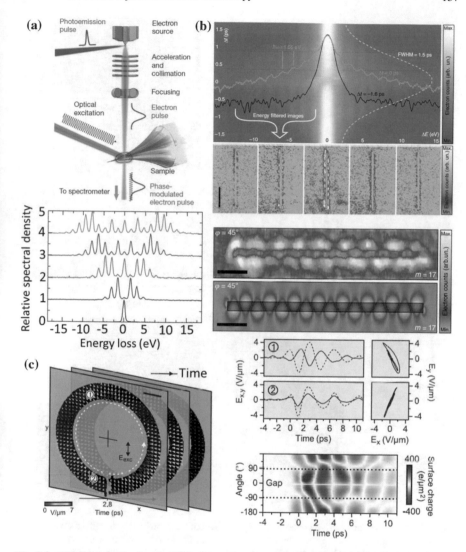

Fig. 7.2 PINEM activities in several groups. **a** Top: By precisely controlling the phase and polarization of the incident light, the PINEM setup can be used to coherently manipulate the free-electron states. Bottom: Electron energy spectra for several field amplitudes of 0, 0.023, 0.040, 0.053, and 0.068 V nm-1 [10]. **b** Top: Time-energy phase-space PINEM map of silver nanowires. Bottom: Experimental and numerical PINEM images of the field distribution on an isolated nanowire with light excitation polarized at the angle of 45∘ with respect to its longitudinal axis [11]. **c** Left: Vector distribution of the lateral components of the electric field for a split-ring resonator excited with a THz laser pulse and probed with electron pulses with only an 80 fs duration. Right, top: Temporal distribution of the excitations at two distinct test points marked by A and B in the left panel and the polarization state of the electric field. Right, bottom: Temporal distribution of the electric field along the inner circumference shown by the dashed line in the left panel [36]. Re-printed by permission from [22] under the CC BY license

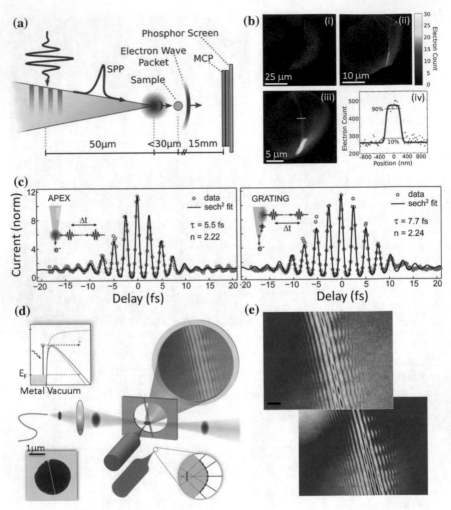

Fig. 7.3 PPM with laser-driven nanotips. **a** By adiabatically nanofocusing the plasmons from the grating to the apex, the sample can be positioned arbitrarily close to the tip, which helps achieve higher magnifications [41]. **b** (i–iii) PPM images from a Ag nanowire recorded by nanofocusing at three different magnifications and (iv) cross section along the white line in the left lower panel [41]. **c** Comparison of the interferometric autocorrelation of photoelectrons emitted from the sample using either adiabatic nanofocusing or direct illumination of the tip apex, as shown in the insets [37]. **d** Recording an interference pattern of electron beams exciting a carbon nanotube using PPM [14]. **e** Several interference fringes are observed, both for a laser-driven gun (upper panel) and for a field-emission gun (lower panel) [14]. Re-printed by permission from [22] under the CC BY license

directly illuminating the tip apex [14, 38, 39] (see Fig. 7.3c and d). The former setup has advantages of bringing the sample closer to the tip, which by itself is beneficial for shorter electron pulses, and achieving higher magnification (see Fig. 7.3b). Using adiabatic nanofocusing, the realization of electron pulses as short as 8 fs has been reported [37]. However, as noted by Mueller et al., dispersion may cause broadening of the emitted electron pulse by a few femtoseconds (see Fig. 7.3c, which shows the interferometric autocorrelation of the photoelectrons from the tip) [37]. A certain advantage of PPM is the ability to perform inline electron holography. Electron holography was suggested by Gabor in 1948 [40] as a new technique to overcome the problems of lens aberration and to improve the electron microscopy techniques. Still, holography in PPM is used to improve the spatial resolution by an order of magnitude [37] thanks to the transverse interference patterns that are observed in the PPM images. Moreover, the comparison between the interference fringes for the case of field-driven and laser-driven tungsten nanotips demonstrates that the effective size of the source is not very different for the two cases. In other words, the promise of utilizing point sources in PPM holds true for a laser-driven source as well as for a field-emission source (see Fig. 7.3e) [14].

7.4 Time and Energy Resolutions in Ultrafast Electron Microscopy

Improving the time resolution of ultrafast techniques in general and ultrafast electron microscopy in particular is highly demanded for elucidating the charge-transfer dynamics and the time evolution of the electromagnetic oscillations in nanostructures and understanding the behaviour of correlated materials [42]. However, it has been increasingly noticed that reaching an attosecond time resolution demands the utilization of sophisticated techniques such as highly precise synchronization with microwave cavities [16, 43] or THz cavities [44] to chirp the electron pulse distribution in time-energy phase space. Although electron pulses as short as a few fs have been realized, the increase in the time resolution is concomitant with the broadening of the pulse in the energy domain, which covers a large bandwidth and is beneficial for spectroscopy and the investigation of resonant phenomena as well. In addition to the synchronization cavities, the inclusion of an appropriate low-emittance field-emitter electron gun utilizing nanolocalized photoemission from a zirconium oxide covered (100)-oriented single-crystalline tungsten tip has also been reported [45], which significantly improves the beam specifications. The ultrafast electron microscope facilitated by this tip has been shown to sustain a spatial resolution of 0.9 Å, an energy broadening of 0.6 eV, and a temporal resolution of 200 fs.

7.5 Photon-Assisted Domain and Spectral Interferometry

Considering electron beams and laser excitations as two individual probes of sample excitation, two different domains have already been discussed, namely, the electron-induced domain and the photon-induced domain. Members of the former domains include EELS and CL techniques, whereas in the latter domain, PINEM and ultra-fast PPM are obvious examples. Within the electron-induced domain, the electron interacts with the optical modes of the sample that are initially in the ground state. In the classical picture, one can assume that the evanescent field of the electron probe polarizes the sample, and the electron is hence scattered by its induced field (see Chap. 3). Interestingly, multiple photon emission processes might also occur, which can result in multiple loss peaks. This effect has been observed in experiments carried out by Powel and Swan as early as 1959 [46, 47] and has also been observed in simulations considering the retardation effect [48, 49]. Within the photon-induced domain, however, the sample is pumped with a strong laser beam into a superposition of number states. In addition to the loss channels caused by the spontaneous emission, stimulated emission and absorption processes will happen as well and, indeed, in most experimental cases become the only relevant loss and gain channels (see Sect. 7.2). In other words, the electron-induced polarization in the sample can be neglected. One might try to classify these two apparently distinct electron-induced and photon-induced domains versus the intensity of the incident laser beam, where at very low laser intensities, the electron-induced polarization can still be probed using EELS. In contrast, at sufficiently high laser intensities, only PINEM excitations are probed by the electron, as the electron-induced spontaneous emission can be neglected.

At an intermediate intensity, coherent electron-induced radiation should be able to interfere with the laser-induced fields [50] (see Fig. 7.4a). This interference phenomenon can then be probed using EELS (see Fig. 7.4b) and is highly beneficial for recording the spectral phase as an example. The laser field amplitude at which this transition occurs depends on the kinetic energy of the electron beam and its impact parameter. For the plasmonic nanostructure considered here and for an electron at a kinetic energy of 200 keV traversing the near field at 5 nm away from the nanostructure, this field amplitude is approximately 10^5–10^6 V m^{-1}, as determined by simulation. Interestingly, conventional EELS in the absence of laser excitation is a more useful spectroscopic tool with the ability to probe all the optical modes that the structure sustains, including, here, the resonances at $E = 0.25$, 1.0, and 1.55 eV. The laser pulse will indeed peak only the mode at $E = 1$ eV. However, the introduced laser-electron pump-probe approach lets us determine the temporal evolution of the selected plasmon excitations. To record such an interference map, ultrashort sub-fs electron pulses are required, which are perfectly synchronized with the laser field [51]. This is because the duration of the electron pulses should be shorter than a cycle of the laser excitation to be able to appropriately probe the gain and loss oscillations along the delay axis [50]. Additionally, only coherent radiation such as transition radiation and coherent bremsstrahlung can interfere with the incident photon pulses.

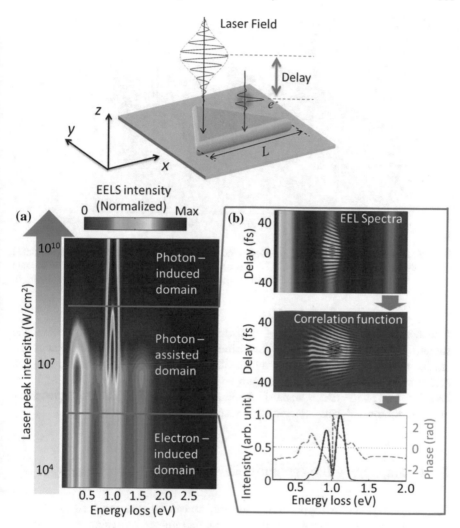

Fig. 7.4 Distinguished domains in the interaction of single-electron pulses and laser pulses with nanostructures. An electron with a kinetic energy of 200 keV interacts with a triangular gold nanostructure with an edge length of $L = 400$ nm and a thickness of 30 nm, which is positioned upon a Si_3N_4 substrate [50]. The structure is pumped by a y-polarized laser pulse with a temporal broadening of 10 fs and a carrier photon energy of 0.98 eV. The electron pulse has a temporal broadening of only 48 as. **a** Electron-induced, photon-assisted, and photon-induced domains are categorized versus the intensity of the laser pulse. **b** Within the photon-assisted domain, the strength of the induced laser illumination is at the level of the electron-induced polarization and can further interfere with it. Additionally, interference patterns can be recorded using EELS and be utilized for recovering the spectral phase. Re-printed by permission from [22] under the CC BY license

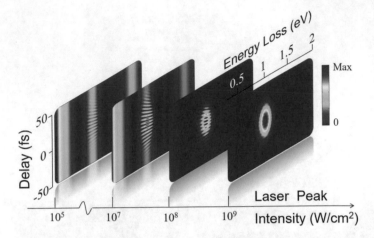

Fig. 7.5 EELS map in the time-energy phase space at different laser peak intensities (Re-printed by permission from [23] under the CC BY license), ranging from the electron-induced domain at a laser peak intensity of 10^5 Wcm^{-2} to the photon-induced domain at a laser peak intensity of 10^9 Wcm^{-2}

Incoherent CL in the interaction of electrons with dielectrics, semiconductors, and semimetals will, however, hamper the visibility of the fringes. Assuming that coherent interference between the energy loss and gain channels is possible, one can record such interference only at certain laser peak intensities (Fig. 7.5). At lower laser intensities, the EELS signal is dominant, whereas at higher laser intensities, only the photon-induced polarization is asserted.

7.5.1 Improving the Temporal Resolution

To observe the interference phenomenon noted above, we have proposed a system for improving the synchronization of the electron and photon pulses upon their arrival at the sample. It is indeed not necessary to trigger electrons with photons. Another approach that is proposed here is to trigger the photon trajectories by electrons. Relativistic electron beams support various mechanisms of radiation, e.g., Cherenkov radiation [52], Smith-Purcell radiation [53], and transition radiation [54]. These radiation mechanisms were fully discussed in Chap. 2. The Smith-Purcell effect is commonly utilized to design free-electron lasers in the THz frequency range [55, 56]. Due to the interaction of the electrons with gratings rather than with a single scatterer, coherent electromagnetic radiation is formed in the far-field region as a result of the interference of the transition radiation from each individual element in the grating in a phase-matched manner. Considering that each scatterer re-irradiates a single photon upon interaction with the electron, certainly more than two photons, and hence a certain number of grating elements, is required to form an interference pattern in the far-field region [57].

Considering the abovementioned mechanisms of radiation, an electron–based spectral interferometry technique is proposed here that can simply be integrated into current state-of-the-art electron microscopes. The idea is to use coherent transition radiation from the interaction of the electron with a structure in order to enforce each single electron to produce its own conjugate photon pulse and hence its own time reference. It should be mentioned that although time-resolved cathodoluminescence experiments with picosecond time resolutions have already been demonstrated [58], there has still been no report on an attosecond time resolution and the reconstruction of the spectral phase of the samples using electron microscopes.

An electron-driven photon source (EDPHS) is considered here. To incorporate this technique into an electron microscope, a mesoscopic metamaterial-based EDPHS is designed. Recently, it has been demonstrated that metamaterials in interaction with relativistic electrons can radiate coherent optical beams [59]. However, the metamaterial-based EDPHS proposed here is designed in such a way as to generate coherent and focused transition radiation along the electron trajectory, as shown in Fig. 7.6a. The emission should be unidirectional so that the generated photons are focused upon the sample instead of being scattered over a very large angular range in

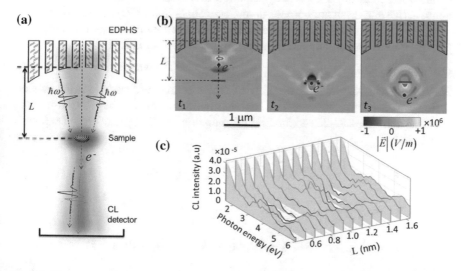

Fig. 7.6 The concept of spectral interferometry with electron microscopes. **a** The setup is composed of an EDPHS, which in interaction with a swift electron generates photons that are scattered in a unidirectional way towards the sample and focused on it. **b** The EDPHS excitation interferes with the electron-induced CL emission of the sample, either destructively or constructively, depending on the distance L between the sample and EDPHS. The sample is positioned at the focal point of the EDPHS. **c** The calculated CL spectra versus distance and energy, where the shift of resonances in the energy–distance map due to the interference is apparent. The sample here is a silicon disc with a thickness of 40 nm and a radius of 150 nm. Re-printed by permission from [24] under the CC BY license

the far-field region, as is schematically shown in Fig. 7.6b. The criteria for the design of this EDPHS and the functionality and characteristic behaviours of the proposed EDPHS will be presented later.

The emission from the EDPHS propagates at the speed of light, which is faster than the electron speed. However, due to retardation, it takes a few femtoseconds for the generated photons to leave the EDPHS. We use this fact to control the arrival time of the electron and the EDPHS radiation at the sample position by changing the distance between the sample and the EDPHS (Fig. 7.6b). A characteristic delay is thus assigned to the photon pulse with respect to the electron as $\tau = (1 - \beta)Lv_{el}^{-1}$, where v_{el} is the electron speed and $\beta = v_{el}/c$ is the speed of the electron normalized to the speed of light (c). L is the distance between the EDPHS and the sample.

It is well known that transition radiation is mutually coherent with the relativistic electron that created it [60]. Moreover, since the dissipative loss is well controlled in the EDPHS, the generated photons are converted into a radiation continuum and focused upon the sample. In such a design, the electron-induced polarization in the sample interferes with the EDPHS radiation, and this interference phenomenon can be either constructive or destructive, depending on the distance L between the EDPHS and the sample. This interference pattern becomes evident in both the EELS and CL spectra. The focus in this paper will be on the CL detection, since the role of interference is more apparently captured in the far-field radiation because the nonradiative and dissipation losses do not contribute to the CL spectra [61, 62].

Theoretically, the angle-resolved CL spectrum is interpreted as the number of photons emitted per unit of solid angle emission Θ and per unit of photon frequency ω to the far field, as $\Gamma_{CL}(\omega, \Theta) = \frac{1}{4\pi^2 \hbar k_0} \left| \vec{E}_0(\omega, \Theta) \right|^2$, where $k_0 = \omega/c$, $\hbar$ is the Planck constant, and $\vec{E}_0$ is the amplitude of the scattered electric field in the far-field region [60]. Since the scattered field has contributions from both the electron-induced polarization in the sample and the EDPHS radiation, the CL spectra can be explicitly written as:

$$
\begin{aligned}
\Gamma_{CL}(\omega, \Theta) &= \frac{1}{4\pi^2 \hbar k_0} \left| \vec{E}_{el}^{ind}(\omega, \Theta) + e^{i\omega\tau} \vec{E}_{EDPHS}(\omega, \Theta) \right|^2 \\
&= \frac{1}{4\pi^2 \hbar k_0} \left| \vec{E}_{el}^{ind}(\omega, \Theta) \right|^2 + \frac{1}{4\pi^2 \hbar k_0} \left| \vec{E}_{EDPHS}(\omega, \Theta) \right|^2 \\
&\quad + \frac{1}{2\pi^2 \hbar k_0} \left| \vec{E}_{el}^{ind}(\omega, \Theta) \right| \cdot \left| \vec{E}_{EDPHS}(\omega, \Theta) \right| \cos(\varphi_{el} - \varphi_{EDPHS} - \omega\tau)
\end{aligned}
\tag{7.2}
$$

where $\vec{E}_{el}^{ind}$ and $\vec{E}_{EDPHS}$ are the electric-field contributions induced by the electron and EDPHS excitations, respectively. Equation (7.1) clearly indicates that, similar to any interference phenomenon, the spectral phases φ_{el} and φ_{EDPHS} due to the electron and EDPHS excitations are encoded in the overall intensity. We observe a maximum intensity whenever $\varphi_{el} - \varphi_{EDPHS} - \omega\tau = 2n\pi$, where n is an integer. Moreover, by controlling the delay (τ) by simply changing the distance, we can record a full interference pattern in the time–energy ($\tau - E$) phase space (Fig. 7.6c).

The fundamental requirements for designing a highly efficient EDPHS that can be used for the application considered here can be outlined as 1st—the ability to couple the electron-induced polarization to the far-field radiation, 2nd—the ability to provide a unidirectional focused radiation upon the sample, and 3rd—the ability to manifest a broad-band far-field spectrum. To meet these criteria, the concomitant incorporation of the photonic crystal and metamaterial concepts is considered here. First, in order to propose a mesoscopic EDPHS with the ability to focus the transition radiation, an inverted superlens is introduced. This multi-layered structure is composed of Ag/Al_2O_3 thin films, each with a thickness of 30 nm, positioned upon a spherical plano–concave silica lens (see Fig. 7.7a), with overall thicknesses of $h_1 = 400$ nm and $h_2 = 600$ nm and a radius of curvature of 1000 nm. In an effective medium approximation, the multi-layered metal/dielectric structure demonstrates an electromagnetic hyperbolic behaviour. This behaviour exhibits a very large photonic density of states and hence an enhanced coupling efficiency of the quantum emitters

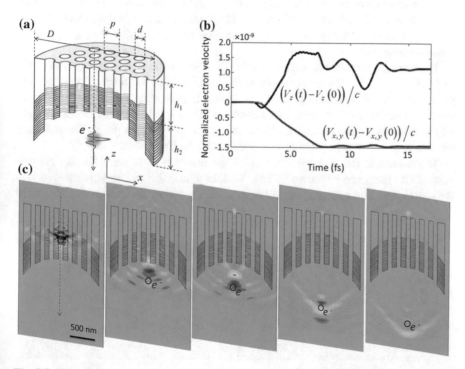

Fig. 7.7 Time-domain response of the EDPHS. **a** Topology of the EDPHS, which is composed of an inverted superlens with an incorporated void hexagonal photonic crystal. $h_1 = 400$ nm, $h_2 = 600$ nm, $D = 1.6\,\mu m$, $d = 100$ nm, and $p = 200$ nm. **b** Computed modulation in the relative electron velocity $((\vec{v}_{el}(t) - \vec{v}_{el}(0))/c$, where t is time, $\vec{v}_{el}$ is the electron velocity, and **c** is the speed of light) due to the interaction with the EDPHS. **c** z-component of the electron-induced electric field at the times depicted in each frame. Re-printed by permission from [24] under the CC BY license

to the optical modes. Moreover, a void hexagonal photonic crystal is also incorporated into the lens. The collective excitation of a photonic crystal lattice can generate a better coupling of the excited beams to the radiation continuum, as will be demonstrated later. This behaviour is especially appealing here to prohibit the activation of non-radiative de-excitation processes.

To understand the interaction of the swift electron at a kinetic energy of 200 keV with the EDPHS, a fully self-consistent numerical approach within the framework of the combined

Maxwell and Lorentz equations has been applied, with the details provided in Chap. 3. The interaction of the relativistic electrons with nanostructures and thin films can be well explained by the undepleted pump approximation_ENREF_35. However, the amount of recoil that the electron receives is illustrative (Fig. 7.7b) and demonstrates well the intensive radiation from the EDPHS. The time reference is set here as the time at which the electron enters the simulation domain. Within the time interval of 2.4–4.4 fs, the electron is traversing the silica substrate, whereas at $t = 7.0$ fs, it leaves the EDPHS near-field region. Interestingly, the modulation of the electron velocity is even more apparent during the time interval of 7–14 fs, when the electron is still in the wake of the EDPHS emission. The strongest recoil that the electron receives is at $t = 11.3$ fs, when the electron reaches the focal point of the plano-concave lens.

The time domain simulations perfectly demonstrate the temporal response of the EDPHS (Fig. 7.7c). The emission from the EDPHS is in the form of an ultrashort positively chirped few-cycle pulse. The radiation is quite intense near the focal point, which is approximately 900 nm away from the EDPHS and along the electron trajectory and, interestingly, catches up to the electron exactly at this point. The radiation spreads very fast by overtaking the electron after the focal point.

To understand the spectral features of the emission from the EDPHS, the calculated CL spectrum is shown in Fig. 7.8a, in which the emission is divided into the transmitted and reflected contributions. This splitting, as well as the computed

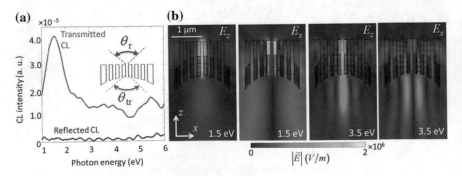

Fig. 7.8 Frequency-domain response of the EDPHS. **a** Calculated CL spectra at the detector positions above and below the EDPHS. The solid angles of the detectors are taken as 1.84 sr. $\theta_{tr} = \theta_r = \pi/2$. **b** Magnitude of the Fourier-transformed electric field components at the energies depicted in each frame. Re-printed by permission from [24] under the CC BY license

Fourier transform of the electric field components (see Fig. 7.8b), demonstrates the unidirectional behaviour of the emission and the focusing capability of the superlens. The polarization state of the EDPHS emission can also be understood from these figures. The field profiles very much resemble the first-order (LG11) and second-order (LG21) transverse magnetic Laguerre-Gaussian optical beams at $E = 1.5\,\text{eV}$ and $E = 4.0\,\text{eV}$, respectively. Moreover, the peak intensity of the EDPHS emission in the focal plane is $2.65\,\text{W}\,\text{cm}^{-1}$.

7.5.2 Spectral Interferometry Technique

A sample positioned along the electron trajectory will interact with both the electron and the EDPHS excitations. To investigate the functionality of the proposed method for analysing the ultrafast responses of different samples, a Ag disc and a Si disc, both with a 300 nm diameter and 40 nm thickness, are used here as samples. The CL spectra of the individual Ag sample and EDPHS structure are shown in Fig. 7.9a by the grey and green shadowed regions, respectively. Due to the excitations of plasmons in the silver disc, the CL emission from the sample is intense, and a clear peak is observed at $E = 2.9$ eV. When both the EDPHS and the sample are positioned along the electron trajectory, they form coupled structures with an overall CL spectrum that is not only the sum over the CL spectra of the different elements. Since the emissions from the EDPHS and the sample are mutually coherent, the interference between the EDPHS and the electron-induced emissions of the sample can alter the CL spectra significantly with respect to the distance L between the sample and EDPHS. For the case of the Ag disc chosen as the sample, the changes in the CL spectra are

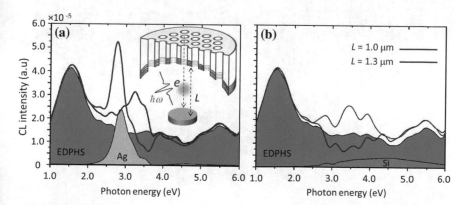

Fig. 7.9 Calculated CL spectra for an isolated EDPHS (teal shadowed region) and a sample coupled to the EDPHS at distances $L = 1.0\,\mu\text{m}$ (violet line) and $L = 1.3\,\mu\text{m}$ (red line) between the sample and EDPHS for **a** a silver disc and **b** a silicon disc as the sample. The CL spectra for the individual Ag and silicon discs are shown by the grey and olive shadowed regions, respectively. Re-printed by permission from [24] under the CC BY license

limited to the energy band of 2.0–3.0 eV. This is due to the fact the plasmon-induced cathodoluminescence is rather narrowband. This is quite different from the case when the sample is a Si disc, for which the CL emission of the sample is very broadband (Fig. 7.9b). Hence, fluctuations in the overall CL spectra of the coupled EDPHS and sample structure also cover a broad energy range, from 2.0 to 6.0 eV.

A full interference pattern in the CL energy-distance map becomes apparent when a series of CL spectra are acquired at preassigned distance steps. A fine sweeping of the distance can be well accomplished by state-of-the-art piezoelectric actuators. Considering this, the proposed methodology can be used to assess the distance/delay axis with enough sampling points to precisely characterize the temporal evolution of the responses according to the Shannon sampling theorem.

It will be shown in the following how the proposed methodology can be used to retrieve the spectral phase. The calculated CL spectra for the coupled system of the EDPHS and silver disc are demonstrated in Fig. 7.10a, versus distance and energy. The distance axis is sampled at intervals of 50 nm, and the whole range is from 0.42 to 1.62 μm. The spectral intensity that is measured in the first step, without using the phase-retrieval algorithm, is the CL from the whole setup, which includes the CL from the individual EDPHS and samples, as well as the interference between the emissions from these two structures. This fact can be comprehended from (7.2). The purpose of the exploited spectral interferometry is to recover the intensity and phase of the CL spectra from the sample. Moreover, Fig. 7.10a clearly demonstrates that the whole CL spectrum is not only slightly different from the EDPHS spectrum but also an accumulative spectrum of the individual EDPHS and sample spectra. Both

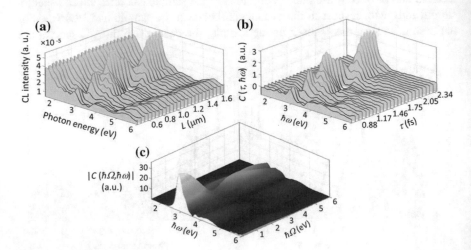

Fig. 7.10 Spectral interferometry using the energy-distance CL map. **a** Series of calculated CL spectra of the coupled system of the EDPHS and a silver disc position at a distance L from the EDPHS, where the distance is swept in steps of 50 nm. **b** The extracted energy-time correlation function [see (7.2)]. **c** The Fourier-transformed two-frequency correlation function. Re-printed by permission from [24] under the CC BY license

the EDPHS and sample transition radiations, as well as the interference between those radiations at the CL detector, are responsible for the overall CL spectra.

To retrieve the spectral phase, a reference CL spectrum is required, which is taken here as the CL spectrum of the EDPHS structure. According to Fourier transform spectral interferometry [63], even a single CL spectrum of the coupled EDPHS/sample system at a precise distance/delay would be enough to retrieve the spectral phase. However, using the full energy-distance map helps to reduce the noise [64]. A method is proposed here that is suited to retrieve the spectral phase of an ultra-broadband electron-induced signal within a 2–6 eV energy range.

The first step is to extract a correlation function from the CL map. Taking the EDPHS CL spectrum as the reference, the energy-distance and its corresponding energy-time correlation map is given by:

$$
\begin{aligned}
C(L, \hbar\omega) &= \frac{\Gamma_{\mathrm{CL}}(L, \omega)}{\Gamma_{\mathrm{CL}}^{\mathrm{EDPHS}}(\omega)} - 1 \\
&= \left|\vec{E}_{\mathrm{el}}^{\mathrm{ind}}(\omega)\right|^2 \Big/ \left|\vec{E}_{\mathrm{EDPHS}}(\omega)\right|^2 \\
&\quad + 2\left|\vec{E}_{\mathrm{el}}^{\mathrm{ind}}(\omega)\right| \Big/ \left|\vec{E}_{\mathrm{EDPHS}}(\omega)\right| \cos(\varphi_{\mathrm{el}} - \varphi_{\mathrm{EDPHS}} - \omega\tau)
\end{aligned}
\tag{7.3}
$$

where $\Gamma_{\mathrm{CL}}(L, \omega)$ is the energy–distance CL map shown in Fig. 7.10a and $\Gamma_{\mathrm{CL}}^{\mathrm{EDPHS}}(\omega)$ is the CL spectrum of the EDPHS. This correlation function is different from the generally used degree of first-order coherence [65] and is only introduced here to acquire the spectral phase with high accuracy. To do so, we first map $C(L, \hbar\omega)$ to the time–energy correlation ($C(\tau, \hbar\omega)$) by simply using the delay-distance relation $\tau = (1 - \beta)L / v_{\mathrm{el}}$ (see Fig. 7.10b). $C(\tau, \hbar\omega)$ can then be transformed into the energy-energy correlation function ($C(\hbar\Omega, \hbar\omega)$) as:

$$
\begin{aligned}
C(\hbar\Omega, \hbar\omega) &= \Im\{C(\tau, \hbar\omega)\} \\
&= \left|\vec{E}_{\mathrm{el}}^{\mathrm{ind}}(\omega)\right|^2 \Big/ \left|\vec{E}_{\mathrm{EDPHS}}(\omega)\right|^2 \delta(\Omega) \\
&\quad + 2e^{+i(\varphi_{\mathrm{el}} - \varphi_{\mathrm{EDPHS}})}\delta(\Omega - \omega)\left|\vec{E}_{\mathrm{el}}^{\mathrm{ind}}(\omega)\right| \Big/ \left|\vec{E}_{\mathrm{EDPHS}}(\omega)\right| \\
&\quad + 2e^{-i(\varphi_{\mathrm{el}} - \varphi_{\mathrm{EDPHS}})}\delta(\Omega + \omega)\left|\vec{E}_{\mathrm{el}}^{\mathrm{ind}}(\omega)\right| \Big/ \left|\vec{E}_{\mathrm{EDPHS}}(\omega)\right|
\end{aligned}
\tag{7.4}
$$

where $\delta(\cdot)$ is the Dirac delta function. It is already apparent here that the fine sweeping of the delay axis and covering at least a few oscillation cycles are required to be able to take the

Fourier transform. $C(\hbar\Omega, \hbar\omega)$ (see Fig. 7.10c) is readily used to retrieve the spectral intensity and phase by integrating it over the positive frequencies. The retrieved electric field spectra will be obtained as:

$$
E^{\mathrm{ret}}(\omega) = e^{+i(\varphi_{\mathrm{el}} - \varphi_{\mathrm{EDPHS}})}\left|\vec{E}_{\mathrm{el}}^{\mathrm{ind}}(\omega)\right| \Big/ \left|\vec{E}_{\mathrm{EDPHS}}(\omega)\right|
\tag{7.5}
$$

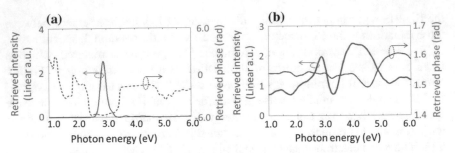

Fig. 7.11 Reconstructed spectral intensity and phase. The retrieved electron-induced electric field **a** in a silver disc and **b** in a silicon disc for an electron at a kinetic energy of 200 keV penetrating through the centre of the disc. The disc has a diameter of 300 nm and a thickness of 40 nm. Re-printed by permission from [24] under the CC BY license

It is apparent from (7.5) that the abovementioned technique can provide us with a direct electric field reconstruction. However, it is the ratio of the electron-induced field to the EDPHS field that is actually retrieved. In other words, the experimental characterization of the EDPHS electric field with an optical interferometry technique such as SPIDER would be highly beneficial. In doing so, the retrieval algorithm presented here will provide us with direct access to $\vec{E}_{el}^{ind}(\omega)$. Hereafter, only the results of the phase retrieval algorithm that lead to (7.5) are presented. The retrieved spectral phase and intensity are shown in Fig. 7.11a and b for the Ag and Si discs, respectively. Within the frequency range of 2.0–3.0 eV, the intensity and phase of the electron-induced electric field in a silver disc can be retrieved with high accuracy thanks to the rather flat spectral intensity of the EDPHS in this frequency range, as can be understood by comparing Fig. 7.11a to Fig. 7.9a. Even the ultrabroadband electron-induced resonances of the silicon disc are captured with good accuracy.

Although the presented EDPHS could serve as a highly efficient photon source for the presented interferometry technique, other electron-driven photon sources that directly employ quantum emitters and active regions can also be used for this purpose. In fact, the collective responses of atoms and quantum dots to the electron beam might also produce a superradiance continuum [66, 67], whereas the combination of these with metamaterials leads to the enhancement of the quantum efficiency and unidirectional emission.

Furthermore, there is no fundamental reason why the interference pattern should not be captured by the EELS detector other than the fact that the EEL spectrum has a direct contribution from loss channels such as dissipation and nonradiative de-excitations. A suitable design for the EDPHS can also overcome these problems by incorporating superradiance combined with metamaterials to control the directionality of the emission.

Within an adiabatic Wolkow approximation [68], the electromagnetic vector potential is accumulated in the phase of the electron wave-function. An EELS or CL spectrum provides us with knowledge about the probability of the electron to emit certain photons or the intensity of the emitted photons in the far field. In this regard, the phase of the electron wave-function, and therefore the accumulated vector potential, is neglected. A spatial holography technique can retrieve the spatial phase of the electron wave-function [69]. The proposed spectral interferometry technique here can be used to characterize the vector potential in time, energy, and space. As an example, it provides direct access to the spatiotemporal multipolar resonances of nanostructures, which is not accessible from the routinely exploited EELS or CL measurements. Although the spectral features of such resonances are captured in CL or EELS measurements, electrodynamic simulation methods are always needed to clarify the physics behind each resonance by retrieving the spatiotemporal distribution of the excitations [70, 71]. Exploiting a pulse characterization technique such as the spectral interferometry technique would be necessary to experimentally retrieve the spatiotemporal phase and oscillations of the electron-induced excitations in the sample.

The EDPHS structure proposed here emits coherent radiation within the energy range of 1–6 eV. In this regard, the time-frequency responses of metamaterial elements, plasmons, and photonic crystals and the quantum optical fluctuations of quantum dots, nanoemitters, atoms and molecules can be probed. In principle, depending on the EDPHS radiation, this technique may also be applied to higher photon energies, by incorporating coherent bremsstrahlung as an example [72]. Moreover, state-of-the-art piezo–stages combined with hybrid manipulators can offer perfect control of the distance between the EDPHS and sample, and hence a wide dynamic range for characterizing the temporal behaviours will be addressable. In fact, a hybrid 3-axis manipulation system and a precisely designed sample holder should be considered as an additional counterpart for a normal electron microscope in order to perform the proposed interferometry experiment [73].

In summary, the demonstrated spectral interferometry technique uses electron excitation and CL detection to acquire an energy–distance CL map, which can be used directly to extract the time–energy correlation function. The time–energy correlation function is then Fourier–transformed to obtain the two-energy correlation and retrieve the spectral phase and intensity for direct electric field reconstruction. This is achieved by a precise design of a photon source, which uses interaction with the electron beam to generate a focused transition radiation that is mutually coherent with the electron itself. Due to the nature of the designed metamaterial-based photon source, the emitted photon beam is ultrabroadband and, interestingly, demonstrates optical excitations that mimic the Laguerre-Gaussian transverse magnetic optical beams. The presented methodology offers vast applications in the fields of ultrafast science and paves the way towards attosecond electron-based spectroscopic techniques. The presented EDPHS design by itself has applications in the full coherent control of electron-induced emissions. In such a way, even few-photon sources would be enough to perform spectral interferometry. There is no apparent need for ultrahigh intensive laser radiation, as the applications are restricted to linear manipulations of the optical density of states.

In the following, we propose, practically realize, and experimentally characterize a planar electron-driven photon source that is suitable for direct insertion inside an electron microscope.

7.6   Merging Transformation Optics with Electron-Driven Photon Sources

In Sect. 7.5, it was shown that a swift electron interacting with a superlens composed of a hexagonal two-dimensional photonic crystal incorporated into an inverted hemispherical geometry will emit focused transition radiation. A simpler planar structure, which in an effective medium approximation exhibits the same behaviour, can be realized by projecting the location of the photonic crystal elements from the curved spherical geometry into the desired plane, which is located directly above the hemisphere (Fig. 7.12a and b). Mathematically, the transformation from the spherical plane to the cylindrical plane leads to the modification of the permittivity as $\varepsilon' = |\Lambda|^{-1} \Lambda \varepsilon \Lambda^T$, where Λ is the Jacobian of the transformation matrix and ε is the initial permittivity of the material in the spherical coordinate system. The inhomogeneous planar refractive index of the plane by projection from the spherical

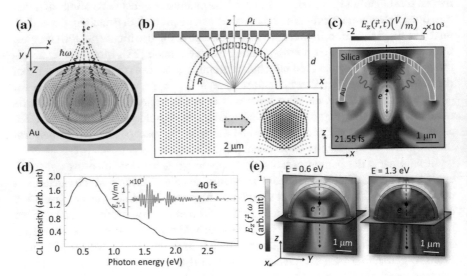

Fig. 7.12 Electron-induced radiation from curved geometries and designing a flat lens to mimic this behaviour. **a** Formation of focused broadband transition radiation in the interaction of an electron beam with an engineered planar lens. The structure is designed to effectively mimic a porous hemispherical geometry. **b** Schematic of the design principles that include the projection of the incorporated holes in the hemispherical lens to a planar gold film. **c** Interaction of a 30 keV electron (incident from the top) with a mesoscopic hemispherical porous film positioned on a silica substrate. The electron passes through one of the holes at the centre. Depicted in the figure is a snapshot of the simulated z-component of the electric field 22.55 fs after impact on the lens surface, which indicates a clear focus. **d** CL spectra of the hemispherical structure computed at a plane positioned 5 micrometres below the structure. Inset: Temporal representation of the z-component of the electric field at the focal point in (**c**). **e** Time-averaged electron-induced electromagnetic wave packet at the energies indicated on each frame. Depicted is the magnitude of the Fourier transform of the z-component of the electric field in the xy and yz planes. Re-printed by permission from [25] under the CC BY license

system to the cylindrical coordinate system is obtained as $\varepsilon'_{\rho\rho} = \varepsilon\rho^2 / (\rho^2 + d^2)$, where $\rho = \sqrt{x^2 + y^2}$. For $\rho \gg d$, $\varepsilon'_{\rho\rho} = \varepsilon$. We also assume that $\mu = \mu' = 1$ for the simplicity of the fabrication processes. A geometrically discrete coordinate transformation is applied. A hexagonal lattice of point defects (i.e., holes) is mapped from spherical coordinates to the cylindrical system, which leads to a distribution of points as illustrated in the bottom plane in Fig. 7.12b (grey dots). To keep the system cylindrically symmetric, holes are only formed inside a certain ring diameter. We use the finite-difference time-domain (FDTD) method for calculating the electron-induced radiation. A snapshot of the z-component of the electron-induced radiation after the interaction of a 30 keV electron with a hemispherical gold film (with dimensions described below) is illustrated in Fig. 7.12c. A hexagonal lattice of holes with a diameter of 100 nm and a rim-to-rim distance of 50 nm is drilled into the spherical film. The radiation is in the form of a focused ultrafast electromagnetic wave packet with the focal point positioned at the centre of the sphere. It originates from the radiation damping of the surface plasmon polaritons (SPPs) in our spherical film by the incorporated lattice of holes, in addition to the transition radiation. While the radiation damping of the SPPs in a planar gold film is negligible, for a thin film with a lattice of holes, the effective permittivity is anisotropic, and as a result, the SPP dispersion is mapped into the light cone. Hence, radiation damping is observed (see Sect. 7.6.1). This behaviour, in addition to the spherical curvature of the film in Fig. 7.12c, causes enhanced radiation that is focused at the centre of the sphere.

The calculated radiation spectrum (cathodoluminescence, CL) for a plane located 5 microns below the hemispherical lens is shown in Fig. 7.12d and extends from 0.4 to 2.6 eV. The spectrum is calculated using the Poynting vector $\Gamma^{\mathrm{CL}}(\omega) = (1 / 2\hbar\omega) \times$ Re $\iint_s \vec{E}(\vec{r}, \omega) \times \vec{H}^*(\vec{r}, \omega) \cdot \vec{ds}$, where $\vec{E}$ and $\vec{H}$ are the scattered components of the electron-induced electric and magnetic fields, respectively, ω is the angular frequency of the emitted light, and the plane s is located in the far field. This formulation allows for a direct comparison with electron energy loss spectroscopy (EELS) using Poynting's theorem [74]. Thus, only the components normal to the plane of the detector (transverse components) are relevant to the CL spectrum. The fact that the emitted light constitutes many frequency components is also reflected by the computed temporal distribution of the z-component of the electric field at the focal point (centre of the sphere; see the inset in Fig. 7.12d). The time-averaged z-components of the electric field magnitude in the xy and yz planes at energies of $E = 0.6\,\mathrm{eV}$ and $E = 1.3\,\mathrm{eV}$ are shown in Fig. 7.12e, indicating focused radiation at the centre of the spherical geometry at both photon energies.

7.6.1   Structure

Our planar metasurface lens has a spectral response that peaks in the visible-near-infrared spectral range (0.5–2.0 eV). We use a hexagonal lattice of holes, each with a diameter of 100 nm and a centre-to-centre distance of 150 nm, projected onto a planar Au disc with a thickness of 50 nm using the abovementioned algorithm. The whole structure is placed on a 20 nm thick Si_3N_4 membrane. The radius of the original hemisphere is $R = 5.7\,\mu m$, and $d = R$, which leads to a focal length $f = R$ for the planar lens. Scanning transmission electron microscopy (STEM) high-angle annular dark-field images of the realized planar lens are shown in Fig. 7.13a and b. A void ring is incorporated on the circumference of the lens in order to facilitate the reflection of the propagating SPPs and hence increase the generation of far-field radiation, as shown below. The inserted void ring has a strong effect on the photon emission capability of the lens, which is understood by comparing theoretically and experimentally the responses of structures including and excluding the ring.

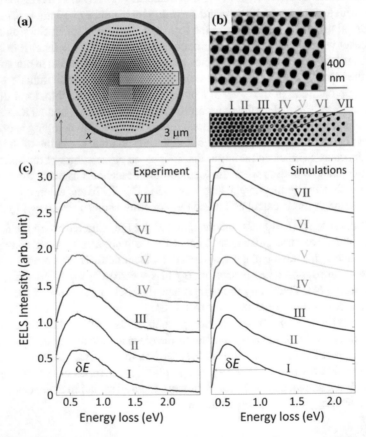

Fig. 7.13 Characterization of a fabricated planar lens. **a** Dark-field STEM image of the fabricated lens. **b** The same image at a higher magnification from the green and red boxes in (**a**). **c** Experimental and simulated EEL spectra (200 keV primary electron beam energy) at the positions depicted by coloured spots in (**b**). Re-printed by permission from [25] under the CC BY license

7.6.2 EELS Investigations

We first investigate the near-field behaviour of the fabricated planar lens using electron energy-loss spectroscopy (EELS) with the unique Zeiss SESAM transmission electron microscope in STEM mode. In EELS, due to the inelastic interaction of the electron beam with the optical modes of the nanostructure, the electron will lose energy, where the amount of energy loss is measured by using an energy-loss spectrometer. The EELS signal reflects the probability of the electron beam exciting optical modes during its interaction with the sample and thus senses the photonic local density of states projected onto the trajectory of the electron from a few tens of meV to hundreds of eV (as discussed in Chap. 2). For this reason, EELS is ideally suited for probing ultra-broadband excitations. The near-field resonances should exhibit similarly broad features to the far-field radiation. The measured and simulated EEL spectra, taken at different locations on the lens, are shown in Fig. 7.13c and show a broad spectral feature at a central energy of $E = 0.6\,\text{eV}$ and a bandwidth of $\delta E = 0.8\,\text{eV}$. An asymmetric resonance is observed due to the high-order chirping of the electron-induced polarizations. The chirping effect is understood by a gradual change in the frequency of the light-field oscillations, as will be described later. Interestingly, the EEL spectra at positions closer to the rim of the structure exhibit broader features than those near the centre, yet the highest intensity for the EELS signal occurs at the centre of the holes, as shown in Fig. 7.14. We simulate the EEL spectra using the non-recoil approximation as $\Gamma^{\text{EELS}}(x_0, y_0, \omega) = (e/\pi\hbar\omega)\times\text{Re}\,\tilde{E}_z(x_0, y_0, k_z = \omega/v_{\text{el}}; \omega)$, where (x_0, y_0) is the impact location of the electron beam in the transverse direction, e is the elementary charge, k_z is the projected wavenumber of the scattered field along the electron

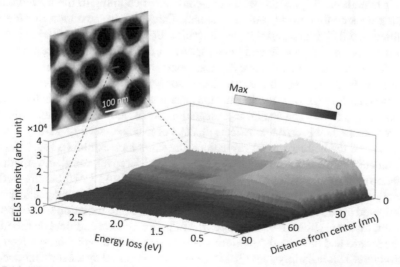

Fig. 7.14 EEL spectra along the radius of the central hole. Colour scale is the EELS intensity in arbitrary units. Re-printed by permission from [25] under the CC BY license

trajectory, $\tilde{E}_z$ is the Fourier transform of the electric field in both the spectral and spatio-spectral domains, and v_{el} is the velocity of the electron. The simulated EEL spectra, as shown in the right column of Fig. 7.13c, shows good agreement with the measured spectra.

7.6.3 Radiation Mechanism

To obtain further insight into the radiation mechanism of the planar lens interacting with a moving electron, we investigate the temporal dependence of the field evolution [Fig. 3(a)]. We use 30 keV electrons, as in the CL measurements described further below. The electron excites the planar lens when it reaches the near-field region of the structure. The interaction of the electron with the thin film takes place only within a sub-fs time scale; however, this interaction induces relatively long-lasting oscillations in the sample, with a relaxation time of $\tau \approx 20$ fs. The contribution of the induced polarization to the radiation is not gradual over time. The first ultrafast pulse, due to transition radiation, leaves the structure directly upon the electron impact in the form of a sub-cycle spherical wave front, overtaking the electron at only a sub-fs time scale (Fig. 7.15b). It should be noted, however, that the duration of the transition radiation is very much dependent on the thickness of the film (two dipoles created and annihilated at the upper and lower surfaces) and the presence of the substrate. The amplitude and position of the image charge are also altered by the presence of the holes.

Simultaneously, the electron excites propagating SPPs of a tailored dispersion due to the inhomogeneous lattice of holes. The excited SPPs gradually scatter due to their hyperbolic dispersion, with the z-component coupling to the far field, thus building up a continuum of radiation modes. Due to their separation in time, the transition and SPP radiations will not interfere for our structure, unlike the observations for the grating-facilitated out-coupling of SPPs [75, 76]. Interestingly, the lower-frequency components leave the lens sooner, followed by a gradual chirping of the entire radiation towards higher-frequency components (see Fig. 7.15b). The time lag between the transition radiation and the plasmon-induced radiation in the present case is $\Delta t = 15.6$ fs. The spatial representation of the field components in the frequency domain, however, demonstrates this fact even better (see Fig. 7.15c). Similar to the case of the hemispherical resonator, our planar lens exhibits a focal waist in the frequency range from 0.6 to 1.4 eV (Fig. 7.15c). Moreover, the electron-induced excitation in the forward direction comprises approximately 43% of the entire electromagnetic energy delivered into the system, where 37% of the electromagnetic energy is dissipated inside the gold film, and 20% is converted into radiation in the upward direction (data not shown). Interestingly, the radiation spectra and field distributions in the forward and backward directions are not symmetric, despite the symmetry of the structure. This is mainly due to the Doppler effect imposed by the moving electron source. The calculated time-averaged amplitude of the electric field in the focal plane is shown in Fig. 7.16, left column. The $\vec{k}-$ space distribution is obtained by mapping

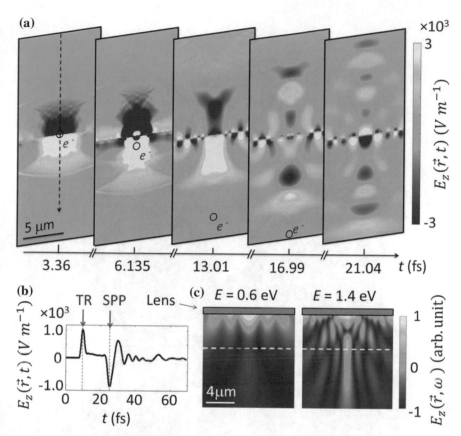

Fig. 7.15 Simulated spatio-temporal response of the planar lens interacting with a 30 keV electron. **a** Snapshots of the z-component of the electric field amplitude at depicted times for the indicated electron positions, which demonstrate that the electromagnetic wave packet is focused into a 1.5 μm waist at a 5.7 μm distance from the structure. **b** Temporal representation of the z-component of the electric field at the focal point. **c** Fourier transform of the temporal data, giving the z-component of the electric field at the depicted energies. The white dashed line marks the focal plane. TR stands for the transition radiation. Re-printed by permission from [25] under the CC BY license

the field components from the space-time $(\vec{r}, t)$ to the $(\vec{k}, \omega)$ domain using a spatial Fourier transformation. However, the fact that an FDTD simulation is performed in a closed domain imposes some restrictions over the integration range of the Fourier transformation. To minimize these errors, we multiplied the field components in the $(\vec{r}, t)$ domain by a cylindrically symmetric step window $\Theta(\rho/a)$, where a is the radius of the cylindrical window that was estimated by an ad hoc procedure to obtain smooth functions with the best signal-to-noise ratio in the $(\vec{k}, \omega)$ domain while also maintaining all the important spatial features in the space-time domain. The amplitude of the electric field as well as the total electromagnetic energy (Fig. 7.16 right column) show a high-intensity signal at the centre of the momentum space in the

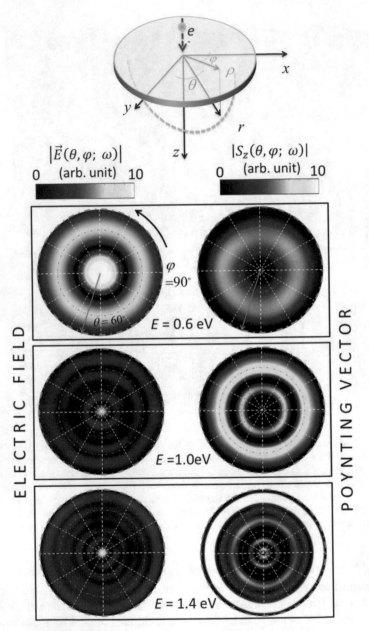

Fig. 7.16 Simulated *k*-space amplitude of the electric field in the focal plane (left column) and cal-culated Poynting vectors (corresponding to angular CL profile) at different energies (right column). Re-printed by permission from [25] under the CC BY license

focal plane. The size of this focus spot is the largest at $E = 0.4\,\text{eV}$; it drastically decreases upon increasing the frequency. The maximum of the total electromagnetic energy remains located at the centre of the reciprocal space at photon energies in the range 0.2–1.6 eV. However, the transverse components of the electromagnetic fields that determine the Poynting vector and hence the CL angular profile for the radiation to the far field exhibit the opposite behaviour (see Fig. 7.16, right column): a high intensity is observed for the large angle, in particular for the higher energies. This is to be expected, as the longitudinal components of a field always demonstrate a different symmetry than the transverse components. For example, for transverse magnetic fields (for the TM$_z$ modes) in a cylindrical coordinate system, $E_\rho \propto \partial^2 E_z / \partial \rho \, \partial z$ and $E_\varphi \propto \partial^2 E_z / \rho \, \partial\varphi \, \partial z$. In other words, the $\varphi-$ component of the electric field should vanish, as the symmetry of the excitation and the lens requires $\partial / \partial\varphi = 0$. E_ρ is, however, present and exhibits a different symmetry from E_z, which explains the vanishing central bright spot for the CL map for high energies. For computing the transverse components of the electron-induced radiation at the far-field region, the effective current distributions have been calculated along a fictitious plane 50 nm below the sample using the electromagnetic equivalence principle. Those current distributions have then been used to calculate the radiation at the far-field region.

7.6.4 SPP Dispersion and Radiation Damping in Porous Thin Films

An electron interacting with a thin metal film creates SPPs and loses momentum and energy accordingly. We have calculated the MREELS using a previously described method (see Chap. 3). The MREELS from a 50-nm thick gold film and 30 keV electrons clearly demonstrates the dispersion of SPPs, as well as a bright and broad peak inside the light cone (Fig. 7.17a). The latter is due to transition radiation, which is the dominant mechanism of radiation for electrons interacting with a thin gold film (Fig. 7.17b). As the dimensions of the holes inside the spherical and planar lenses are much smaller than the wavelength of the SPPs, an effective medium theory can be used to understand the qualitative behaviour of the SPPs in such structures. The effective permittivity becomes anisotropic, with $\varepsilon_{||} = \varepsilon_{xx} = \varepsilon_{yy} = (d_1 + d_2)\varepsilon_{\text{Au}}\varepsilon_{\text{d}} / (d_1 \varepsilon_{\text{Au}} + d_2 \varepsilon_{\text{d}})$ and $\varepsilon_\perp = (d_1 \varepsilon_{\text{d}} + d_2 \varepsilon_{\text{Au}}) / (d_1 + d_2)$, where d_1 and d_2 are the diameter of the holes and the rim-to-rim distance between the holes, respectively [77]. Here, we assumed that the permittivity components parallel to the surface of the thin film are the same; hence, the bulk permittivity is modelled with that of a uniaxial crystal. Moreover, the frequency-dependent propagation constant of the ordinary and extraordinary SPP waves propagating at the surface of such materials in general are given as $\gamma_{\text{o}} = \beta_{\text{o}} + i\alpha_{\text{o}} = k_0\sqrt{\varepsilon_{||}(\varepsilon_{||} + 1)^{-1}}$ and $\gamma_{\text{e}} = \beta_{\text{e}} + i\alpha_{\text{e}} = k_0\sqrt{(\varepsilon_{||}\varepsilon_\perp - \varepsilon_{||})(\varepsilon_{||}\varepsilon_\perp - 1)^{-1}}$, respectively, where β is the phase constant and α is the attenuation constant [78]. For a gold thin film with an inhibited

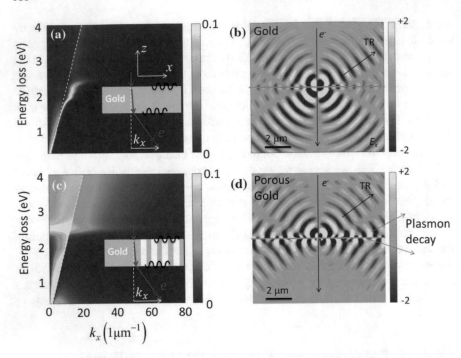

Fig. 7.17 Comparing electron-induced SPPs in pure and porous gold thin films. Momentum-resolved EELS and electron-induced radiation (including transition radiation and radiation damping of launched SPPs) for the interaction of 30 keV electrons with a uniform 50 nm thick **a, b** gold film and **c, d** gold film with an inhibited hexagonal lattice of holes. Shown in the **b** and **d** panels are the spatial profiles in the xz plane of the x-component of the scattered electric field at a photon energy of $E = 1.55$ eV. The colour bar is the EELS intensity in arbitrary units. **a** and **c** panels show the momentum-resolved EELS map. Re-printed by permission from [25] under the CC BY license

hexagonal lattice of holes, we assume $d_1 = 100$ nm and $d_2 = 50$ nm. An electron interacting with this film excites extraordinary SPP waves, with a bright intensity of the EELS signal inside the light cone (Fig. 7.17c). Moreover, the intensity of the TR radiation is less than that of the pure gold film, whereas the SPP radiation damping is greatly enhanced (Fig. 7.17d). Obviously, the inclination angle of the radiation for a uniform gold film without any discontinuity is towards a direction that is out of reach of the focus point. However, hyperbolic SPPs reaching the void ring will be reflected and hence radiate into the inclination angle required for focusing.

7.6.5   Cathodoluminescence Investigations

We measure the angle-resolved cathodoluminescence emission of the lens using an SEM equipped with a Schottky field-emission electron source operated at 30 keV. The emitted light is collected using an aluminium paraboloid mirror with a focal distance

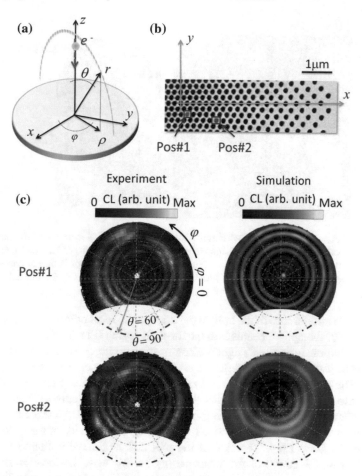

Fig. 7.18 CL investigations. **a** Detection geometry specified by the hemisphere positioned above the structure. **b** Impact positions for the electron beam, and **c** Experimental and simulated cathodoluminescence angular profiles at $E = 1.9$ eV at two different electron impact positions for the incident electron marked by red and blue colours in panel (**b**), respectively. Re-printed by permission from [25] under the CC BY license

of 0.5 mm, which provides a collection solid angle of 1.46 π sr[20]. The mirror is mounted above the sample, hence collecting the photons emitted in the backward direction, as depicted at Fig. 7.17. The CL measurements have been performed at a photon energy of $E = 1.9$ eV (wavelength $\lambda = 650$ nm). For calculating the CL map, however, the Poynting vector along the r—direction (S_r) has been considered. At this energy, the CL map exhibits 8 centrosymmetric bright rings when the electron traverses the structure closed to the centre. Additional measurements (Fig. 7.19) show that the number of rings decreases with the decreasing energy. This behaviour points at a geometrical interference effect, caused by the reflection of the SPPs from the inhibited void ring surrounding the lens. This behaviour can be understood by the

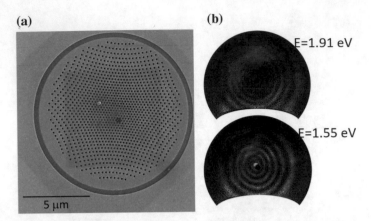

Fig. 7.19 Measured CL map at two different photon energies. **a** SEM image of a fabricated planar lens with a focal length of 2 μm. **b** CL map at two different wavelengths of 1.55 and 1.91 eV, with the electron beam exciting the structure at the blue spot. The grey spot is due to contamination during the measurement. Re-printed by permission from [25] under the CC BY license

constructive interference of the optical radiation at the focal point and the subsequent divergence of the optical beams into the far-field region (Fig. 7.18).

The radiation from the designed electron-driven photon source is first focused at the focal plane and then diverges, causing an interference pattern in the far-field region. The number of maxima and minima in the detector plane (CL detector) is calculated here. We first note that the plasmons are gradually radiated from the structure, caused by the engineering of the refractive index. An effective medium theory is used to derive the effective (anisotropic) permittivity of the gold film, as discussed in Sect. 7.6.4. The size of the unit cell is gradually altered versus the distance from the centre of the structure (r). In this way, the location of the nth projected hole from the centre of the lens is $R_n = nf S \left(f^2 - (nS)^2\right)^{-0.5}$, where $S = 150$ nm is the lattice size of the initial hexagonal lattice and f is the focal length. We calculate $d_1 = 100$ nm and $d_{2n} = R_n - R_{n-1} - d$. The plasmon dispersion for $d_2 = d_1 = 100$ nm for both ordinary and extraordinary waves for a thin film made of an anisotropic material with permittivity values described in Sect. 7.6.4 is calculated. It is apparent that the SPP dispersion is significantly altered by the presence of the holes and that the plasmon dispersion is shifted into the light cone. Comparing Figs. 7.17 and 7.20 shows that the swift electrons are coupling to the extraordinary plasmon waves, and hence the radiation damping is significantly enhanced.

Extraordinary plasmon waves are plasmon polaritons propagating in a uniaxial anisotropic material in the direction normal to the optic axis [79]. In other words, the relevant effective mode index is that of the extraordinary plasmons with

$$n_{\text{eff}} = \sqrt{\left(\varepsilon_{||}\varepsilon_{\perp} - \varepsilon_{||}\right)\left(\varepsilon_{||}\varepsilon_{\perp} - 1\right)^{-1}}, \qquad (7.6)$$

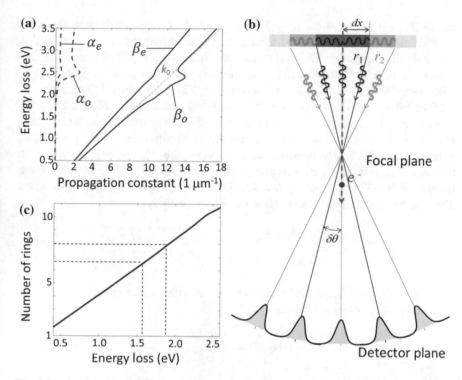

Fig. 7.20 Number of far-field rings versus the photon energy (Re-printed by permission from [25] under the CC BY license). **a** SPP dispersion in a thin film made of gold and a hexagonal lattice of air holes, modelled using an effective medium approach. β is the phase constant, and α is the attenuation constant. **b** A geometrical consideration for describing the number of intensity maxima observed in the measured CL maps. **c** Calculated number of rings versus photon energy for a lens with radius $D = 5.6\,\mu$m

where n_{eff} is the mode index. We further derive a simple geometrical optic formulation that describes the focusing ability of the lens as well as the number of rings m in the CL map (Fig. 7.20b). To obtain constructive interference at the focal plane, we should have

$$n_{\text{eff}}k_0\,r + k_0\sqrt{f^2 + r^2} = k_0 f \pm 2m\pi, \tag{7.7}$$

where f is the focal length and r the distance of the propagation, which is measured from the centre of the structure. This simple phase relation can further be integrated over r to derive m as

$$m(\lambda) = \left| \frac{1}{\lambda D} \int_{r=0}^{D} \left(r\, n_{\text{eff}}(r, \lambda) + \sqrt{f^2 + r^2} - f \right) \mathrm{d}r \right|, \tag{7.8}$$

where D is the radius of the lens (5.6 μm). Note that n_{eff} is itself depends on the wavelength λ and distance. The number of rings is approximately linearly dependent on the photon energy (inversely related to the wavelength) (Fig. 7.20c) and in perfect agreement with the experimental data, showing $m = 7.94$ at $E = 1.9\,\mathrm{eV}$ ($\lambda = 652.5\,\mathrm{nm}$) and $m = 6.63$ at $E = 1.55\,\mathrm{eV}$ ($\lambda = 729.3\,\mathrm{nm}$) (Fig. 7.20c). Here, we used the following parameters: $D = 5.6\,\mu\mathrm{m}$, $f = 2\,\mu\mathrm{m}$, and $R = 50\,\mathrm{nm}$. The linear approximation is understood by the dominance of the third term inside the integrand in (7.8), whereas the first and second terms add up to corrections to the phase of the radiation in such a way to facilitate focusing. When the electron beam traverses the lens at a certain distance with respect to the symmetry axis of the lens, the mode indexes of the SPPs propagating towards the $+x$ and $-x$ directions will be different. The criterion for constructive interference will be given by

$$k_0 f + 2n\pi = n_{\mathrm{eff}}^+(r, \lambda)k_0 r + k_0\sqrt{f^2 + r^2 - 2f r\,\cos\theta}$$
$$= n_{\mathrm{eff}}^-(r, \lambda)k_0 r + k_0\sqrt{f^2 + r^2 + 2f r\,\cos\theta} \qquad (7.9)$$

where θ is the inclination angle of the radiation with respect to the lens axis. By assuming $r \ll f$, (7.9) is recast as

$$\cos(\theta) = \frac{1}{2}\Delta n_{\mathrm{eff}}, \qquad (7.10)$$

where $\Delta n_{\mathrm{eff}} = n_{\mathrm{eff}}^+ - n_{\mathrm{eff}}^-$ is the difference in the effective refractive index of plasmons propagating in the $\pm x$ directions.

As a result of the discussions above, the maximum number of rings observed with a certain angular range is then directly related to the focal length (f), the effective refractive index experienced by the SPPs (n_{eff}), and the wavelength.

When the electron is placed 300 nm away from the centre, the radiation is no longer normal to the structure but is slightly inclined (pos#1 at Fig. 7.18c). The inclination angle of the radiation is increased by further moving the electron beam away from the centre of the structure; however, the radial symmetry of the radiation at the plane normal to the wave vector is generally preserved (pos#2 at Fig. 7.18c). This behaviour has been also confirmed by simulations and gives excellent tunability of the polarization state and direction of emissions of the electron-driven light source (Fig. 7.21). We have also investigated the effect of electron impact parameters for lenses with different focal lengths, which demonstrate the same sensitivity of the CL map to the electron impact position (see Fig. 7.22). Finally, we emphasize that the diameter of the holes in our lens design is chosen to be much smaller than the wavelength of the light so that the diffraction from the holes is negligible, contrary to the case for Fresnel lenses and photon sieves [80]. This is confirmed by the fact that spectrally broadband behaviour is observed. The lens thus operates by focusing the radiation as a result of the hyperbolic dispersion of the SPPs, which is due to the engineering of the spatial distribution of the holes. Additionally, the contribution of the SPPs to the focused radiation is most prominent for the backward propagating

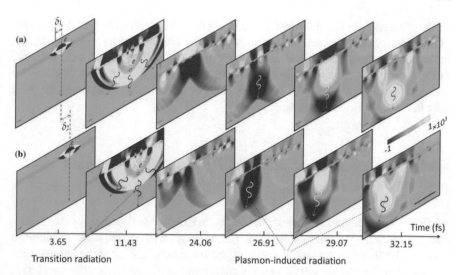

Transition radiation   Plasmon-induced radiation

Fig. 7.21 Dependence of the angle of the radiation on the distance of the electron beam trajectory from the centre of the lens (Re-printed by permission from [25] under the CC BY license). Snapshots of the z-component of the electric field at the depicted times for the electron impact position at **a** 400 nm and **b** 1000 nm away from the centre ($\delta_1 = 400$ nm, $\delta_2 = 600$ nm). The simulations here have been performed for a lens with only a 5 μm radius, and therefore the dependence of the radiation angle upon the distance from the centre is further enhanced in comparison with the lens structures proposed in the main manuscript. Colour bar is the electric field amplitude in units of $V\,m^{-1}$. Scale bar is 2 μm

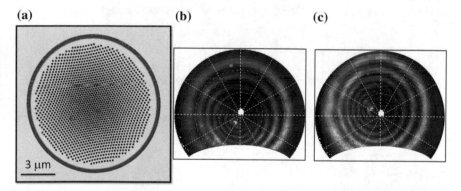

Fig. 7.22 Controlling the directionality of the radiation by electron impact. **a** A fabricated planar lens with a focal length of 10 μm. **b** and **c** CL maps at $E = 1.6$ eV at the depicted positions (red spot corresponds to CL image **b**, and blue spot corresponds to CL image **c**). Re-printed by permission from [25] under the CC BY license

SPPs reflected from the outer void rim and not for the outgoing SPPs that are propagating directly after the excitations towards the rim. We attribute this behaviour to the inclination angle of the radiation from the hyperbolic SPPs, which for the reflected SPPs from the void ring is directed towards the focal plane. In the following, we compare the design principles based on transformation optics and compare them with those of a photon sieves lens.

7.6.6 Transformation Optics and Differences with Photon Sieves

We first discuss the possibilities for designing an EDPHS that operates by the diffraction of propagating plasmons from incorporated holes. There are a few similarities between the structure discussed here and photon sieves. The latter are based on the diffraction of free-space waves by embedded holes in dielectric thin films and have been established as an efficient way for focusing X-rays (Fig. 7.23). In the notation of Fig. 7.23a, radiation from a point source S is focused at point P by means of a structured thin film consisting of a distribution of holes. The size of these transmissive pinholes and their distribution are both chosen in such a way to allow for the constructive interference of the diffracted rays at the focal point P. To allow for this, the distribution of the pinholes should satisfy $\sqrt{r_n^2 + S^2} + \sqrt{r_n^2 + f^2} = S + f + n\lambda$. Here, r_n is the distance of the nth hole from the origin, λ is the wavelength of the source, and n is an integer. Obviously, the focusing ability strongly depends on the

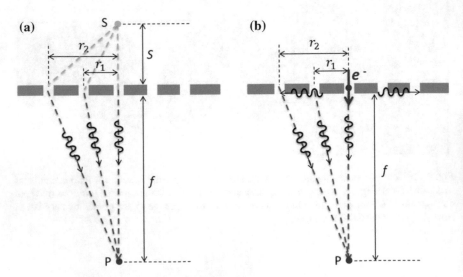

Fig. 7.23 Comparing the functionality of photon sieves for far-field light and electron-induced polaritons. Schematic of the focusing capability of a photon sieve for **a** X-rays, and **b** electron-induced plasmon polaritons. Re-printed by permission from [25] under the CC BY license

wavelength of the source, which hinders photon sieves from offering a broadband response.

In a similar photon-sieve geometry for a moving electron interacting with a metallic thin film, we consider the excitation of plasmon polaritons at the impact position of the electron (Fig. 7.23). A distribution of pinholes (or another diffractive centres such as metamaterial elements or ribs) is embedded to allow for the outcoupling of the propagating plasmons, creating a radiation continuum. For the constructive interference of the outcoupled beams in the focal plane, the distribution of the diffractive elements (here pinholes) should satisfy

$$n_{\text{eff}}r_n + \sqrt{r_n^2 + f^2} = f + n\lambda \tag{7.11}$$

where n_{eff} is the effective mode index of the plasmons in the structured thin film. It should be noted that for both cases (photon sieves for either X-rays or plasmons), the size of the pinholes should be large enough to allow for either a transmissive response or an efficient outcoupling of the plasmon polaritons, i.e., the diameter of the holes should be comparable to the wavelength of the source or excited plasmons.

In the following, we show how the photon sieve design principle can be used to focus the electron-induced radiation. We consider here an 80 nm-thick Al film and use (7.11) to calculate the required distribution of the pinholes to facilitate focusing at ($E = 6\,\text{eV}$). At this wavelength, the 80-nm Al film sustains plasmon polaritons with a mode index of $n_{\text{eff}} = 1.106 + i0.015$ (for this thickness, the plasmon polaritons at the upper and lower surfaces are uncoupled, and the even and odd modes are degenerate). Furthermore, the diameter of the pinholes is varied between $D = 60\,\text{nm}$ and $D = 120\,\text{nm}$ (Fig. 7.24a). The radiation from a non-structured Al thin film interacting with a moving electron covers a broad angular range (Fig. 7.24b), though by imbedding the pinholes with the desired distribution, we achieve focusing at $f = 900\,\text{nm}$ away from the structured film (Fig. 7.24c). However, at a slightly different energy of $E = 6.3\,\text{eV}$, the radiation becomes completely unfocused (Fig. 7.24d).

Given the unwanted sensitivity of the photon sieves to the wavelength, we exploit other possibilities that avoid such sensitivity from the design principle. As discussed in the main text, focusing capabilities based on geometrical refraction, such as focusing by a hemispherical thin film, are inherently broadband in nature. We therefore here exploit the principles of transformation optics to generate a distribution of non-transmissive pinholes, with the aim of engineering the effective mode index of plasmon polaritons to mimic the response of a hemisphere. The general assumption is based on adiabatically tuning the periodicity of the lattice of holes to achieve an engineered mode index as a function of the distance from the origin. Using the Jacobian of the transformation matrix Λ for mapping a function from a spherical coordinate system with coordinates (r, θ, φ) to a cylindrical system with coordinates (ρ, φ, z) (with φ and θ the zenithal and azimuthal angles, respectively), the permittivity of the structure in the cylindrical system $\varepsilon' = |\Lambda|^{-1} \Lambda \hat{\varepsilon} \Lambda^T$ is derived from the permittivity of the material in spherical coordinates ($\hat{\varepsilon} = \varepsilon_{\text{Au}} \mathbf{I}$, where ε_{Au} is the permittivity of gold and $\mathbf{I}$ is the identity matrix). Specifically, we take

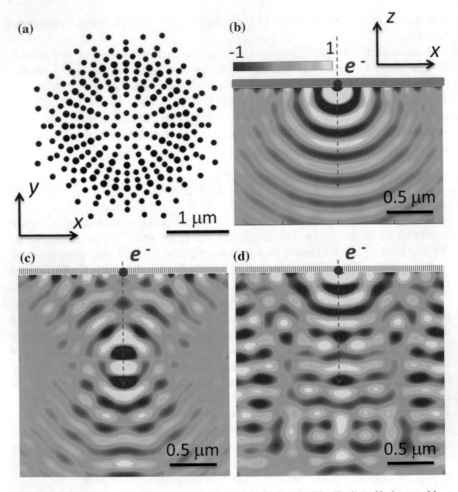

Fig. 7.24 Photon sieve for electron-induced plasmonpolaritons. **a** Distribution of holes to achieve focused radiation at $f = 900$ nm and $E = 6$ eV for an Al thin film. The z-component of the electric field at a given time for **b** a non-structured and **c** a structured Al film at $E = 6$ eV and **d** for a structured Al film at $E = 6.3$ eV. Colour bar is the electric field amplitude in arbitrary units. Re-printed by permission from [25] under the CC BY license

$\varepsilon'_{\rho\rho} = \varepsilon_{Au} \rho^2 / (\rho^2 + d^2)$, $\varepsilon'_{\rho\varphi} = \varepsilon'_{zz} = 0$, and $\varepsilon'_{\varphi\varphi} = \varepsilon_{Au}$. Formally, such a structure seems practically out of reach, considering that $\varepsilon'_{zz} = 0$. Nevertheless, $\varepsilon'_{\rho\rho}$ dominates the propagation of polaritons and can be suitably engineered. For a hemisphere with the focal point at the centre, we set for our mapping purpose $d = R$, so that $\varepsilon'_{\rho\rho} = \varepsilon_{Au}(\rho/R)^2 ((\rho/R)^2 + 1)^{-1} = \varepsilon_{Au} \sin^2 \theta$. At $\rho \gg f$, $\varepsilon'_{\rho\rho} \simeq \varepsilon_{Au}$. In other words, the distance between the holes should be adiabatically increased with increasing distance from the origin to obtain the pure gold permittivity at $\rho \gg f$.

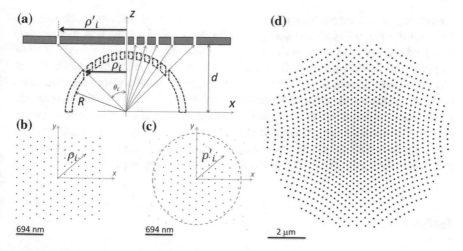

Fig. 7.25 Projection algorithm for the designed EDPHS. **a** Mapping principle for projecting the lattice of pinholes from a spherical domain to a planar thin film. The 2-dimensional distribution of the lattice points at the **b** original coordination and **c** final coordinates. **d** The final distribution of lattice points for the fabricated EDPHS. Re-printed by permission from [25] under the CC BY license

A geometric projection algorithm that maps the position of the holes on the gold hemispherical film onto a thin film located at $z = d = R$ is used to obtain such a distribution of pinholes (see Fig. 7.25a). We assume a hexagonal lattice of holes with centres located at (x_i, y_i), as shown in Fig. 7.25b, and use the mapping algorithm to obtain the new distribution for the holes in the thin film as $\rho_i' = R \tan \theta_i$ and $\tan \theta_i = \rho_i / \sqrt{R^2 - \rho_i^2}$ (Fig. 7.25b). To maintain an azimuthally symmetric radiation pattern, only the projected holes within a certain radius are included in the design.

For $R \to \infty$, the original lattice will be unambiguously maintained in the projected domain as well (a hemispherical film with $R \to \infty$ corresponds to a flat film). Finally, for the EDPHS investigated here, we assumed $R = 5.7 \, \mu m$, under which the lattice in Fig. 7.25d is obtained. At each lattice point, a pinhole with a diameter of 100 nm is considered. The overall lattice is incorporated onto an Au thin film with a thickness of 50 nm.

7.7 Summary

In this chapter, we briefly discussed the photon-induced domain with an emphasis on two different techniques that involve either fast electrons or slow electrons. We also proposed that the transition from the electron-induced domain to the photon-induced domain is formed by another distinguished domain, i.e., the photon-assisted domain.

Depending on the broadening of the electron pulses, an interference phenomenon between the electron-induced and photon-induced polarizations was theoretically observed and opened new directions for interferometry. We also proposed a method to practically realize such interference maps in the CL spectra by introducing an electron-driven photon source. This source in interaction with electrons generates a collimated light beam that is phase-matched and mutually coherent with the electron itself, thus significantly improving the temporal resolution of ultrafast electron microscopes. We also discussed and experimentally realized an appropriate photon source by using concepts from transformation optics and characterized the functionality of this source by means of EELS and CL angle-resolved mapping.

References

1. S.T. Park, M.M. Lin, A.H. Zewail, Photon-induced near-field electron microscopy (PINEM): theoretical and experimental (in English). New J. Phys. **12**, 123028 (2010). https://doi.org/10.1088/1367-2630/12/12/123028
2. B.J. Siwick, J.R. Dwyer, R.E. Jordan, R.J.D. Miller, An atomic-level view of melting using femtosecond electron diffraction (in English). Science **302**(5649), 1382–1385 (2003). https://doi.org/10.1126/science.1090052
3. B. Barwick, D.J. Flannigan, A.H. Zewail, Photon-induced near-field electron microscopy (in English). Nature **462**(7275), 902–906 (2009). https://doi.org/10.1038/nature08662
4. A.H. Zewail, J.M. Thomas, *4D electron microscopy, imaging in space and time* (Imperial College Press, Singapore, 2010)
5. A.H. Zewail, Four-dimensional electron microscopy (in English). Science **328**(5975), 187–193 (2010). https://doi.org/10.1126/science.1166135
6. G. Sciaini, R.J.D. Miller, Femtosecond electron diffraction: heralding the era of atomically resolved dynamics (in English). Rep. Prog. Phys. **74**(9), 096101 (2011). https://doi.org/10.1088/0034-4885/74/9/096101
7. R.J.D. Miller, Mapping atomic motions with ultrabright electrons: the chemists' Gedanken experiment enters the lab frame (in English). Annu. Rev. Phys. Chem. **65**, 583–604 (2014). https://doi.org/10.1146/annurev-physchem-040412-110117
8. G.M. Vanacore, A.W.P. Fitzpatrick, A.H. Zewail, Four-dimensional electron microscopy: ultrafast imaging, diffraction and spectroscopy in materials science and biology (in English). Nano Today **11**(2), 228–249 (2016). https://doi.org/10.1016/j.nantod.2016.04.009
9. Y.M. Lee, Y.J. Kim, Y.J. Kim, O.H. Kwon, Ultrafast electron microscopy integrated with a direct electron detection camera (in English). Struct. Dynam-Us **4**(4), 044023 (2017). https://doi.org/10.1063/1.4983226
10. A. Feist, K.E. Echternkamp, J. Schauss, S.V. Yalunin, S. Schafer, C. Ropers, Quantum coherent optical phase modulation in an ultrafast transmission electron microscope (in English). Nature **521**(7551), 200–203 (2015). https://doi.org/10.1038/nature14463
11. L. Piazza et al., Simultaneous observation of the quantization and the interference pattern of a plasmonic near-field (in English). Nat. Commun. **6**, 6407 (2015). https://doi.org/10.1038/ncomms7407
12. K. Bücker et al., Ultrafast transmission electron microscopy reveals electron dynamics and trajectories in a thermionic gun setup, in *European Microscopy Congress 2016: Proceedings*. Wiley-VCH Verlag GmbH & Co. KGaA (2016)
13. G.L. Cao, S.S. Sun, Z.W. Li, H.F. Tian, H.X. Yang, J.Q. Li, Clocking the anisotropic lattice dynamics of multi-walled carbon nanotubes by four-dimensional ultrafast transmission electron microscopy (in English). Sci. Rep. **5**, 8404 (2015). https://doi.org/10.1038/srep08404

14. D. Ehberger et al., Highly coherent electron beam from a laser-triggered tungsten needle tip (in English). Phys. Rev. Lett. **114**(22), 227601 (2015). https://doi.org/10.1103/physrevlett.114.227601
15. M. Walbran, A. Gliserin, K. Jung, J. Kim, P. Baum, 5-femtosecond laser-electron synchronization for pump-probe crystallography and diffraction (in English). Phys. Rev. Appl. **4**(4), 044013 (2015). https://doi.org/10.1103/physrevapplied.4.044013
16. W. Verhoeven, J.F.M. van Rens, E.R. Kieft, P.H.A. Mutsaers, O.J. Luiten, High quality ultrafast transmission electron microscopy using resonant microwave cavities. Ultramicroscopy **188**, 85–89 (2018). https://doi.org/10.1016/j.ultramic.2018.03.012
17. W. Verhoeven et al., Time-of-flight electron energy loss spectroscopy using TM110 deflection cavities. Struct. Dynam-Us **3**(5), 054303 (2016). https://doi.org/10.1063/1.4962698
18. J. Maxson, D. Cesar, G. Calmasini, A. Ody, P. Musumeci, D. Alesini, Direct measurement of sub-10 fs relativistic electron beams with ultralow emittance. Phys. Rev. Lett. **118** (15), 154802 (2017) [Online]. Available: https://link.aps.org/doi/10.1103/PhysRevLett.118.154802
19. J. Yang et al., Diffractive imaging of a rotational wavepacket in nitrogen molecules with femtosecond megaelectronvolt electron pulses. Nat. Commun. **7**, 11232 (2016). https://doi.org/10.1038/ncomms11232
20. S.P. Weathersby et al., Mega-electron-volt ultrafast electron diffraction at SLAC National Accelerator Laboratory. Rev. Sci. Instrum. **86**(7), 073702 (2015). https://doi.org/10.1063/1.4926994
21. K.E. Priebe et al., Attosecond electron pulse trains and quantum state reconstruction in ultrafast transmission electron microscopy (in English). Nat. Photonics. **11**(12), 793–797 (2017). https://doi.org/10.1038/s41566-017-0045-8
22. N. Talebi, Electron-light interactions beyond the adiabatic approximation: recoil engineering and spectral interferometry AU—Talebi, Nahid. Adv. Phys. X **3**(1), 1499438 (2018). https://doi.org/10.1080/23746149.2018.1499438
23. N. Talebi, W. Sigle, R. Vogelgesang, P. van Aken, Numerical simulations of interference effects in photon-assisted electron energy-loss spectroscopy. New J. Phys. **15**(5), 053013 (2013). https://doi.org/10.1088/1367-2630/15/5/053013
24. N. Talebi, Spectral Interferometry with Electron Microscopes. Sci. Rep. **6**, 33874 (2016). https://doi.org/10.1038/srep33874, https://www.nature.com/articles/srep33874#supplementary-information
25. N. Talebi et al., Merging transformation optics with electron-driven photon sources. Nat. Commun. (2019)
26. F.J.G.d. Abajo, M. Kociak, Electron energy-gain spectroscopy. New J. Phys. **10** (7), 073035 (2008). Available: http://stacks.iop.org/1367-2630/10/i=7/a=073035
27. S.T. Park, A.H. Zewail, Photon-induced near field electron microscopy (in English). Proc. Spie. **8845**, 884506 (2013). https://doi.org/10.1117/12.2023082
28. A.H. Zewail, 4D ultrafast electron diffraction, crystallography, and microscopy (in English). Annu. Rev. Phys. Chem. **57**, 65–103 (2006). https://doi.org/10.1146/annurev.physchem.57.032905.104748
29. B.K. Yoo, Z.X. Su, J.M. Thomas, A.H. Zewail, On the dynamical nature of the active center in a single-site photocatalyst visualized by 4D ultrafast electron microscopy (in English). Proc Natl Acad Sci USA **113**(3), 503–508 (2016). https://doi.org/10.1073/pnas.1522869113
30. B.W. Shore, J.H. Eberly, Analytic approximations in multilevel excitation theory (in English). B Am. Phys. Soc. **23** (1), 34–34 (1978). Available: <Go to ISI>://WOS:A1978EG45500157
31. S. Aaronson, A. Arkhipov, The computational complexity of linear optics (in English). ACM Sympos. Theory Comput. 333–342 (2011) [Online]. Available: <Go to ISI>://WOS:000297656800035
32. A. Crespi et al., Integrated multimode interferometers with arbitrary designs for photonic boson sampling (in English). Nat. Photonics **7**(7), 545–549 (2013). https://doi.org/10.1038/Nphoton.2013.112
33. K.E. Echternkamp, A. Feist, S. Schafer, C. Ropers, Ramsey-type phase control of free-electron beams (in English). Nat. Phys. **12**(11), 1000–1004 (2016). https://doi.org/10.1038/nphys3844

34. Y. Morimoto, P. Baum, Diffraction and microscopy with attosecond electron pulse train (in English). Nat. Phys. **14**(3), 252–256 (2018). https://doi.org/10.1038/s41567-017-0007-6
35. A. Losquin, T.T.A. Lummen, Electron microscopy methods for space-, energy-, and time-resolved plasmonics (in English). Front. Phys-Beijing **12**(1), 127301 (2017). https://doi.org/10.1007/s11467-016-0605-2
36. A. Ryabov, P. Baum, Electron microscopy of electromagnetic waveforms (in English). Science **353**(6297), 374–377 (2016). https://doi.org/10.1126/science.aaf8589
37. M. Muller, V. Kravtsov, A. Paarmann, M.B. Raschke, R. Ernstorfer, Nanofocused plasmon-driven sub-10 fs electron point source (in English). ACS Photonics **3**(4), 611–619 (2016). https://doi.org/10.1021/acsphotonics.5b00710
38. A.R. Bainbridge, C.W.B. Myers, W.A. Bryan, Femtosecond few-to single-electron point-projection microscopy for nanoscale dynamic imaging (in English). Struct. Dynam-Us **3**(2), 023612 (2016). https://doi.org/10.1063/1.4947098
39. E. Quinonez, J. Handali, B. Barwick, Femtosecond photoelectron point projection microscope (in English). Rev. Sci. Instr. **84**(10), 103710 (2013). https://doi.org/10.1063/1.4827035
40. D. Gabor, A new microscopic principle. Nature **161**(4098), 777–778 (1948). https://doi.org/10.1038/161777a0
41. J. Vogelsang et al., Ultrafast electron emission from a sharp metal nanotaper driven by adiabatic nanofocusing of surface plasmons (in English). Nano Lett. **15**(7), 4685–4691 (2015). https://doi.org/10.1021/acs.nanolett.5b01513
42. P. Baum, On the physics of ultrashort single-electron pulses for time-resolved microscopy and diffraction. Chem. Phys. **423**, 55–61 (2013). https://doi.org/10.1016/j.chemphys.2013.06.012
43. A. Gliserin, A. Apolonski, F. Krausz, P. Baum, Compression of single-electron pulses with a microwave cavity (in English). New J. Phys. **14**, 073055 (2012). https://doi.org/10.1088/1367-2630/14/7/073055
44. C. Kealhofer, W. Schneider, D. Ehberger, A. Ryabov, F. Krausz, P. Baum, All-optical control and metrology of electron pulses. Science **352**(6284), 429–433 (2016). https://doi.org/10.1126/science.aae0003
45. A. Feist et al., Ultrafast transmission electron microscopy using a laser-driven field emitter: femtosecond resolution with a high coherence electron beam (in English). Ultramicroscopy **176**, 63–73 (2017). https://doi.org/10.1016/j.ultramic.2016.12.005
46. C.J. Powell, J.B. Swan, Origin of the characteristic electron energy losses in aluminum (in English). Phys. Rev. **115**(4), 869–875 (1959). https://doi.org/10.1103/physrev.115.869
47. C.J. Powell, J.B. Swan, Origin of the characteristic electron energy losses in magnesium (in English). Phys. Rev. **116**(1), 81–83 (1959). https://doi.org/10.1103/physrev.116.81
48. F.J. García de Abajo, Multiple excitation of confined graphene plasmons by single free electrons. ACS Nano **7**(12), 11409–11419 (2013). https://doi.org/10.1021/nn405367e
49. R. Garciamolina, A. Grasmarti, A. Howie, R.H. Ritchie, Retardation effects in the interaction of charged-particle beams with bounded condensed media (in English). J. Phys. C Solid State **18**(27), 5335–5345 (1985). https://doi.org/10.1088/0022-3719/18/27/019
50. N. Talebi, W. Sigle, R. Vogelgesang, P. van Aken, Numerical simulations of interference effects in photon-assisted electron energy-loss spectroscopy (in English). New J. Phys. **15**, 053013 (2013) (10.1088/1367-2630/15/5/053013)
51. A. Howie, Stimulated excitation electron microscopy and spectroscopy (in English). Ultramicroscopy **151**, 116–121 (2015). https://doi.org/10.1016/j.ultramic.2014.09.006
52. V.L. Ginzburg, Radiation by uniformly moving sources—Vavilov-Cherenkov Effect, Doppler-Effect in a Medium, Transition Radiation and Associated Phenomena (in English), in *Progress in Optics, Vol. Xxxii*, vol. 32 (1993), pp. 269–312 [Online]. Available: <Go to ISI>://WOS:A1993BA26A00005
53. J. Urata, M. Goldstein, M.F. Kimmitt, A. Naumov, C. Platt, J.E. Walsh, Superradiant smith-purcell emission (in English). Phys. Rev. Lett. **80**(3), 516–519 (1998). https://doi.org/10.1103/physrevlett.80.516
54. F.J. García de Abajo, A. Rivacoba, N. Zabala, N. Yamamoto, Boundary effects in Cherenkov radiation. Phys. Rev. B **69** (15), 155420 (2004). Available: http://link.aps.org/doi/10.1103/PhysRevB.69.155420

55. P. Zhang, L.K. Ang, A. Gover, Enhancement of coherent Smith-Purcell radiation at terahertz frequency by optimized grating, prebunched beams, and open cavity. Phys. Rev. Spec. Top. Accelerators Beams **18** (2), 020702 (2015). Available: http://link.aps.org/doi/10.1103/PhysRevSTAB.18.020702

56. S.E. Korbly, A.S. Kesar, J.R. Sirigiri, R.J. Temkin, Observation of frequency-locked coherent terahertz Smith-Purcell radiation (in English). Phys. Rev. Lett. **94**(5), 054803 (2005). https://doi.org/10.1103/physrevlett.94.054803

57. F.J.G. de Abajo, Smith-purcell radiation emission in aligned nanoparticles (in English). Phys. Rev. E **61**(5), 5743–5752 (2000). https://doi.org/10.1103/physreve.61.5743

58. M. Merano et al., Probing carrier dynamics in nanostructures by picosecond cathodoluminescence (in English). Nature **438**(7067), 479–482 (2005). https://doi.org/10.1038/nature04298

59. G. Adamo et al., Electron-beam-driven collective-mode metamaterial light source (in English). Phys. Rev. Lett. **109**(21), 217401 (2012). https://doi.org/10.1103/physrevlett.109.217401

60. F.J.G. de Abajo, Optical excitations in electron microscopy (in English). Rev. Mod. Phys. **82**(1), 209–275 (2010). https://doi.org/10.1103/revmodphys.82.209

61. A. Losquin, M. Kociak, Link between cathodoluminescence and electron energy loss spectroscopy and the radiative and full electromagnetic local density of states (in English). ACS Photonics **2**(11), 1619–1627 (2015). https://doi.org/10.1021/acsphotonics.5b00416

62. N. Talebi, A directional, ultrafast and integrated few-photon source utilizing the interaction of electron beams and plasmonic nanoantennas (in English). New J. Phys. **16**, 053021 (2014). https://doi.org/10.1088/1367-2630/16/5/053021

63. I.A. Walmsley, C. Dorrer, Characterization of ultrashort electromagnetic pulses (in English). Adv Opt Photonics **1**(2), 308–437 (2009). https://doi.org/10.1364/aop.1.000308

64. E.M. Kosik, A.S. Radunsky, I.A. Walmsley, C. Dorrer, Interferometric technique for measuring broadband ultrashort pulses at the sampling limit (in English). Opt Lett **30**(3), 326–328 (2005). https://doi.org/10.1364/ol.30.000326

65. L. Mandel, E. Wolf, Coherence properties of optical fields (in English). Rev. Mod. Phys. **37**(2), 231 (1965). https://doi.org/10.1103/revmodphys.37.231

66. R.H. Dicke, Coherence in spontaneous radiation processes (in English). Phys. Rev. **93**(1), 99–110 (1954). https://doi.org/10.1103/physrev.93.99

67. M.O. Scully, A.A. Svidzinsky, The super of superradiance (in English). Science **325**(5947), 1510–1511 (2009). https://doi.org/10.1126/science.1176695

68. D.M. Wolkow, Über eine Klasse von Lösungen der Diracschen Gleichung. Zeitschrift für Physik, journal article **94**(3), 250–260 (1935). https://doi.org/10.1007/bf01331022

69. P.A. Midgley, R.E. Dunin-Borkowski, Electron tomography and holography in materials science. Nat. Mater. **8**(4), 271–280 (2009). https://doi.org/10.1038/nmat2406

70. N. Talebi et al., Excitation of mesoscopic plasmonic tapers by relativistic electrons: phase matching versus eigenmode resonances (in English). ACS Nano **9**(7), 7641–7648 (2015). https://doi.org/10.1021/acsnano.5b03024

71. N. Talebi, B. Ögüt, W. Sigle, R. Vogelgesang, P.A. van Aken, On the symmetry and topology of plasmonic eigenmodes in heptamer and hexamer nanocavities. Appl. Phys. A **116**(3), 947–954 (2014). https://doi.org/10.1007/s00339-014-8532-y

72. U. Timm, Coherent Bremsstrahlung of electrons in crystals (in English). Fortschr Physik **17**(12), 765–808 (1969). https://doi.org/10.1002/prop.19690171202

73. M. Nakajima, F. Arai, D. Lixin, T. Fukuda, A hybrid nanorobotic manipulation system integrated with nanorobotic manipulators inside scanning and transmission electron microscopes, in *4th IEEE Conference on Nanotechnology*, 16–19 Aug 2004 (2004), pp. 462–464. https://doi.org/10.1109/nano.2004.1392386

74. N. Talebi, A directional, ultrafast and integrated few-photon source utilizing the interaction of electron beams and plasmonic nanoantennas. New J. Phys. **16**(5), 053021 (2014). https://doi.org/10.1088/1367-2630/16/5/053021

75. M. Kuttge et al., Local density of states, spectrum, and far-field interference of surface plasmon polaritons probed by cathodoluminescence. Phys. Rev. B **79**(11), 113405 (2009). https://doi.org/10.1103/physrevb.79.113405

76. J.T. van Wijngaarden, E. Verhagen, A. Polman, C.E. Ross, H.J. Lezec, H.A. Atwater, Direct imaging of propagation and damping of near-resonance surface plasmon polaritons using cathodoluminescence spectroscopy. Appl Phys Lett. **88**(22), 221111 (2006). https://doi.org/10.1063/1.2208556
77. O. Kidwai, S.V. Zhukovsky, J.E. Sipe, Effective-medium approach to planar multilayer hyperbolic metamaterials: strengths and limitations. Phys. Rev. A **85**(5), 053842 (2012). https://doi.org/10.1103/physreva.85.053842
78. M. Liscidini, J.E. Sipe, Quasiguided surface plasmon excitations in anisotropic materials. Phys. Rev. B **81**(11), 115335 (2010). https://doi.org/10.1103/physrevb.81.115335
79. N. Talebi, Optical modes in slab waveguides with magnetoelectric effect (in English). J. Optics-UK **18**(5), 055607 (2016). https://doi.org/10.1088/2040-8978/18/5/055607
80. L. Kipp et al., Sharper images by focusing soft X-rays with photon sieves. Nature **414**, 184 (2001). https://doi.org/10.1038/35102526

Chapter 8
Electron-Light Interactions Beyond Adiabatic Approximation

Abstract Aligned with the technological developments of electron-based characterization techniques, our theoretical frameworks are yet to be adapted to the strong-laser and slow-electron regimes. More specifically, there exist certain domains where our adiabatic approximations might break down. This is practically important from several viewpoints: (i) in PPM, the shape and amplitude of electron beams are both strongly manipulated, in addition to their phase; (ii) even in free-space electron-light interactions, purely elastic approximations might appear to be a mere over-simplification (Kozak et al. in Nat Phys Lett, 2017 [1]); (iii) during the interaction of electron beams with gratings and light, electron bunching appears to be an additional mechanism to the electron acceleration, where both the acceleration and bunching mechanisms are controlled by the longitudinal broadening of the electron beam relative to the grating period (Talebi in New J Phys 18:123006, 2016 [2]); and (iv) shaped electron beams interacting with matter have different selection rules and might offer approaches for manipulating the electron-induced radiations (Sergeeva et al. in Opt Express 25:26310–26328, 2017 [3]; Tsesses et al. in Phys Rev A 95:013832, 2017 [4]; Kaminer et al. in Phys Rev X 6:011006, 2016 [5]). The last point is fundamentally important, as even for a single electron wave packet, when the electron beam is in a superposition of at least two momentum states, interferences between different quantum paths in the interaction of photons with the electron may occur (Peatross et al. in Phys Rev Lett 100:153601, 2008 [6]). As noted by Keitel and co-workers, the quantum eigenstates of electrons in a nonplanar laser beam or in general shaped light waves are unknown (Peatross et al. in Phys Rev Lett 100:153601, 2008 [6]). For this reason, the development of self-consistent numerical methods may facilitate a better understanding of the outcomes of experiments (Talebi in New J Phys 18:123006, 2016 [2]; White et al. in Phys Rev B 86:205324, 2012 [7]; Kohn et al. in Phys Rev 140:1133, 1965 [8]) and stimulate the design of new experiments.

© Springer Nature Switzerland AG 2019
N. Talebi, *Near-Field-Mediated Photon–Electron Interactions*,
Springer Series in Optical Sciences 228,
https://doi.org/10.1007/978-3-030-33816-9_8

The present chapter provides an overview of electron-light interactions from a new perspective, i.e., non-adiabatic analysis and recoil engineering within the semiclassical quantum mechanical approaches, by combining the time-dependent numerical solutions of the Maxwell and Schrödinger equations in a single toolbox.

8.1 Simulations Beyond Adiabatic Assumptions

The majority of the strong-field effects in the light-matter interaction can be qualitatively understood using adiabatic approximations [11]. These approximations most often treat the electron gas as being unbound to the ions, where the evolution of their wave function in interaction with the laser field is represented by Wolkow states [12]. This approach treats the change in the electron wave function via its propagation in the laser field with a special case of the eikonal approximation, where the change in the amplitude is neglected. The Wolkow wave function is given by

$$\psi(\vec{r}, t) = \frac{1}{(2\pi)^{\frac{3}{2}}} \exp\left(i\vec{k}_{\text{el}} \cdot \vec{r} - i\omega_e t\right)$$

$$\times \exp\left(-i\frac{e}{\hbar m_0} \int_{-\infty}^{t} \left[\hbar \vec{k}_{\text{el}} \cdot \vec{A}(\tau) + \frac{e}{2}\left|\vec{A}(\tau)\right|^2 - m_0\varphi(\tau)\right] d\tau\right) \quad (8.1)$$

where m_0 is the electron mass, $\vec{k}_{\text{el}} = m_0 v_{\text{el}}/\hbar$ is the electron wave vector, $\omega_e = \hbar k_{\text{el}}^2/2m_0$, and $\vec{A}(t)$ is the vector potential, which is considered to be only dependent on time, as the wavelength of the photons is assumed to be much larger than the extent of the whole wave function. The first term in the integrand underpins the acceleration and deceleration of the electron in its interaction with light, within the dipole approximation. The second term describes the ponderomotive force on the electron, and the last term is correlated with the change in the phase of the electron via its propagation in the effective electrostatic potential of the light.

8.1.1 Slowly Varying Approximation

In general, however, the time-dependent Schrödinger equation including electromagnetic interactions is given by

$$-\frac{\hbar^2}{2m_0}\nabla^2\psi + \frac{-i\hbar e}{m_0}\vec{A} \cdot \vec{\nabla}\psi + \frac{e^2}{2m_0}\left|\vec{A}\right|^2\psi - e\varphi\psi = i\hbar\frac{\partial\psi}{\partial t} \quad (8.2)$$

where the Coulomb gauge has been implied and φ is the scalar potential. Note that despite the fact that the Schrödinger equation is nonrelativistic, relativistic corrections can still be applied to be able to employ (8.8) to elucidate the dynamics of relativistic electrons. This is achieved by simplifying the Dirac equation into a scalar form by ignoring the electron spin [13], which is most often unimportant in electron microscopy. Moreover, the Dirac equation introduces formidable complications in electron-beam physics such as zitterbewegung behaviour [14] and spontaneous electron-positron pair creation.

A common practice in electron microscopy, especially when high-energy electron beams are involved, is to write the wave function as

$$\psi(\vec{r}, t) = \psi_0(\vec{r}, t) \exp\left(i\vec{k}_{\text{el}} \cdot \vec{r} - i\omega_{\text{el}} t\right) \tag{8.3}$$

where $\psi_0(\vec{r}, t)$ is the slowly varying amplitude. Using this approximation, we recast (8.8) into a more practical form as

$$-\frac{\hbar^2}{2m_0}\nabla^2\psi_0(\vec{r}, t) - \frac{i\hbar^2}{m_0}\vec{k}_{\text{el}} \cdot \vec{\nabla}\psi_0(\vec{r}, t)$$
$$+ \frac{-i\hbar e}{m_0}\vec{A} \cdot \left[\vec{\nabla}\psi_0(\vec{r}, t) + i\vec{k}_{\text{el}}\psi_0(\vec{r}, t)\right]$$
$$+ \frac{e^2}{2m_0}\left|\vec{A}\right|^2\psi_0(\vec{r}, t) - e\,\varphi\psi_0(\vec{r}, t) = i\hbar\frac{\partial\psi_0(\vec{r}, t)}{\partial t} \tag{8.4}$$

By assuming $\vec{\nabla}\psi_0 \ll i\vec{k}_{\text{el}}\psi_0$ [slowly varying approximation (SVA)], (8.4) is further simplified into the form

$$-\frac{\hbar^2}{2m_0}\nabla^2\psi_0(\vec{r}, t) - e\vec{A} \cdot \vec{v}_{\text{el}}\psi_0(\vec{r}, t) + \left(\frac{e^2}{2m_0}\left|\vec{A}\right|^2 - e\varphi\right)\psi_0(\vec{r}, t) = i\hbar\frac{\partial\psi_0(\vec{r}, t)}{\partial t} \tag{8.5}$$

where, in the strong-field approximation, the second term in (8.4) has been neglected. To derive the Wolkow wave functions from (8.5), we assume that $\psi_0(\vec{r}, t) = \psi_0(t)$, which, by insertion into (8.5) and appropriate normalization, leads to (8.1).

It has already been noted that in the case of energetic and relativistic electron beams, the Wolkow states and SVA provide us with an analytical treatment of many experimental observations, from PINEM [15, 16] to electron holography [17]. However, certainly Wolkow wave functions will not provide us with enough insight into more advanced methodologies based on the shaping of the electron wave functions and point projection microscopy with slow electrons. This is because the Wolkow states are outcomes of an adiabatic approximation in which the amplitude of the electron wave function is neglected. Moreover, implicit to (8.5) is the dipole approximation, which might break down in some realistic situations.

8.1.2 Full-Wave Analysis and Self-consistent Approach

Indeed, the most well-known self-consistent approach is the Hartree approximation. Hartree's method is based on the simplification of the many-body Hamiltonian of the Schrödinger equation, as

$$\frac{-\hbar^2}{2m_0}\nabla^2\psi(\vec{r}) + U^{\text{ion}}(\vec{r})\psi_i(\vec{r}) + U^{\text{el}}(\vec{r})\psi_i(\vec{r}) = E\,\psi(\vec{r}) \tag{8.6}$$

in the energy-domain representation; $U^{\text{ion}}(\vec{r})$ is the potential of the ions given by

$$U^{\text{ion}}(\vec{r}) = -Ze^2\sum_{\vec{R}}\frac{1}{4\pi\varepsilon_0\left|\vec{r} - \vec{R}\right|} \tag{8.7}$$

and the electron-electron interaction potential is averaged over the density of the electron charges as

$$U^{\text{el}}(\vec{r}) = -e\int \mathrm{d}r'\frac{\rho(\vec{r}')}{4\pi\varepsilon_0|\vec{r} - \vec{r}'|} \tag{8.8}$$

where the charge density distribution would be

$$\rho(\vec{r}) = -e\sum_{i}|\psi_i(\vec{r})|^2 \tag{8.9}$$

and the sum is over all occupied one-electron levels. The Hartree approximation hence fails to consider the effect of the shape of the other electron wave functions over the single-electron wave function of interest, as only an average over the charge distributions is implied. In other words, it is only the average density that appears in (8.8), regardless of the shape of the involved electronic orbitals. Despite the simplicity of the approximation, (8.6) is still quite demanding from the numerical point of view. Additional improvements to the Hartree approximation are provided by the Slater determinant and the Hartree-Fock exchange potentials. The further introduction of self-consistent density-functional theory [8, 18, 19] and inclusion of pseudopotentials [20, 21] have added to the accuracy of the single-electron approximations, albeit with an increasing level of numerical complexity. Moreover, the generalization of density-functional theory to the time-dependent Schrödinger equation has led to the introduction of the action integral, compared to the energy vibrational, which is to be minimized [22]. In practice, however, the electron-electron interactions in (8.8) can be treated based on a time-dependent variation of the Hartree-Fock approximation [23].

8.1.3 Self-consistent Maxwell–Schrödinger Approximation

Equations (8.12)–(8.15) should be solved self-consistently. The iteration includes the insertion of initial approximations of the electron wave functions in the material under investigation, the computation of the potentials and the insertion of the potentials into (8.12), the calculation of the new set of wave functions, and the repetition of the cycle until $U^{el}(\vec{r})$ (or the wave functions) does not change any more. Indeed, the solution of $U^{el}(\vec{r})$ is often given by the Poisson equation ($\nabla^2 U^{el}(\vec{r}) = -e\rho(\vec{r})$); hence the problem is considered static. For swift electron beams in electron microscopes however, a common practice is to include an additional current density distribution given by [24]

$$\vec{J}_{el}(\vec{r}, t) = \frac{e\,\hbar}{m_0}\text{Im}\left\{\sum_i \psi_i(\vec{r}, t)\vec{\nabla}\psi_i^*(\vec{r}, t)\right\} - \frac{e^2\rho(\vec{r}, t)}{m_0}\vec{A}(\vec{r}, t) \qquad (8.10)$$

This equation should be inserted into the Maxwell's equations to account for a self-consistent theory of field approximation, including retardation. It is noted that (8.10) is necessary in order to satisfy the continuity of the electric charges and especially important to model the electron-electron interactions similar to the Hartree approximation, though also considering screening effects. In particular, from the Maxwell's equations, one can obtain $\vec{\nabla} \cdot \vec{D}(\vec{r}, t) = \rho(\vec{r}, t)$, which in turn leads to the Poisson equation in a dielectric medium

$$\nabla^2\varphi(\vec{r}, t) = -\frac{\rho(\vec{r}, t)}{\varepsilon_0\varepsilon_r} \qquad (8.11)$$

whenever the Coulomb gauge is considered (accounting for the Lorentz gauge leads to the Helmholtz equation for the scalar potential). ε_r is the dielectric function of the material, or the environment surrounding the electron wavefunction. In vacuum, the electron potential $U^{el}(\vec{r}, t) = -e \int d\vec{r}'\rho(\vec{r}', t)/4\pi\varepsilon_0|\vec{r} - \vec{r}'|$ is simply reproduced, whereas in a material, the inclusion of the dielectric function accounts for the screening effects [25]. In other words, introducing the current density distribution into the combined Maxwell–Schrödinger system of equations allows for the calculation of the magnetic vector potential; the latter is crucial for self-consistent field theory including retardation.

We use the abovementioned retarded self-consistent field theory to simulate the dynamics of electron wave packets interacting with nanostructures and samples. Our crude assumption here is that the samples can be modelled using the dielectric theory, and hence we do not account for the electron diffraction by the ionic potentials. The modelling of the latter effect is achievable by including appropriate pseudopotentials, but this is not the subject of the present review. Our model, however, considers collective electron-electron interactions representable by the dielectric function.

The following self–consistent steps form the current state of our developed Maxwell–Schrödinger numerical toolbox. We consider two different simulation domains of arbitrary sizes and arbitrary grids, where the connections between them are held by accurately mapping the current distribution and potentials. In each time step, we

(i)   simulate the electron wave function by solving (8.8) using a pseudospectral Fourier method [2, 26], which conserves the norm of the wave function,
(ii) calculate the current-density distribution and project it from the Schrödinger domain into the Maxwell domain using an accurate mapping technique,
(iii) solve the different field components using an FDTD algorithm,
(iv) calculate the potentials from the field components by considering the Coulomb gauge theory, using $\nabla^2 \varphi = -\vec{\nabla} \cdot \vec{E}$ and $\partial \vec{A}/\partial t = -\vec{\nabla}\varphi - \vec{E}$, and
(v)   map the potentials from the Maxwell domain into the Schrödinger domain.

Additionally, appropriate absorbing boundary conditions are satisfied for both the wave function and the field components in the Schrödinger and Maxwell domains, individually. It is finally noted here that for the sake of simplicity, the samples in the present simulations included in this review are all modelled only with their bulk permittivity.

In the following, we first investigate the electron-light interactions in the free space by means of both the Wolkow approximation (8.1) and full-wave analysis stated above. We particularly examine carefully a specific case, i.e., the Kapitza-Dirac effect [10].

8.2   Interference Between Quantum Paths in the Coherent Kapitza-Dirac Effect

In the Kapitza-Dirac effect, atoms, molecules, or swift electrons are diffracted off a standing wave grating of the light intensity created by two counter-propagating laser fields. In ultrafast electron optics, such a coherent beam splitter offers interesting perspectives for ultrafast beam shaping. Here, we study, both analytically and numerically, the effect of the inclination angle between two laser fields on the diffraction of pulsed, low-energy electron beams. For sufficiently high light intensities, we observe a rich variety of complex diffraction patterns. These not only reflect interferences between electrons scattered off intensity gratings that are formed by different vector components of the laser field but may also result, for certain light intensities and electron velocities, from interferences between these ponderomotive scattering and direct light absorption and stimulated emission processes, which are usually forbidden for far-field light. Our findings may open up perspectives for the coherent manipulation and control of ultrafast electron beams by free-space light.

The coherent control of the shape of quantum wave functions has set the way towards bond-selective chemistry [27], quantum computing [28, 29], and the ultrafast control of plasmons [30]. Quantum coherent control originates from the ability to manipulate the interference between quantum paths towards realizing a desired shape of a target wave function by means of shaped, coherent, and/or strong laser excitation. Theoretically, such quantum paths may intuitively be studied using Feynman's path integral approach [31]. This has been instrumental in interpreting observations of above-threshold ionization [32, 33] and high-harmonic generation [34] but also for the selective control of the quantum paths using circular and elliptical polarizations [35]. The success of path integrals in understanding coherent control lies in the offered ease of selecting a few physically significant paths from a wealth of mathematically available options. Moreover, important concepts such as the *action* are easily related to classical quantities.

To date, coherent control has mostly been used to shape the wave function of bound-electron states [36], while applications in controlling free electron waves have emerged only recently [37]. Specifically, the inelastic interaction of electrons with optical near-fields can cause attosecond longitudinal electron bunching. The population amplitudes of certain electron states can be controlled precisely by the laser phase in a Ramsey-type experiment [38]. It has been learned that the inelastic processes involved in electron-near-field scattering require moderate laser intensities and are well understood in a minimal coupling Hamiltonian, neglecting ponderomotive forces [15].

For future progress in ultrafast electron microscopy, it appears desirable to avoid the need for matter-based near-field interactions and to simply use light waves in free space to control and shape pulsed electron beams. As a step in this direction, we show here how to use the elastic interaction of free-electron waves with focused freely propagating laser fields to coherently control the transversal distribution of electron wave packets. This is achieved by a generalization of the Kapitza-Dirac (KD) effect [39, 40] to the concomitant utilization of standing-wave and travelling-wave light patterns. In the normal KD effect, electron waves travelling through a standing-wave pattern of light are diffracted to transversely populated momentum states at multiples of twice the momentum of free-space light. We show that the KD effect can be generalized to the realization of arbitrary momentum states of the electron wave packet by controlling the interference between quantum pathways originating from distinctly different parts, absorptive and ponderomotive, of the interaction Hamiltonian. This offers fundamentally new degrees of freedom for designing light-controlled phase masks for free-space electron pulses.

8.2.1 Results

In a normal KD effect, electrons are propagating through a standing wave pattern of the light intensity and in a direction perpendicular to the momentum of the light. We assume here that the standing wave pattern of light is formed by two counter-propagating light waves with the wave vectors $\vec{k}_1 = +k_{ph}\hat{y}$ and $\vec{k}_2 = -k_{ph}\hat{y}$. Consequently, the electrons are transversely diffracted into distinct orders that are displaced

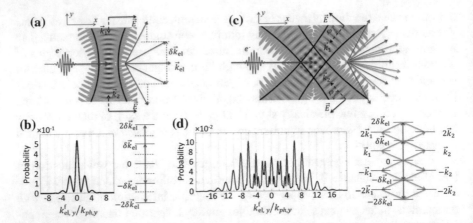

Fig. 8.1 Quantum coherent interference paths in the interaction of electron beams with free-space light patterns. **a** When electron beams interact with a standing-wave light pattern in free space, only pondermotive forces are effective, resulting in **b** transverse recoils to the harmonics of $k_y^{el} = n\delta k$. **c** For pulsed electrons synchronized with the laser excitation, interference paths between various parts of the Hamiltonian will result in **d** the formation of an exotic diffraction pattern. Re-printed by permission from [10] under the CC BY license

by $\delta k_{el} = \left| \vec{k}_2 - \vec{k}_1 \right| = 2k_{ph}$, where k_{el} and k_{ph} are the electron and light wavenumbers, respectively (Fig. 8.1a and b). The probability distribution of the diffraction orders has been shown to be dominated by Bessel functions as $J_n(\kappa t)$, where t is the time of propagation of electrons through the standing wave pattern, and the coupling strength $\kappa = e^2 E_0^2 / 8m_0\hbar\omega^2$ scales with the square of the electric field amplitude E_0 [41, 42]. Here, e and m_0 are the elementary charge and mass, respectively, $\hbar$ the Planck constant, and ω the angular frequency of the light wave. Essentially, the spatially varying light intensity introduces a periodic ponderomotive potential along the y-axis that results in a bunching of the electron density in regions of low light intensity. After leaving the interaction zone, this electron wave packet transforms into a coherent superposition of plane waves propagating in the direction $\vec{k}_{el}^f = \vec{k}_{el} \pm n\delta k_{el}\hat{y}$.

Theoretically, the dynamics of such free-space electron wave packets in KD gratings are commonly described in terms of the Wolkow representation of the electron wavefunction propagating through a vector potential $\overrightarrow{A}(\vec{r}, t)$ (see Sect. 8.1.1). Then, the wavefunction can be represented as [12]

$$\psi(\vec{r}, t) = \psi_0(\vec{r}, t \to 0) \exp\left(-i\frac{e^2}{2\hbar m_0} \int_0^t \vec{A}(\vec{r}, \tau)^2 d\tau\right)$$

$$\exp\left(-i\frac{ek_{el}}{m_0} \int_0^t A_x(\vec{r}, \tau) d\tau\right) \tag{8.12}$$

where $\psi_0(\vec{r}, t \to 0)$ is the initial state of the electron wave packet, which for a plane wave electron is given by $\psi_0(\vec{r}, t \to 0) = (2\pi)^{-1.5} \exp(ik_{el}x - i\Omega t)$. Here, $\hbar\Omega = 0.5 m_0 v_{el}^2$ is the kinetic energy of the electron, propagating with velocity v_{el} along the x-axis. As usual, the vector potential $\vec{A}$ is related to the electric field, by neglecting the scalar potential (this approximation is justified because the electromagnetic field in free space is purely transverse), as $\vec{E} = -\dot{\vec{A}}$. Furthermore, by $\vec{A}^2$, we mean $\vec{A} \cdot \vec{A}$. To describe the KD effect, it is sufficient to write the vector potential as $\vec{A} = A_0 \hat{x} \cos(\omega t) \cos(k_{ph} y)$, which by direct substitution in (8.12) can be cast in the form [41]

$$
\psi(\vec{r}, t) = \frac{1}{(2\pi)^{\frac{3}{2}}} \exp(ik_{el}x - i\Omega t) \exp\left\{-i \frac{e^2 E_0^2}{2k_{el}\hbar^2\omega^2} x\right\}
$$
$$
\times \sum_n i^n J_n\left(\frac{e^2 E_0^2}{2k_{el}\hbar^2\omega^2} x\right) \exp(i2nk_{ph}y)
$$

(8.13)

For this, we have used $E_0 = \omega A_0$ and $k_{ph} = \omega/c$ and have related the delay τ to the propagation length x as $x = v_{el}\tau$, with v_{el} being the phase velocity of the electrons [16, 43]. To derive (8.13), the temporal oscillation of $A^2 \cos^2(\omega t)$, giving highly oscillatory phase variations at twice the light frequency, has been replaced by its average value $A_0^2/2$. Then, the **ponderomotive phase modulation**, [the first exponential in (8.12)] shows a phase variation $\varphi \propto \cos^2(k_{ph}y) = \frac{1}{2}(1 + \cos(2k_{ph}y))$ along the transverse direction. Using the Jacobi-Anger transform [44], this can be expressed as a series of Bessel functions corresponding to the different diffraction orders of the electron beam. This is the last term in (8.13). The probability for populating the n-th diffraction order is then given as $P_n^{A^2} = J_n^2\left(\frac{e^2 E_0^2}{2m_0\hbar\omega^2} t\right)$. Moreover, the argument of the exponential inside the summation in (8.13) can be written as $i2nk_{ph}y = in(\vec{k}_2 - \vec{k}_1) \cdot \vec{r}$. This specifies that the final electron wave function can be regarded as a superposition of the plane wave electrons occupying the momentum states $|n, -n\rangle$ in the complete basis specified by the momentum states of the two incident light beams, leading to the final momentum of $\hbar\vec{k}_{el}^f = \hbar\vec{k}_{el} + n\hbar\vec{k}_2 - n\hbar\vec{k}_1$.

Importantly, the phase modulation introduced by the second exponential in (8.12) is usually discarded in the description of the Kapitza-Dirac effect. As shown in more detail in the Supporting Information, this term causes an ultrafast oscillation of the phase, which for plane-wave electrons averages to zero. Hence, the second exponential in (8.12) is simply replaced by a multiplicative factor of unity in (8.13).

This shows that in conventional treatments of the Kapitza-Dirac effect, only the ponderomotive forces on the electrons, expressed by the Hamiltonian $H_1 = e^2 A^2/2\hbar m_0$, are responsible for the phase modulations and the resulting electron diffraction. In contrast, the absorptive part $H_2 = ek_{el}A_x/m_0$ of the Hamiltonian in (8.12) induces only highly oscillatory phase terms, which—in a phase-cycling approach—are neglected. This is justified whenever considering sufficiently long interaction times between the electron and field. However, this interaction can be shortened dramatically to a few or even less than a single cycle of the light field by letting the electrons interact with spatially confined optical near-fields. In this case, it is the absorptive part H_2 of the Hamiltonian that becomes dominant and causes inelastic-electron photon interactions, as, for example, in photon-induced near-field electron microscopy [15, 16, 41, 45].

The abovementioned approximations lead to the general assumption that free electrons and light waves cannot inelastically interact in free space. In contrast, for restricted electron-light interaction times achieved, e.g., by ultrafast laser excitations and/or employing slow-electron pulses, at energies below 100 eV, none of the previous assumptions necessarily holds true. This has recently been evidenced, e.g., by the direct acceleration and bunching of electrons with laser pulses in free space [46, 47]. In this case, H_1 may not only contribute to the inelastic but also to the elastic interaction, which will be demonstrated below. As a result, one might observe, even in free space, interferences between the quantum paths arising from both parts of the Hamiltonian. We show that, for light gratings formed by optical beams with finite inclination angles, two different paths can reach the momentum recoil of $k_{\text{el}, y} = 2nk_{\text{ph}}$ provided by the H_1 and H_2 parts of the Hamiltonian, respectively (see Fig. 8.1c and d). These quantum path interferences appear as modulations in the diffraction pattern of the electron waves, different from the diffraction orders observed in the normal KD effect, and hence can—in principle—be observed by a regular position-sensitive detector. To verify these conclusions, we present comprehensive analytical and numerical studies directly from first principles.

For the sake of simplicity of our analytical model, we first consider the interaction of an electron wave packet with two plane waves of light propagating at angles φ and $-\varphi$ with respect to the direction of propagation of the electron (see Fig. 8.1c). In this way, we construct a standing wave pattern along the y-axis and perpendicular to the initial propagation direction of the electron, whereas along the x-axis, the superposition of the light beams imposes a travelling wave. The x- and y-components of the vector potential will be given by $A_x = -2A_0 \sin \varphi \cos(k_{\text{ph}} \sin \varphi \, y) \cos(\omega t + k_{\text{ph}} \cos \varphi \, x)$ and $A_y = -2A_0 \cos \varphi \sin(k_{\text{ph}} \sin \varphi \, y) \times \sin(\omega t + k_{\text{ph}} \cos \varphi \, x)$, respectively. By inserting these components into the Wolkow propagator (8.12) and including only the H_1 (pondermotive) Hamiltonian, we obtain

$$\psi(\vec{r}, t) = \frac{1}{(2\pi)^{\frac{3}{2}}} \exp(ik_{\text{el}}x - i\Omega t) \exp\left(-i\frac{e^2 E_0^2}{2k_{\text{el}}\hbar^2\omega^2}\{1 + 2\cos^2(\varphi)\}x\right)$$

$$\times \sum_n i^n J_n\left(\cos(2\varphi)\frac{e^2 E_0^2}{2k_{\text{el}}\hbar^2\omega^2}x\right) \exp(2ink_{\text{ph}}\sin(\varphi)y) \qquad (8.14)$$

Evidently, for $\varphi = 90°$, (8.14) will recast (8.13). Interestingly we observe that the probabilities for coupling to the n-th diffraction order, with a transverse momentum $n\delta\vec{k} = n(\vec{k}_1 - \vec{k}_2)$, now depends sensitively on the inclination angle φ. Of particular interest is that for $\varphi = 45°$, only the 0-th diffraction order is populated. In other words, for $\varphi = 45°$, the plane wave electrons will be not diffracted by the light beams. We explain this by the competition between the $H_1^x = e^2 A_x^2/2\hbar m_0$ and $H_1^y = e^2 A_y^2/2\hbar m_0$ parts of the Hamiltonian, being related to the $\sin^2 \varphi$ and $\cos^2 \varphi$

terms, respectively. Hence, at exactly $\varphi = 45°$, destructive interference between the two paths of the Hamiltonian cancels the diffraction. This cancellation at $\varphi = 45°$ will also occur for incoherent electron beams since their interaction with the pondermotive potential is phase-insensitive.

To further investigate the formation of KD diffraction orders under the pure effect of ponderomotive forces, we simulated the interaction of a pulsed electron wave packet with two inclined continuous optical beams with $E_0 = 60\,\mathrm{GV\,m^{-1}}$, $\omega = 62.79\,\mathrm{P\,rad\,s^{-1}}$ ($\lambda_{ph} = 30\,\mathrm{nm}$), and $\varphi = 50°$. Such shortwave and intense fields, difficult to realize in the laboratory with current state-of-the-art technology, have been chosen for reducing the numerical complexity of the simulations. A scaling to less intense, longer wavelength pulses will be discussed in more detail below. In these simulations, the interaction Hamiltonian is restricted to $H^{int} = e^2 A^2 / 2m_0$; i.e., we neglect the part linearly related to the vector potential. The initial longitudinal and transverse broadenings of the incident electron wave packet are $W_x = 2.4\,\mathrm{nm}$ and $W_y = 20\,\mathrm{nm}$, respectively, and its initial carrier velocity is $v_{el} = 0.03c$. At this velocity, an electron experiences many ponderomotive diffraction orders up to $n = 3$ (see Fig. 8.2a and b for the lower-order diffraction probabilities). This electron wave packet, which has a pulse duration of about 2.67 fs, propagates along the x-axis, and its overall interaction time with the focused light field is roughly 6 fs, corresponding to about 26.7 oscillation cycles of the light field (Fig. 8.2c). In the simulations, we assume a fixed relative phase between the light field and electron wave packet, as is commonly the case in ultrafast laser-driven electron generation schemes. At the exit of the interaction zone, we notice a gradual formation of symmetric and transverse electron bunching effects with up to 12 distinct lobes (Fig. 8.2c). This generation of KD diffraction orders is better visualized by the momentum representation of the electron wave function (Fig. 8.2d). Longitudinal electron bunching along the direction of the propagation of the electron is not observed, which confirms that inelastic scattering processes are not effective. The generation of up to 28 KD diffraction orders, however, is nicely seen. The number of diffraction orders in the momentum space is significantly larger than the observed number of lobes in real space. This is related to the fact that many of the observed diffraction orders in momentum space may result from a pure modulation of the phase of the wave function in real space; thus, they do not become apparent in the amplitude of the wave function in real space. Considering that the detector is not energy-selective, the integration of the probability amplitude along the k_x-axis as $P(k_y) = \int dk_x |\psi(k_x, k_y)|^2$ gives the probability distribution of finding the electron in the transverse momentum state k_y (Fig. 8.2e). The peaks of the probability distribution are found at $k_{el,y} = nk_{ph,y} = n\delta\vec{k}_{el} = nk_{ph}\sin\varphi$, in agreement with the predictions of our previous eikonal approximation (8.13).

Noticeable changes in the diffraction spectrum are observed when considering the full interaction Hamiltonian as $H^t = -i\hbar e\, m_0^{-1} \vec{A} \cdot \vec{\nabla} + 0.5\, e^2 m_0^{-1} A^2$. In the Wolkow approximation, the complete wave function is then formulated as the product of two terms as $\psi(\vec{r}, t) = \psi_{A^2}(\vec{r}, t)\psi_{\vec{A}\cdot\vec{k}}(\vec{r}, t)$, where ψ_{A^2} is given by (8.14) and $\psi_{\vec{A}\cdot\vec{k}}$ is obtained as

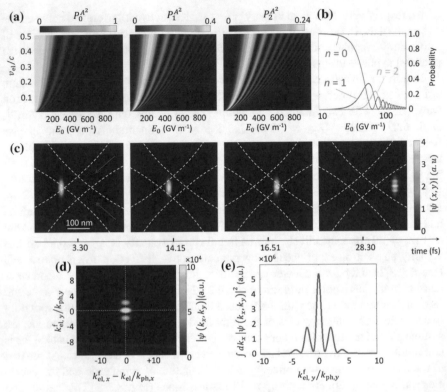

Fig. 8.2 Diffraction of electron pulses by the ponderomotive potential of a standing wave light pattern at the wavelength of $\lambda_0 = 30$ nm (Re-printed by permission from [10] under the CC BY license). **a** Population densities in the lowest diffraction orders as a function of electron velocity and electric field amplitude of the light beams. **b** Probability of finding the electron in one of the three lowest diffraction orders, calculated for a plane wave electron at a velocity of $v_{el} = 0.03c$. Amplitude of the electron wave function for an electron moving at $v_{el} = 0.03c$ in **c** real space and **d** momentum space, at selected times during the interaction with the standing-wave light field. The electric field amplitude of each optical beam is $E_0 = 60$ GVm^{-1} and the wavelength is 30 nm. **e** Transverse momentum distribution of the electron pulse before and after the interaction with the standing-wave field, demonstrating the diffraction of the electron wave into orders $k_{el,y}(n) = n\delta k = n2k_{ph,y}$

$$\psi_{\vec{A}\cdot\vec{k}} = \exp\left(i\frac{ek_{el}}{m_0}2A_0\sin(\varphi)\cos(k_{ph}\sin(\varphi)y)\int_0^{\delta t}\cos(\omega\tau + k_{ph}\cos(\varphi)x)d\tau\right). \tag{8.15}$$

This can be further recast into

$$\psi_{\vec{A}\cdot\vec{k}}(\vec{r},t) = \sum_n\sum_l\sum_m\sum_o i^{-(l+m+n+o)} \times \exp\left(i\left\{l\vec{k}_1 + o\vec{k}_2\right\}\cdot\vec{r}\right)$$
$$\times J_{n-l}(\alpha_s)J_{m-o}(\alpha_s)J_n(\alpha_c)J_m(\alpha_c) \tag{8.16}$$

where $\alpha_c = (ek_{el}E_0/\omega^2 m_0)\sin(\varphi)(1 - \cos(\omega\delta t))$ and $\alpha_s = (ek_{el}E_0/\omega^2 m_0)\sin(\varphi)\sin(\omega\,\delta t)$. Here, $\delta t = x/v_{el}$ is the interaction time. More precisely, it is the ratio of the overall interaction time to the temporal period of the laser oscillations, i.e., $\delta t/T$, where $T = 2\pi/\omega$, which causes oscillations in the coupling strengths, i.e., $\alpha_c \propto \sin^2(\pi\,\delta t/T)$ and $\alpha_s \propto \sin(2\pi\,\delta t/T)$. This cycling behaviour is akin to the Rabi oscillation of the electronic population densities between the photonic states of the momentum ladder set by the free-space electromagnetic excitations [37].

The arguments of the Bessel functions, denoted as $\alpha_{s,\,c}$, are functions of the parameters of the laser excitation (E_0 and ω) as well as the parameters of the electron such as its mass, charge and initial momentum. They scale linearly with the product of the initial momentum and field amplitude, $k_{el}E_0$, in contrast to the arguments obtained for the ponderomotive Hamiltonian, H_2, scaling quadratically with the field amplitude. Additionally, the arguments are well related to the harmonics of the interaction time δt relative to the period of the oscillation of the laser excitation and the angle φ. Obviously for $\varphi = 0$, $\alpha_c = \alpha_s = 0$ and $\psi_{\vec{A}\cdot\vec{k}} \equiv 1$, regardless of the laser intensity and the interaction time. This occurs because for $\varphi = 0$, there exists no projection of the vector potential over the momentum of the moving electron. Moreover, for interaction times $\delta t = nT$ with n being an integer, no diffraction that can be correlated to the H_2 part of the Hamiltonian will occur.

It is of particular interest to note that, in contrast to the action of the ponderomotive potential, causing the generation of diffraction orders only at harmonics of $\delta\vec{k} = \vec{k}_1 - \vec{k}_2$, the action of the absorptive part $H_1 = -i\hbar e\,m_0^{-1}\vec{A}\cdot\vec{\nabla}$ will result in harmonics of the form $l\vec{k}_1 + o\vec{k}_2$, where l and o can now be any arbitrary integer—note that within the eikonal approximation, H_1 can be simplified to $H_1 = \hbar e\,m_0^{-1}\vec{A}\cdot\vec{k}_{el}$, as introduced indirectly in the Wolkow solution (8.12). Thus, this interaction can, in principle, prepare the electron wave packet in states $|l, o\rangle$ with momentum $\hbar\vec{k}_{el}^f = \hbar\vec{k}_{el} + l\hbar\vec{k}_1 + o\hbar\vec{k}_2$. In contrast, only states $|l, -l\rangle$ can be accessed by ponderomotive scattering. The probability of finding the electron after the interaction in state $|l, o\rangle$ can be written as

$$P_{l,o}^{\vec{A}\cdot\vec{k}} = \left|\sum_m \sum_n i^{-(m+n)} J_{n-l}(\alpha_s) J_{m-o}(\alpha_s) J_n(\alpha_c) J_m(\alpha_c)\right|^2. \qquad (8.17)$$

Interestingly, for fixed values of δt and φ, we note a different dependence of the population densities $P_{l,o}^{\vec{A}\cdot\vec{k}}$ on the electron velocity and electric field amplitude than for $P_n^{A^2}$ in the case of purely ponderomotive scattering (compare Fig. 8.3 with Fig. 8.2a). Even at very high field amplitudes, it is still possible to observe the 0-th diffraction order, albeit only for electrons slower than $0.03c$. The locations of the maxima, in general, follow a hyperbolic curve in both (E_0, v_{el}) and (E_0, λ_{ph}) spaces. This is different for the curves formed under the action of the H_1 Hamiltonian, which follow a parabolic shape. Interestingly, for fast electrons, the electron wave function populates higher momentum states already for a much lower field amplitude since

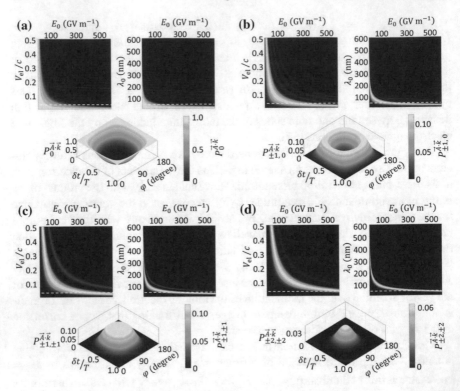

Fig. 8.3 Population densities in different diffraction orders resulting from the interaction of electron beams with a propagating light field created by the interference of two inclined light beams under the effect of the $\vec{A} \cdot \vec{k}$ Hamiltonian (Re-printed by permission from [10] under the CC BY license). **a** $|0, 0\rangle$, **b** $|0, 1\rangle$ and $|1, 0\rangle$, **c** $|\pm 1, \mp 1\rangle$, and **d** $|\pm 2, \mp 2\rangle$ final states. Shown in each panel is $P_{l,o}^{\vec{A} \cdot \vec{k}}$ (top left): as a function of the electron velocity (v_{el}) and electric field amplitude (E_0) at $\delta t = 0.3T$, $\varphi = 50°$, $\lambda_{ph} = 30$ nm, (top right): as a function of the photon wavelength and electric field amplitude at $\delta t = 0.3T$, $\varphi = 50°$, and $v_{el} = 0.03c$, and (bottom): as a function of $\delta t / T$ and φ, at $\lambda_{ph} = 30$ nm and $v_{el} = 0.03c$

now the coupling scales linearly with the field amplitude and electron momentum. This behaviour is in distinct contrast with the action of the pondermotive potential on the electron, for which larger field amplitudes will be necessary to observe higher-order diffractions of the electron by the light. At a specific photon energy and electron velocity, two other parameters, i.e., δt and φ, might be used to control the diffraction orders (Fig. 8.3, bottom panels). For interaction times given by $\delta t = nT$, the Rabi oscillations between the photonic states will cause the population densities to average back to the initial $|0, 0\rangle$ state. For this reason, controlling the interaction time appears to be a precise way of tuning the population densities. Additionally, the excitation angle φ of the photon beams with respect to the direction of the propagation of the electron affects the diffraction orders correlated with both the H_1 and the H_2 parts of the Hamiltonian, although in a quite different way. Interestingly, at a given interaction time, which is set by the longitudinal broadening of the electron wave packet, the angle φ can still be used to suppress or release certain diffraction orders.

Importantly, in the presence of large field amplitudes and slow electrons, both parts of the Hamiltonian, i.e., the H_1 and H_2 parts may contribute to the diffraction of the electron beam. This offers a new degree of freedom for manipulating the temporal and spatial structure of the electron pulse by controlling the quantum path interferences resulting from the action of both H_1 and H_2 on the transition of the electron beam from the $|0, 0\rangle$ initial state to the final state $|l, o\rangle$. To better understand this, we derive the final electron wave packet using (8.12), (8.13), and (8.16) as

$$
\begin{aligned}
\psi(\vec{r}, t) = {} & \frac{1}{(2\pi)^{\frac{3}{2}}} \exp(-i\Omega t) \exp\left\{ i\left(k_{\text{el}} - \frac{e^2 E_0^2}{2\hbar^2 \omega^2 k_{\text{el}}} \cos^2(\varphi) \right) x \right\} \\
& \sum_p i^p J_p\left(\frac{e^2 E_0^2}{2\hbar^2 \omega^2 k_{\text{el}}} \cos(2\varphi) x \right) \\
& \times \sum_n \sum_l \sum_m \sum_o i^{-(l+n)} i^{-(o+m)} J_n(\alpha_c) J_m(\alpha_c) J_{n+p-l}(\alpha_s) J_{m-o-p}(\alpha_s) \\
& \times \exp\left(i\left\{ l\vec{k}_1 + o\vec{k}_2 \right\} \cdot \vec{r} \right).
\end{aligned}
\tag{8.18}
$$

As a result, the probability of finding the electron at the final $|l, o\rangle$ momentum state will be given as

$$
\begin{aligned}
P_{l,o} = {} & \left| \sum_p \sum_n \sum_m i^{p-(l+n+o+m)} J_p\left(\frac{e^2 E_0^2}{2\hbar^2 \omega^2 k_{\text{el},x}} \cos(2\varphi) x \right) \right. \\
& \left. J_n(\alpha_c) J_m(\alpha_c) J_{n+p-l}(\alpha_s) J_{m-o-p}(\alpha_s) \right|^2.
\end{aligned}
\tag{8.19}
$$

This establishes an interesting interference pattern for each given final $|l, o\rangle$ momentum state (Fig. 8.4). One might observe, in specific ranges of the field amplitude and electron velocities, quantum coherent interference paths in the diffraction orders. In contrast to previously reported Rabi oscillations in photon-induced near-field electron microscopy [37, 38], the quantum interferences discussed here will occur in free space and may be utilized as a neat way of controlling the transverse distribution of the electron wave packets, without changing their energy. Interestingly, even in the presence of the H_2 interaction, the probability of finding the electrons in the final $|l, -l\rangle$ states is still dominant over that of finding the electrons in the $|l, 0\rangle$ states. More specifically, the interferences between the H_1 and H_2 interactions are only observed at specific electron velocities. At higher velocities, it is the H_2 interaction that dominates the final diffraction orders of the electron wave packet, whereas at lower electron velocities, the pondermotive interaction is more pronounced (compare Fig. 8.4 with Figs. 8.2a and 8.3). This behaviour implies that there is a range of electron velocities that allows for the abovementioned interference phenomenon to

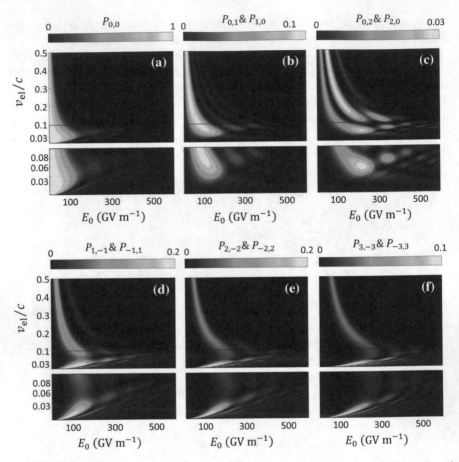

Fig. 8.4 Population densities for different diffraction orders (**a–d**) resulting from the interaction of an electron beam with a propagating light field created by the interference of two inclined Gaussian light beams. The electrons interact with both the ponderomotive A^2 and absorptive $\vec{A} \cdot \vec{k}$ parts of the Hamiltonian. Here, $\varphi = 50°$, $\delta t = 0.3T$, and $\lambda_{ph} = 30$ nm. Quantum path interferences resulting from the interaction with both terms of the Hamiltonian result in an oscillatory dependence of the electron density on the electric field amplitude. Re-printed by permission from [10] under the CC BY license

take place between the H_1 and H_2 interaction paths. This range of electron velocities strongly depends on the applied light frequencies and interaction time as well.

For example, for $\lambda_{ph} = 30$ nm, $v_{el} = 0.03c$, and at proper laser field amplitudes, it should be possible to observe an interference between the $|0, 0\rangle \xrightarrow{H_1} |l, -l\rangle$ and $|0, 0\rangle \xrightarrow{H_2} |l, -l\rangle$ paths and furthermore to coherently control it by means of the incidence angle φ and the laser intensity.

The above mentioned theory is further benchmarked by directly employing a time-dependent Schrödinger solver based on first principles to analyse the electron-light interaction in the regime where we expect interference between the two terms of the Hamiltonian. The electron wave packet has initial longitudinal and transverse broadenings of $W_x = 20$ nm and $W_y = 25$ nm, respectively, and its kinetic velocity is $v_{el} = 0.03c$. The laser carrier wavelength and electric-field amplitude are set to $\lambda_{ph} = 30$ nm and $E_0 = 200$ GV m^{-1}, respectively, and $\varphi = 50°$. By including the complete Hamiltonian, the interaction of the electron wave packet with the laser beams (Fig. 8.5a) is very different from the case of purely pondermotive scattering (Fig. 8.2c). The centre of the wave packet is soon depopulated, and the wave packet breaks up into a sequence of sub-peaks along the transverse direction. These subpeaks become increasingly narrower as the wave packet continues to propagate through the interaction region. The formed bunches of ultrathin wave packets move further away from the interaction region along a direction preassigned by the wave vector of the

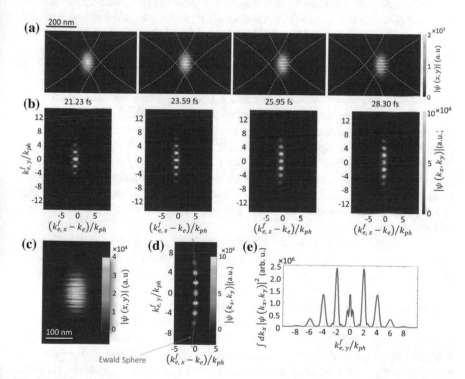

Fig. 8.5 Study of the interaction of a pulsed electron beam with two Gaussian optical beams from first principles. Snapshots of the amplitude of the electron wave function are shown in **a** real space and **b** momentum space at the interaction time indicated in each frame. The dashed lines are inserted at the locations where the electric field intensity has dropped by a factor of e^{-1}. **c** Final shape of the electron wave packet 20 fs after leaving the interaction region, shown by the dashed lines in (**a**). **d** Final distribution of the electron wave function in momentum space. **e** Probability distribution of the different diffraction orders, as recorded on an electron detector. Re-printed by permission from [10] under the CC BY license

light beams, hence building up the coherent diffraction orders. In the momentum space, we notice the population of the electron beams prominently into the $|l, -l\rangle$ states, as evidenced by the 5 distinguished bright spots along the transverse direction. However, fainter probability maxima located at $\vec{k}_{el} \pm \vec{k}_1$ and $\vec{k}_{el} \pm \vec{k}_2$ unravel from the non-equilibrium population distributions at the $|\pm 1, 0\rangle$ and $|0, \pm 1\rangle$ states, as well as the higher-order $|0, \pm 2\rangle$ and $|\pm 2, 0\rangle$ states, which only last for ultrashort times during the interaction (Fig. 8.5b). Moreover, these states provide means to form the desired interference paths between the various quantum levels (see Fig. 8.1d), resulting in diffraction orders that are not located at $|l, -l\rangle$ states (Fig. 8.5d and e). These diffraction orders appear as ultrathin electron bunches that are caused by the interference between the abovementioned quantum paths (Fig. 8.5c).

Interestingly, all the diffraction levels will form along the circumference of an Ewald-like sphere (marked in Fig. 8.5d), demonstrating the striking similarity between the elastic interaction of electrons with light and crystalline matter, even though the interaction Hamiltonian is completely different (Fig. 8.5 d). The radius of the Ewald sphere is equal to the wavenumber of the electron k_{el}, as expected. This further strengthens the fact that the inelastic light-matter interaction, though temporarily happening during the interaction of an electron wave packet with continuous-wave light in vacuum, cannot give rise to a steady-state outcome. This is because (i) both light waves specified by wave vectors $\vec{k}_1$ and $\vec{k}_2$ have exactly similar field strengths, polarizations, and wave numbers, while the only difference between them is the propagation direction, and (ii) the probability amplitudes for the transition of the electron wave function from the $|0, 0\rangle$ momentum state to the 4 final states $|\pm l, \pm l\rangle$, under the effect of the H_2 interaction and for a given l, are all equal. Nevertheless, the transverse (elastic) scattering to the two $|l, -l\rangle$ and $|-l, l\rangle$ states will be caused by both the H_1 and H_2 parts of the Hamiltonian, and the different paths giving rise to the transverse interaction will interfere with each other, causing the appearance of subpeaks in the detected diffraction pattern, which are not displaced by $2nk_{ph}$ (Fig. 8.5e).

The abovementioned quantum coherent interferences can be controlled by several parameters, including φ and E_0. Of particular interest is the role of the laser electric-field amplitude. For slow electrons and low laser intensities, the role of the pondermotive force will become dominant. This is to be understood particularly by comparing Fig. 8.4d with Figs. 8.3e and 8.2a. As a result, one expects to observe diffraction peaks on the order of only $n\delta\vec{k}_{el}$, which is further confirmed by our first-principle calculations (Fig. 8.6). This ponderomotive scattering remains the dominant interaction at electric field amplitudes $E_0 \leq 100\,\text{GVm}^{-1}$. However, for $E_0 > 100\,\text{GVm}^{-1}$, the diffraction orders correlated with the H_2 part of the Hamiltonian become active as well, and the resulting diffraction pattern is an outcome of the interference between the quantum paths associated with each part of the Hamiltonian. By increasing the field amplitude, the overall number of diffraction orders will also be increased. Interestingly, the highest-order diffraction patterns are all at the orders of only $\vec{k}_{el}^f = \vec{k}_{el} + 2nk_{ph}\hat{y}$, which demonstrates that the pondermotive potential still dominates the overall response. Hence, the distribution of the diffraction orders due

(a)   **(b)**

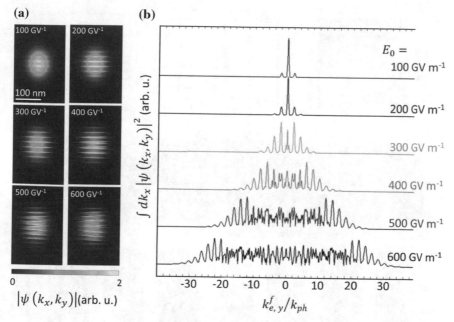

Fig. 8.6 Dependence of the diffraction orders on the electric-field amplitude. **a** Spatial distribution of the amplitude of the electron wave function and **b** integrated probability distributions after the interaction versus the amplitude of the electric-field component of the laser excitation. $\varphi = 50°$, $W_x = 20\,\text{nm}$, $W_y = 25\,\text{nm}$, and $v_{\text{el}} = 0.03c$. See the text for descriptions of the parameters. Re-printed by permission from [10] under the CC BY license

to the H_1 part of the Hamiltonian is within a smaller angular range compared to the total angular distribution of the diffraction patterns.

8.2.2   Phase Cycling in Interaction with a Single Laser Beam

The discussions above demonstrated that the interaction of an electron wave packet with two laser beams will result in an ultrafast phase modulation and the diffraction of the electron wave packet by the generated standing wave pattern of the light. It will be illustrative to investigate such effects when only one laser path (a travelling wave pattern) is considered.

Here, we consider the interaction of an electron wave packet at a kinetic energy of $10\,\text{keV}$ and transverse and longitudinal broadenings of $60\,\text{nm}$ and $20\,\text{nm}$, respectively, with an x-polarized pulsed laser excitation at a temporal broadening of $3\,\text{fs}$ (Fig. 8.7a). The laser pulse has a carrier wavelength of $\lambda_0 = 300\,\text{nm}$ and electric field amplitude of $10^9\,\text{Vm}^{-1}$. Figure 8.7b and c show the distribution of the electron wave packet in the real space and momentum space at certain times during the interaction. In particular, we observe a gradual occupation of momentum states by the electron wave

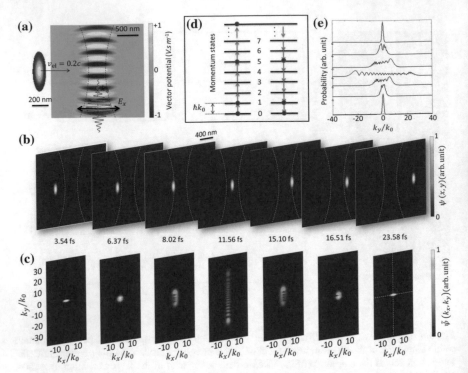

Fig. 8.7 Ultrafast phase modulation and phase cycling of an electron wave packet at the kinetic energy of 10 keV with a pulsed laser at the carrier wavelength of 300 nm and temporal broadening of 3 fs. **a** Schematic of the system under investigation. Snapshots of the **b** real-space and **c** momentum-space representations of the electron wave packet at the depicted time during the interaction with the laser pulse. **d** Marching of the electron wavepacket upon the photonic momentum ladder versus time. **e** Probability distribution of the electron wave packet versus transverse recoil

packet, where the spacing within the momentum ladder is specified by $\hbar k_y = \hbar k_0$, where k_0 is the photon wave number in the free space (Fig. 8.7d and e). The electron continues moving upward along the momentum ladder until it reaches the highest state of $\hbar k_y = 22 \, \hbar k_0$, when it is spatially positioned at the peak of the laser intensity, at $t = 11.56$ fs. After that the electron starts marching down the ladder, until it again reaches the initial state after the interaction. The ultrafast phase modulations that the electron experiences during the interaction with the laser, thus is averaged out in time, in such a way that the electron does not experience any change in its momentum after the interaction. However, such fast phase modulations might be captured out in the radiation from the electron wave packet.

8.2.3 Discussions

The short wavelength choice of $\lambda_{\text{ph}} = 30$ nm and ultrahigh field amplitudes employed above were only introduced to meet our simulation resources. These parameters are much beyond what practically can currently be realized in the laboratory. Here, we briefly outline practical approaches towards the realization of quantum path interferences with field strengths and laser wavelengths routinely employed in light-matter interaction experiments. The initial condition for the formation of interferences in quantum paths is to have similar levels of interaction strengths associated with the H_1 and H_2 parts of the Hamiltonian. Since H_1 is quadratic and H_2 is linear in $\vec{A}$, there will be always two values for the vector potential amplitudes where H_1 and H_2 meet each other—one point is $A_0 = 0$, and the other point is related to the electron velocity as $A_0 = 2m_0 v_{\text{el}}/e$. Thus, the lower the electron velocity, the lower the required field amplitude becomes (Fig. 8.8a). Additionally, the electric field amplitude is related to the vector potential as $E_0 = \omega A_0$. In other words, it will be possible to employ lower field amplitudes by using oscillating fields at lower frequencies (Fig. 8.8b). For example, by assuming a laser wavelength of $\lambda_{\text{ph}} = 1250$ nm, the quantum path interferences between H_1 and H_2 will be observed at electron velocities within $0.01c < v_{\text{el}} < 0.1c$ and electric field amplitudes as low as $0.2\,\text{GVm}^{-1} < E_0 < 1.0\,\text{GVm}^{-1}$ (Fig. 8.8c and d). Such conditions can certainly be created by currently available light sources or by including near-field field-enhancement effects [48]. The dynamics of the sideband modulations in the scattering of single-particle electron wave packets with free-space light can be described as the outcomes of a quantum walk [49] in the discrete momentum states specified by the classical electromagnetic waves. This behaviour is akin to the random walks of photons in classical reliable interferometers with low losses and high stability performances [50]. As a result of the coherent action of the unitary operator specified in (8.1) on the single electron wave packet, at each given time, the electron will be left entangled between different momentum states. As expected, features of this quantum walk include interferences and boson sampling [50] from a sea of many possible states, as, for example, selectively selecting a few elastic or inelastic scatterings as a result of such inferences. The ability to dynamically control the outcome of the random walk by a few parameters such as the polarization, wavelength, intensity, and inclination angle of the incident Gaussian beams, and particularly by avoiding matter and hence electron-electron and electron-core interactions, make the proposed system a promising candidate for bosonic-fermionic random walks and a new generation of boson-sampling devices [49, 51, 52].

As a summary, we have generalized the KD effect to the inclusion of two laser fields that propagate at an inclination angle with respect to the electron trajectory. The spatio-temporal behaviour of the introduced light waves appears as a standing wave pattern transverse to the electron momentum—similar to the normal KD effect. In addition, a travelling wave pattern forms longitudinal to the initial electron velocity—in contrast with the normal KD effect. We have shown, both theoretically and numerically, that the interaction of the electron wave packet with the introduced light

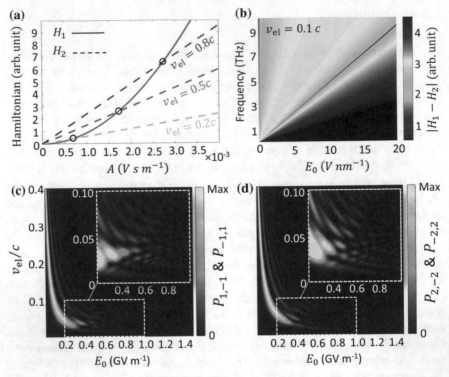

Fig. 8.8 Criterion for observation of quantum path interferences between ponderomotive H_1 and absorptive H_2 parts of the Hamiltonian. **a** Interaction strengths versus the amplitude of the magnetic vector potential for different electron velocities. **b** $|H_1 - H_2|$ as a function of the frequency and electric field amplitude at the electron velocity of $v_{el} = 0.1c$. Population densities for the **c** $|1, -1\rangle$ and $|-1, 1\rangle$ and **d** $|2, -2\rangle$ and $|-2, 2\rangle$ diffraction orders resulting from the interaction of an electron beam with a propagating light field created by two inclined Gaussian light beams. The electrons interact with both the ponderomotive A^2 and the absorptive $\vec{A} \cdot \vec{k}$ parts of the Hamiltonian. Here, $\varphi = 50°$, $\delta t = 0.3T$, and $\lambda_{ph} = 1250$ nm. Much lower electric field amplitudes are required to observe the interference effects compared with in Fig. 8.4. Re-printed by permission from [10] under the CC BY license

beams results in an exotic momentum distribution in the final electron wave packet, which can be described as the quantum path interferences between the two parts of the Hamiltonian, namely, the ponderomotive and absorptive channels. These interference paths and their coupling strengths can be further controlled in general by the shape of the light waves and in particular by the inclination angles and intensity of the two Gaussian beams but also by tuning the interaction time and the electron velocity. We anticipate that the abovementioned interference effects can be proposed as a new boson sampling device for tailoring the random walk of a single quantum wave packet in the discrete levels imposed by the momentum of the light.

Next, we discuss the interaction of electron waves with optical gratings and demonstrate that when near-field interaction zones are employed, inelastic interaction will become possible as well as elastic interaction. By imposing a synchronous condition between the electron and the diffraction orders of the grating, acceleration or bunching effects may also become possible.

8.3 Interaction of Slow Electrons with Dielectric Laser Accelerators

To boost the inelastic interaction and facilitate the necessary momentum and energy transfer, the near-field distribution of the optical modes is usually considered as the interaction medium [43]. In particular, the evanescent field in the vicinity of discontinuities such as an interface [53, 54], plasmonic nanostructures (as described in Chaps. 3–7), photonic crystals [55], and gratings [56] has been so far considered. Interestingly, a classical treatment of the electron–photon interaction demonstrates the necessity of the momentum conservation criterion to be met as $\omega_{ph} = \vec{k}_{ph} \cdot \vec{v}_{el}$ (where ω_{ph} is the angular momentum of the photon, $\vec{k}_{ph}$ is the wave vector of the photon, and $\vec{v}_{el}$ is the velocity of the electron), which agrees perfectly well with the first principles of quantum mechanics (see Chap. 2). When a grating is incorporated as the medium of interaction, the dispersion of the electron-induced polarization is mapped into the optical cone as $\omega_{ph} = \left(\vec{k}_{ph} + \hat{\alpha} 2m\pi/\Lambda \right) \cdot \vec{v}_{el}$, according to Floquet's theorem. Here, $\hat{\alpha}$ is a unitary vector along the grating axis, Λ is the period of the grating, and m is an integer denoting the diffraction order. Indeed, this mapping is the reason behind the Smith–Purcell radiation into the far-field region. When an external laser field is also applied to the grating, an electron travelling adjacent to the grating may gain energy by absorbing photons, a phenomenon referred to as the inverse Smith–Purcell effect [57].

Although classical theory can be routinely applied to investigate the photon emission process in the absence of external radiation, it cannot be applied to photon-induced emissions, as for PINEM. In particular, the probability of the electron emitting or absorbing several quanta of photons in a multiple scattering process can only be treated quantum mechanically [41], which underlines the progress in the description of PINEM experiments through the computation of quantum mechanical propagators by considering either the Lippmann-Schwinger equation [16, 58] or making use of scattering theory [15]. In this regard, quantum mechanics should also be applied to study the inverse Smith–Purcell effect. Although the Lorentz force can be used to understand the linear acceleration in the interaction of electrons with the synchronized mode of a grating, it cannot describe the interferences or the bunching, which happens due to multiple scattering and nonlinear processes within one single-electron wave packet. Therefore, classical mechanics cannot describe the chirping of single-electron wave packets due to such interactions.

The main reason that such quantum mechanical approaches have not yet been employed to study the inverse Smith–Purcell effect is due to the very short de Broglie wavelength of electrons at high energies compared to the optical wavelengths and dimensions of the employed devices, in addition to a lack of an equivalent and stable particle-in-cell numerical approach in quantum mechanics. In fact, self-consistent Maxwell–Lorentz simulations and particle-in-cell numerical approaches are routinely applied to study the interaction of electrons with electromagnetic waves in electron-driven photon sources and accelerators [59–61]. Here, we apply our self-consistent Maxwell–Schrödinger numerical toolbox to the inverse Smith–Purcell effect. Using the proposed Maxwell–Schrödinger numerical toolbox, the interaction of a single-electron wave packet with light and gratings will be studied. It will be shown that a chirped grating would be necessary to avoid the fast dephasing of electrons. Moreover, the effect of the electron pulse broadening on the shape of the electron wave packet after its interaction with the grating will be discussed. It will also be demonstrated that a symmetric grating configuration, though prohibiting the electron from being deflected away from the grating [62, 63], imposes another purely quantum mechanical effect on especially slow single-electron wave packets: in fact the electron waves will be diffracted symmetrically in the transverse direction via a two-photon process because of the Kapitza-Dirac effect.

8.3.1   *Electrons Travelling Parallel to a Single Grating*

As experimentally demonstrated by Furuya and coworkers [57], an electron propagating parallel to a grating can be accelerated once the grating is illuminated with external laser radiation if the synchronicity condition is met. These observations were the driving force behind the development of a new class of linear accelerators (linacs), with acceleration gradients as high as 100 MeV m^{-1} [64]. As has been recently demonstrated, reaching even higher acceleration gradients would be possible by incorporating dielectric gratings [65, 66]. Dielectrics in comparison with metals offer a higher damage threshold at optical frequencies. In addition, ultrafast lasers operating at optical frequencies are so stable that even sub-cycle carrier-envelope phase effects can be investigated. For these reasons, dielectric laser acceleration is emerging as the next-generation linac.

A parallel field of study in wave science is ultrafast electron microscopy using low-energy photoemission electron sources [67–77]. Pulsed electrons from photoemission sources are highly coherent and can be directly exploited in ultrafast imaging and diffraction, and in spectroscopy [76, 78–80]. As will be shown here, low-energy pulsed electrons are also better accelerated and shaped because they are more sensitive to the electromagnetic radiation, and their interaction with the laser excitation will take place in a coherent way. Hereafter, the interaction of low-energy electron pulses with a dielectric grating will be investigated.

A two-dimensional silicon grating composed of silicon nanorods, each with a width of 16 nm and a height of 20 nm, as shown in Fig. 8.9, is considered here as the

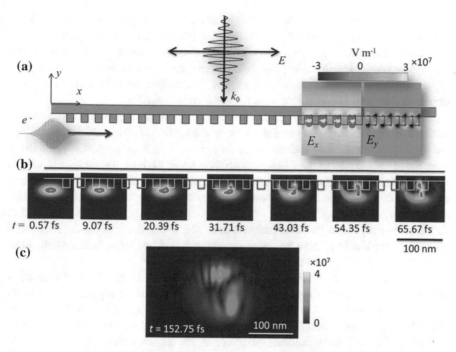

Fig. 8.9 Inverse Smith–Purcell effect. **a** An electron pulse with longitudinal and transverse broadenings of 5 nm and 10 nm, respectively, travels at a distance of 4 nm from and parallel to a silicon grating. The silicon grating has a period of 33.2 nm and is illuminated with a laser pulse at a carrier wavelength of 830 nm and a temporal broadening of 80 fs. The spatial distribution of the electric field components at a given time is depicted in the inset. Snapshots of the spatial distribution of the electron wavefunction **b** at the times depicted on each frame during the interaction with the grating and **c** after the electron has left the grating. Re-printed by permission from [2] under the CC BY license

acceleration medium. The grating is illuminated with a linearly polarized laser pulse at a carrier wavelength $\lambda_0 = 830$ nm, peak electric field amplitude of 0.5 GV/m, and broadening of 80 fs. An electron pulse at an initial carrier energy of 408 eV ($v_{el} = 0.04c$, where c is the speed of light) travelling parallel to the grating axis and at a distance of 4 nm from the grating is considered. The initial distribution of the electron pulse is assumed to be a two-dimensional Gaussian function as:

$$\psi(\vec{r}, t = 0) = \exp\left(ik_x^{el}x\right)\left(\frac{1}{\sqrt{2\pi}\,W_t}\exp\left(-\frac{1}{2}\frac{(x-x_0)^2}{W_t}\right)\right)^{\frac{1}{2}}$$

$$\left(\frac{1}{\sqrt{2\pi}\,W_L}\exp\left(-\frac{1}{2}\frac{(y-y_0)^2}{W_L}\right)\right)^{\frac{1}{2}} \tag{8.20}$$

where W_t and W_L are the transverse and longitudinal broadenings, $k_x^{el} = \hbar^{-1}m_0 v_{el}$ and (x_0, y_0) is the initial position of the pulse centre at $t = 0$. The length of the grating is $960\,\mu$m so that the time of flight of the electron through the acceleration medium becomes approximately 80 fs. Because the laser excitation is perpendicular to the electron trajectory, the synchronicity condition between the electron pulse and the grating near field can be simplified as $\Lambda = m v_{el} \lambda_0 / c$, where, by assuming $m = 1$, $\Lambda = 33.2$ nm is obtained. In this regard, the first diffraction order in the Floquet's expansion term will be considered as the synchronous mode.

The polarization of the laser illumination is along the grating axis. Such an illumination couples efficiently to the TE_z mode of the grating. The spatial distribution of the total electric near field is depicted in the inset of Fig. 8.8a. Apparently, although only the $m = 1$ mode in the Floquet expansion set is desired, the dominant term is $m = 0$, as for all gratings for which the grating period is smaller than the wavelength of the incident light. On the other hand, the synchronicity condition $v_{el}c^{-1} = \Lambda\lambda_0^{-1}$ does not allow for the utilization of diffraction gratings for which $\Lambda > \lambda_0$. This fact implies that, although acceleration is possible, it will come at the cost of severe chirping of the electron wave function. However, it will be shown later that this kind of chirping will be beneficial in electron bunching.

As an electron pulse with initial broadenings of $W_t = 5$ nm and $W_L = 10$ nm travels through the accelerating medium, it is strongly scattered by the grating. Snapshots of the spatial distribution of the electron wavefunction $|\psi(\vec{r}, t_i)|$ (Fig. 8.9b) show that interacting with several elements of the grating will cause the electron pulse to get bunched along the longitudinal direction. Moreover, because of the asymmetry of the grating configuration, the electron wave packet is deflected away transversely in the direction normal to the grating, a condition known as defocusing [62]. The shape of the electron wave after it has left the grating demonstrates a pronounced defocusing in the transverse direction and a severe chirping in the longitudinal direction (Fig. 8.9c).

The E_y component of the electric field is localized at the edges of the grating elements with an evanescent distribution in the vacuum and correspondingly large wave vectors, which imposes severe scattering potentials for the electron wave. Moreover, the sign of the E_y component and also the vector potential component A_y are opposite at the adjacent edges, which causes a part of the wavefunction to be scattered to the $\pm y$ directions. This can be better understood by comparing the momentum distributions of the electron wave as $\tilde{\psi}(k_x^{el}, k_y^{el}; t) = \frac{1}{2\pi} \int_{-\infty}^{+\infty} dx \int_{-\infty}^{+\infty} dy \exp(-ik_x^{el}x) \exp(-ik_y^{el}y) \psi(x, y; t)$ before and after its interaction with the grating, as shown in Fig. 8.10a and b, respectively. It is apparent that the electron wave is mostly decelerated. Considering that all the transversely diffracted signals are detected, the probability of the electron having the longitudinal momentum k_x^{el} (or velocity $m_0^{-1}\hbar k_x^{el}$) is obtained as $\Gamma_L(k_x^{el}; t \rightarrow +\infty) = \int_{-\infty}^{+\infty} dk_y \left|\tilde{\psi}(k_x^{el}, k_y^{el}; t \rightarrow +\infty)\right|^2$, which is shown in Fig. 8.10c. Moreover, several resonances are observed in the momentum distribution of the electron wave function along the longitudinal momentum axis k_x^{el}, with an equidistant spacing of $\delta k_x^{el} = 2\pi\,\Lambda^{-1}$. Considering the phase-matching condition $\omega v_{el}^{-1} = 2\pi\,\Lambda^{-1}$, these resonances are due to the emissions of photons at the energy of $E_{ph} = \hbar v_{el} 2\pi\,\Lambda^{-1}$, which is exactly the laser carrier energy. Because the broadening of the electron

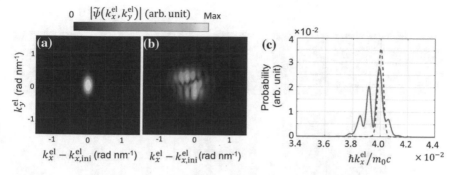

Fig. 8.10 Electron wavefunction in momentum space. **a** Initial distribution of the Gaussian wavefunction and **b** final distribution of the electron wave function before and after its interaction with the grating, respectively. **c** Probability distribution of the electron versus the normalized longitudinal component of the velocity $\hbar k_x^{el}/m_0 c$. $k_\alpha^{el} = p_\alpha/\hbar$ is the electron wavenumber and p_α is the momentum of the electron along the α-axis. Dashed line: initial, and solid line: final probability distribution of the electron wave function. Re-printed by permission from [2] under the CC BY license

provides a sufficiently long time for multiple electron-photon interactions, even up to third-harmonic photon emission and absorption takes place, in a very similar way to the multiple electron–photon scattering in PINEM.

8.3.2 Symmetric Grating

It has been discussed in [62] that a symmetrical force pattern should be employed to circumvent the defocusing of the electrons in interaction with the exponentially decaying near field of the grating, at least on the axis of the accelerator. To examine this proposal numerically, a symmetric configuration is considered by placing two gratings parallel to each other and illuminating the whole structure with two mutually coherent phase-stabilized counter-propagating laser beams with the same specifications as mentioned above (see Fig. 8.11). In this way, only the even mode of the grating is excited, for which the E_x field component has an even distribution, while both the E_y and H_z field components have odd distributions along the y-axis. An initially Gaussian electron wave function at a carrier energy of 408 eV and with $W_t = 8$ nm and $W_L = 15$ nm travels through the grating along the symmetric axis. The gap between the gratings is considered to be 30 nm.

It is anticipated that a symmetric force pattern causes the momentum transferred from the light to the grating to be mostly along the longitudinal direction so that the electron will be trapped in the transverse direction, quite similar to photon trapping in quantum wells. In this way, even after a few grating elements, the electron wavefunction becomes bunched (Fig. 8.11a). The strong chirping of the electron in interaction with the grating is apparent from the final distribution after the electron has left the grating. Interestingly, the simulation results reveal that focusing happens only within a certain spatial region along the electron pulse, especially in the leading part of the wavefunction, whereas a strong defocusing is observed in the trailing part.

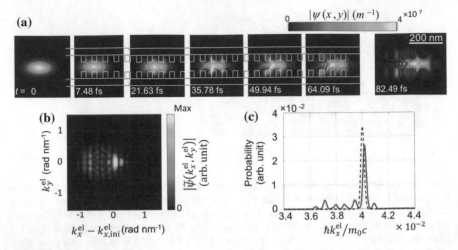

Fig. 8.11 A Gaussian electron wavefunction interacting with a double grating illuminated with light in a symmetric phase-stabilized way. **a** Snapshots of the electron wavefunction at the times depicted on each frame, **b** momentum representation of the electron wavefunction after interaction with the grating, and **c** probability distribution of the electron versus the normalized longitudinal component of the velocity $\hbar k_x^{el} / m_0 c$. $k_\alpha^{el} = p_\alpha / \hbar$ is the electron wavenumber, and p_α is the momentum of the electron along the α-axis. Dashed line: initial, and solid line: final probability distribution of the electron wave function. Re-printed by permission from [2] under the CC BY license

Indeed, the gap between the gratings supports a standing-wave pattern for near-field components with large wave vectors, which causes the electron to diffract transversely in a very similar way to the Kapitza-Dirac effect. However, the Kapitza-Dirac effect is known to be a purely elastic scattering mechanism taking place with standing wave patterns in free space. In contrast, the diffractions that occur here are due to the interaction of the electrons with the near-field of the grating and hence are happening in an inelastic way. Indeed, this inelastic diffraction phenomenon is another source of the defocusing of the electron wave packet after the interaction with the gratings. The representation of the electron wavefunction in momentum space demonstrates that multiple loss and gain peaks occur along the longitudinal axis, whereas several diffraction orders are also observed in momentum space (Fig. 8.11b).

Computing $\Gamma_L(k_x^{el}; t \rightarrow +\infty)$ reveals that this configuration indeed leads to a net acceleration of the electron wave packet, though an acceleration gradient of only 4 MeV m^{-1} is achieved (Fig. 8.11c). For computing the acceleration gradient, the rate of increase of the kinematic energy of the centre of the pulse is considered as $m_0 v_{el} dv_{el}/dx$. Several energy-loss peaks are observed, which prohibit a net acceleration of the electron wave packet. In fact, because the initial electron energy is low, its acceleration takes place quite fast, and after a few periods, the synchronicity condition is not satisfied anymore. This effect is called dephasing [63]. The grating length after which dephasing happens is approximated as being only 280 nm. Therefore, it is not surprising that the electron cannot reach the highest possible acceleration gradient. To circumvent this difficulty, the phase velocity of the acceleration mode of the grating should remain synchronous with the electron. One solution is to introduce a multistage accelerator by incorporating many tapered gratings of different periodicities [66]. Another solution is to introduce a chirped grating, as considered here.

8.3.3 Chirped Grating

The chirped double grating is designed in such a way as to maintain the synchronicity condition with the electron wave function by increasing the period in an adiabatic way. We start with $\Lambda_i = v_{el} \lambda_0 c^{-1}$ and aim to reach $\Lambda_f = 2 v_{el} \lambda_0 c^{-1}$ within 800 nm (Fig. 8.12a). The spatiotemporal behaviour of the electron wavefunction during the interaction with such a grating is shown in Fig. 8.11a. Both the leading and trailing parts of the wave function are defocused, but the centre of the pulse is focused. This is somehow expected, as the synchronized longitudinal forces resulting from the x-component of the vector potential are always concomitant with the transverse components that are out of phase with the longitudinal component. The overall longitudinal dispersion and electron-pulse expansion, however, is much less than in the case of non-chirped symmetrical grating. As a consequence, the final longitudinal broadening of the electron pulse is also less. The momentum representation of the

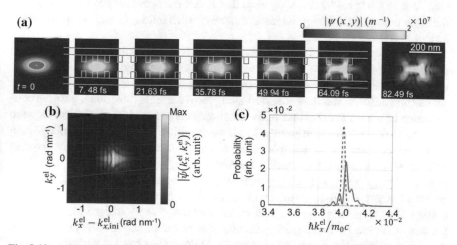

Fig. 8.12 A Gaussian electron wavefunction interacting with a double chirped grating illuminated with light in a symmetric phase-stabilized way. **a** Snapshots of the electron wavefunction at the times depicted on each frame, **b** momentum representation of the electron wavefunction after interaction with the grating, and **c** probability distribution of the electron versus the normalized longitudinal component of the velocity $\hbar k_x^{el} (m_0 c)^{-1}$. $k_\alpha^{el} = p_\alpha / \hbar$ is the electron wavenumber and p_α is the momentum of the electron along the α-axis. Dashed line: initial, and solid line: final probability distribution of the electron wave function. Re-printed by permission from [2] under the CC BY license

electron pulse, as shown in Fig. 8.12b, again reveals several energy loss and gain peaks. However, in contrast to the previous cases, the energy-gain peaks are more pronounced, which is due to the acceleration of the centre of the electron wave-packet in a perfectly synchronized way through the whole interaction length (see Fig. 8.12c). As a consequence, an acceleration gradient of 7.5 MeV m^{-1} is observed.

An interesting aspect of such numerical treatments is the ability to examine the contribution of each individual part of the Hamiltonian to the spatiotemporal behaviour of the electrons in the electromagnetic fields. For a relativistic electron pulse (initial kinetic energy higher than 50 keV), it is well justified to approximate the interaction Hamiltonian by the only leading term $\hat{H}_{\text{int}} = \hbar e\, m_0^{-1} \vec{A} \cdot \vec{k}^{\text{el}}$, as has been considered in developing the theory of PINEM and discussed previously (see Sect. 8.2). However, for slow electrons, this assumption may lead to severe misunderstandings. First, the interaction Hamiltonian stated above leads to energy-loss and energy-gain peaks symmetrically distributed over the loss and gain parts of the spectrum. It is apparent from the abovementioned results that this is not the case here. To understand the reason behind this discrepancy, it is sufficient to assume a slowly varying electron wave packet as in (8.3) and substituting this wave packet into (8.2) to obtain the SVA equation of motion (8.4). Apparently, the terms $-\hbar^2 (2m_0)^{-1} \nabla^2 \psi_0$ and $-i\hbar^2 m_0^{-1} \vec{k}^{\text{el}} \cdot \vec{\nabla} \psi_0$ are responsible for the dispersion, chirping, and expansion of an initially Gaussian wave packet in free space, without considering any electromagnetic interaction. The term $\hbar e m_0^{-1} \vec{A} \cdot \vec{k}^{\text{el}} \psi_0 = \hbar e m_0^{-1} A_x k_x^{\text{el}} \psi_0$ causes the acceleration, as well as the symmetrically distributed energy loss and gain peaks [15]. It is the $-i\hbar m_0^{-1} e \vec{A} \cdot \vec{\nabla} \psi_0$ term that is not time-reversely symmetric and is the main cause of bunching, inelastic transverse diffraction, as well as asymmetric energy loss and gain peaks.

It is already apparent from the results presented above that coherent electron wave packets interact in a different way with optical gratings in comparison with monolithic classical electrons at an equivalent centre of mass energy. Specifically, the distinct diffraction orders observed along the transverse direction are a pure consequence of the wave behaviour of the electrons, very much like an electromagnetic wave traversing a diffraction grating. Moreover, multiple energy-loss and energy-gain peaks, which form due to multiple electron–photon scatterings, are not observed in a classical particle-in-cell numerical method.

Considering the facts stated above, it would be interesting to investigate the interaction of electron pulses at different longitudinal broadenings with the chirped-grating DLA introduced above. It is anticipated that more localized electron wave packets, achieved by making $W_{t,L}$ smaller, can be better accelerated provided that the near synchronicity condition is satisfied. On the other hand, a vacuum is highly dispersive for non-relativistic electrons, and if no lens or collimation system is incorporated [81], a well-localized electron pulse expands rapidly by propagating over distances as short as a few hundred nanometres. In this regard, a compromise between the acceptable spatial broadening of the pulse and the best achievable acceleration gradient is to be searched for. In all the cases considered below, the carrier envelope phase is tuned in such a way that the E_x field component experiences a maximum at the time when the electron reaches the first grating element, whereas the delay of the optical excitation is controlled to manifest the best synchronization with the electron. In other words, the centre of the electron wave packet and the centre of the optical pulse concomitantly reach the middle of the grating. Moreover, the broadening of the optical pulse is considered to be equal to the travelling time of the centre of the electron wave packet along the grating (78 fs). This case, however, is an optimal condition for acceleration, if the dispersion effect could be ignored. In practice, there is no such control over the entrance phase. Furthermore, obtaining off-synchronism may be advantageous for acceleration, at least statistically, even if we ignore the dispersion.

Figure 8.13 shows an electron wave packet at an initial broadening of only 5 nm interacting with the same chirped grating as introduced above. Although the initial longitudinal broadening is smaller than the distance between grating elements, the final broadening is approximately three times larger. The transverse broadening is also significantly affected by the diffraction from the standing wave light pattern captured within the space in between the gratings. As a result of this diffraction phenomenon, the electron is divided into sub-bunches, each having a certain linear momentum normal to the symmetry axis of the grating. This behaviour is better resolved by demonstrating the wave function in the reciprocal space (Fig. 8.13b). Nevertheless, the acceleration gradient is significantly improved in comparison with that of an electron wave packet at an initial broadening larger than the periodicity of the structure (Fig. 8.13c).

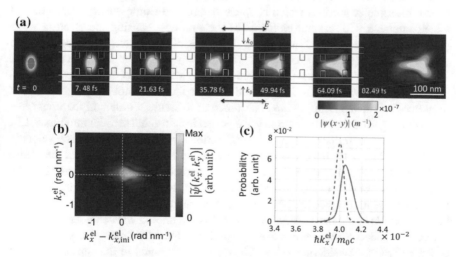

Fig. 8.13 An electron wave packet with a longitudinal broadening of 5 nm interacting with a chirped grating. **a** Snapshots of the electron wavefunction at the times depicted on each frame, **b** momentum representation of the electron wavefunction after interaction with the grating, and **c** probability distribution of the electron versus the normalized longitudinal component of the velocity $\hbar k_x^{el} (m_0 c)^{-1}$. $k_\alpha^{el} = p_\alpha / \hbar$ is the electron wavenumber, and p_α is the momentum of the electron along the α-axis. Dashed line: initial, and solid line: final probability distribution of the electron wave function. Re-printed by permission from [2] under the CC BY license

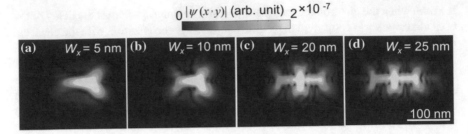

Fig. 8.14 The spatial distribution of the magnitude of the electron wavefunction, after the electron wave packet has left the grating, captured at $t = 82.5$ fs, for initial longitudinal electron broadenings of **a** $W_L = 5$ nm, **b** $W_L = 10$ nm, **c** $W_L = 20$ nm, and **d** $W_L = 25$ nm

Our investigations show that for wave packets with $W_t < 4$ nm, the expansion in the transverse direction is quite severe during the propagation process. We choose to maintain the transverse broadening of the initial wave packet as $W_t = 8$ nm and only discuss the effect of the longitudinal broadening on the final shape of the electron wave function (Fig. 8.14). The broadenings considered here are just trial tests. An interesting consequence is that the larger the electron-wave packet broadening, the better is the bunching effect. This spatiotemporal analysis reveals that electron bunching occurs when the electron wave packet is broad enough to cover several elements of the gratings. As this is the case for electrons emitted from state-of-the-art electron guns, either field emission or photoemission, one can expect multiple scattering, which happens within one single-electron wave packet which interacts with several elements of the gratings, causing the electron wave packet to become bunched. Moreover, the shape of an electron wave packet with only $W_L = 5$ nm (time broadening $W_s = W_L/v_{el} = 417$ as) is much better preserved, and the wavefunction is also not bunched. At the leading part of the wavefunction, defocusing still occurs. The implication of the spatiotemporal chirping behaviour in momentum space unravels the occurrence of diffraction orders in the transverse direction, regardless of the electron-wave packet longitudinal broadening (Fig. 8.15). Because the angular distribution of the diffraction is relevant to the electron velocity, as shown in Fig. 8.15, the effect is not a pure elastic effect. However, for the centre of the wave packet, the scattering can still be considered elastic. According to the Kapitza-Dirac effect, an electron scattered from a standing-wave light undergoes a two-photon-scattering process. As a result, a net transverse-momentum transfer equal to $p^{el} = 2n\hbar k^{ph}$ will be observed, where k^{ph} is the photon wavenumber and n is an integer. Considering this, the electron will be diffracted by an associated wavenumber of $k_y^{el} = 2nk^{ph}$. Considering $\lambda^{ph} = 30$ nm, which is equal to the gap between the gratings, $k_y^{el} = n\,0.419$ rad nm^{-1} will be obtained, which agrees perfectly well with the numerical results obtained. It should be pointed out here that in contrast with the usual Kapitza-Dirac effect, which is a result of the interaction of a plane-wave electron with a free-space standing wave, the effect considered here is a pure near-field effect, happening because of the enhanced E_y component of the electric field excited at the edges of the grating elements (see Fig. 8.9a). The near-field distribution of the

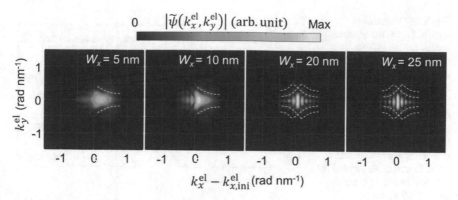

Fig. 8.15 Absolute value of the electron wavefunction in momentum space, after interaction with the grating, for initial longitudinal electron broadenings of **a** $W_L = 5\,$nm, **b** $W_L = 10\,$nm, **c** $W_L = 20\,$nm, and **d** $W_L = 25\,$nm

grating provides a significantly large momentum to facilitate strong electron–photon coupling; a consequence of this is the clear diffraction orders happening here even by incorporating a moderate light intensity, which is not easily addressable in the Kapitza-Dirac effect. However, the momentum transfer here is still a two–photon process, quite similar to the Kapitza-Dirac effect.

The signature of the spatiotemporal bunching in energy-momentum space is the appearance of distinct resonances along the longitudinal momentum axis, causing electron energy-loss and energy-gain peaks (Fig. 8.15c and d). A quantum mechanical interpretation of similar effects has been deliberately developed within the context of PINEM experiments, for which an electron-wave packet interacts with the near field of a sample and loses quanta of photon energy. However, a clear distinction between PINEM and the electron-photon scatterings happening in the inverse Smith–Purcell effect is a net acceleration of the centre of the wave packet due to the satisfaction of the synchronicity condition. This net acceleration causes the resonance peaks to not appear symmetrically in the gain and loss regions.

An interesting result of this investigation is that the more localized the electron wave packet is in the longitudinal direction, the better it is accelerated (Fig. 8.16). This is because the multiple electron photon scatterings do not happen for short electron wave packets due to the short interaction time of the single-electron wave packet with light, which does not allow the quantum mechanical interference to take place. In other words, the light energy will be preserved for a net acceleration of the whole electron wave packet, instead of being transferred to nonlinear and multiple scattering processes.

In summary, we studied the interaction of single-electron pulses with lights and gratings, a phenomenon referred to as the inverse Smith-Purcell effect, which is incorporated regularly in DLAs. Interesting consequences of such an investigation include the bunching of single-electron wave packets in a fully coherent way, which only happens for electron pulses of sufficient longitudinal broadening, in such a way

Fig. 8.16 Probability distribution of the electron versus the normalized longitudinal component of the velocity $\hbar k_x^e / m_0 c$, taken for various initial longitudinal broadenings, as depicted in the figure. Dashed line: initial, and solid line: final probability distributions of the electron wave function. Re-printed by permission from [2] under the CC BY license

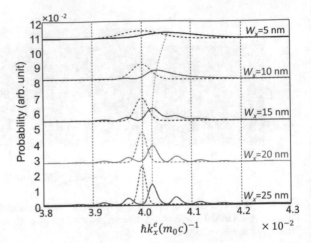

that they can cover a few grating elements. A fully quantum mechanical aspect of the interaction is the occurrence of several diffraction peaks in the transverse direction. Although it has been proposed that a symmetric grating may be incorporated to make the electron wave packet focused, the Kapitza-Dirac effect can violate the assumptions made using Lorentzian mechanics.

Electrons at low energies are easily recoiled and bunched in interaction with gratings and light, which might not be desirable in applications in which a Gaussian shape should be preserved. However, these consequences might be employed for an efficient electron wave packet shaping, such as electron bunching methods to be incorporated in novel spectroscopy techniques.

8.4 Point-Projection Electron Wave Packets Interacting with Near Fields of Nanoparticles

In PPM, divergent electron wave packets emitted from sharp tungsten or gold tips positioned at short distances from the sample interact with the sample, as a result of which a magnified image of the sample is formed in the far field (Fig. 8.17). Of particular interest in the point projection microscopy setup is the formation of transverse interference effects, which can be used for spatial holography (see Fig. 8.3 and discussions therein). Here, we would like to numerically investigate the possibility of controlling and manipulating the transverse and longitudinal interferences by inducing coherent polarizations in the sample by means of laser excitations (Fig. 8.17b). As when slow electrons are involved, the longitudinal interferences might not seem trivial due to the extensive mismatch between the electron and light velocities, which causes difficulties in matching the momentums of the moving electrons and light [see

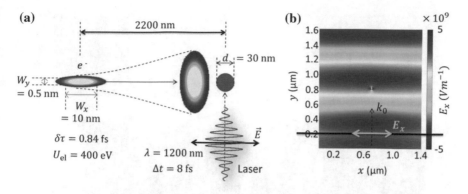

Fig. 8.17 Interaction of divergent electron wavepackets with laser-induced excitations in nanos-tructures. **a** Schematic of the simulation domain with the initial parameters for the laser and electron wavepacket. **b** x-component of the total electric field

Chap. 2, particularly (2.2)]. We will show here, however, in contrast to our expecta-tions, both the longitudinal and transverse modulations are nicely manifested in the electron wave packet after its interaction with the laser-induced near-field excitations of the sample. We describe this effect as the interaction of the electron with coherent ultrafast charge oscillations of the free electrons in the gold nanorod, which can be effectively modelled by the scalar potential.

As the first example, we consider a Gaussian electron wave packet at an initial kinetic energy of 400 eV with transverse and longitudinal broadenings of $W_y = 0.5$ nm and $W_x = 10$ nm (temporal broadening of only $\delta\tau = 84$ fs), interacting with a gold nanorod with a diameter of 30 nm. The nanorod is excited with a femtosecond laser at a carrier wavelength of $\lambda = 1200$ nm, broadening of $\Delta t = 8$ fs and electric-field amplitude of $E_0 = 10^{10}$ Vm^{-1} (Fig. 8.18 a). After propagating over a distance of 2200 nm in vacuum, the electron wave packet experiences dispersion and expansion and approaches a lateral broadening of approximately 30 nm. The sample is modelled by only its dielectric function. The induced optical near field of the nanostructure can be decomposed into transverse and longitudinal terms.

Adopting the Coulomb gauge, as discussed in Sect. 8.1, given by $\vec{\nabla} \cdot \vec{A} = 0$, the vector potential remains purely transverse (solenoidal). This behaviour is also obvi-ous in Fig. 8.18b, where the presence of the particle only slightly affects the incident coming wave. Particularly for a nanostructure with dimensions much smaller than the wavelength of the incident light, the induced electric dipole inside the material is completely manifested by the scalar potential (Fig. 8.18c). More generally, as far as the induced magnetic moments in a material can be neglected, a quasi-static approach will sufficiently model the complete interaction process.

To calculate the electromagnetic potentials, we adopted a pseudopotential approach. We first calculate the field components using the FDTD method. The polarization terms in our method are calculated then by expanding the dielectric function using a Drude model in addition to two critical point functions [82, 83]. In

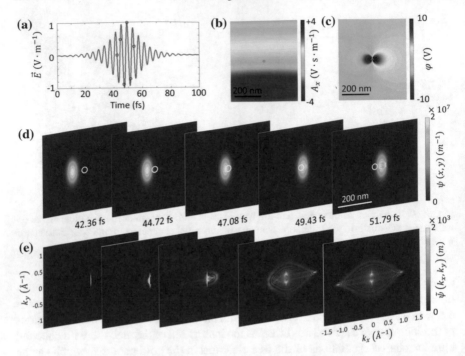

Fig. 8.18 Temporal evolution of the electron wave packet during its interaction with the near-field distribution of the gold nanorod. **a** Temporal distribution of the incident laser field. Spatial distributions of the **b** x-component of the vector potential and **c** scalar potential at a given time. Evolution of the electron wave packet in **d** real space and (**b**) reciprocal space

this way, the displacement vector is described as

$$\vec{D}(\omega) = \varepsilon_0 \vec{E}(\omega) + \vec{P}_D(\omega) + \sum_{n=1,2} \vec{P}_n(\omega) \tag{8.21}$$

Note that we intentionally assumed $\varepsilon_\infty = 0$. In this way, the polarization vector $\vec{P}_D(\omega)$ is the correct Drude-like polarization given by

$$\vec{P}_D = -\varepsilon_0 \frac{\omega_P^2}{\omega^2 + i\gamma_D\omega} \vec{E} \tag{8.22}$$

where ω_P is the plasma frequency and γ_D is the damping ratio associated with the Drude model. The values for the plasma frequency and the damping ratio in energy units are given by $\hbar\omega_P = 8.782\,\text{eV}$ and $\hbar\gamma_D = 0.09\,\text{eV}$, respectively. $\vec{P}_n(\vec{r}, \omega)$ is given by

$$\vec{P}_n(\vec{r}, \omega) = f_n \left[\frac{e^{i\varphi_n}}{\omega_n - i\Gamma_n - \omega} - \frac{e^{-i\varphi_n}}{\omega_n + i\Gamma_n - \omega} \right] \vec{E}(\vec{r}, \omega) \tag{8.23}$$

where the oscillator strengths are $\hbar f_1 = 12.690\,\mathrm{eV}$ and $\hbar f_2 = 0.739\,\mathrm{eV}$, $\hbar\omega_1 = 2.460\,\mathrm{eV}$, $\hbar\omega_2 = 2.469\,\mathrm{eV}$, $\hbar\Gamma_1 = 1.881\,\mathrm{eV}$, $\hbar\Gamma_2 = 0.304\,\mathrm{eV}$, $\varphi_1 = -1.043\,\mathrm{rad}$, and $\varphi_2 = -1.750\,\mathrm{rad}$.

Each term of the polarization vector is individually calculated using the generally exploited recursive formalism within the FDTD approach [84]. The same formalism can be used to adopt nonlinear effects as well, though a general knowledge of the dispersion of the nonlinear susceptibility is required. The Gauss theorem of Maxwell's equations is then used to derive

$$\nabla^2\varphi = -\frac{\vec{\nabla}\cdot\vec{D}(\vec{r},t) - \vec{\nabla}\cdot\vec{P}(\vec{r},t)}{\varepsilon_0} = \rho_E \qquad (8.24)$$

One might adopt a finite-difference formalism to solve (8.24). However, here, we prefer to use a pseudopotential approach. In this way, (8.24) is transformed to the Fourier domain to obtain $\varepsilon_0(k_x^2 + k_y^2)\tilde{\varphi}(k_x, k_y) = \tilde{\rho}_E$. Then, the scalar potential is obtained as $\varphi(x, y) = \mathfrak{I}^{-1}\{\tilde{\varphi}(k_x, k_y)\}$. The vector potential is calculated as $\vec{A}(\vec{r}, t + \delta t) = \vec{A}(\vec{r}, t) - \delta t\left\{\vec{E}(\vec{r}, t) + \vec{\nabla}\varphi(\vec{r}, t)\right\}$ using the discrete-time propagator.

The electron wave packet modulation by the near-field distribution of the nano-object lasts for a duration of 8 fs. This temporal duration is approximately two times the laser oscillation period. The electron-light interaction process causes both amplitude and phase modulations (Fig. 8.17d and e). The amplitude of the electron wave packet is strongly affected by the shape of the nanostructure. This is because it is only the very neighbourhood of the nanostructure that provides such high momentums that can satisfy the energy-momentum conservation criterion for electron-light interaction (2.2). This criterion, given by $\vec{k}_{ph} \cdot \vec{v}_{el} = \omega_{ph}$, is satisfied by the near-field excitations and results in an ultrafine longitudinal phase modulation that is better resolvable in the momentum representation (Fig. 8.18e). By assigning the longitudinal momentum to the kinetic energy of the electron as $U = \hbar^2 k_{x,\,el}^2/2m_0$ and the transverse momentum to the scattering angle as $\theta = \tan^{-1}(k_y^{el}/k_x^{el})$, the angle-resolved energy gain spectrum of the probability distribution will be obtained, as shown in Fig. 8.19a. The longitudinal interference effect causes energy modulations up to the 92nd order of the photon energy ($\hbar\omega_{ph} = 1.033\,\mathrm{eV}$; see Fig. 8.18c). These longitudinal phase modulations are understood as an efficient coupling between the near-field regions of the nanostructure resonating at a frequency of ω_{ph} with the electron wave packets. Hence, the nanostructure acts as a mediator for transferring energy and momentum from light waves to matter waves.

In addition to the longitudinal interferences, transverse interference effects are observed as well, though with different interspacings between the maximum intensities compared to the longitudinal interferences. The reason for the observation of the transverse interference effects is not because of the charging effects in the sample, as is generally observed for DC-driven and laser-driven point-projection electrons interacting with carbon nanotubes [85]. The latter interference patterns for carbon nanotubes are observed because a positively charged carbon nanotube acts as a biprism for the electrons. The interaction of electron beams with the carbon nanotube results in the charging of the nanotube, and hence, it acts as a biprism

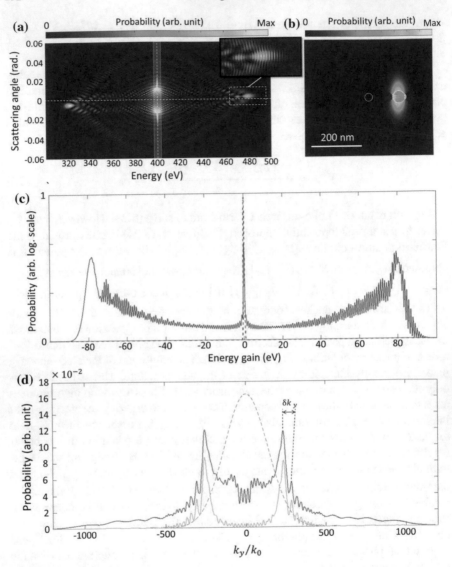

Fig. 8.19 Final distribution of the electron wave packet. **a** Angle-resolved energy map of the probability amplitude and the **b** spatial representation of the electron wave packet after the interaction. **c** Energy-gain spectra of the electron by integrating over the entire angular distribution for the final (blue solid line) and initial (red dashed line) wave packets. **d** Electron distribution along the transverse direction for an electron propagating without interaction (red dashed line) and for the electron after the interaction by integrating over the complete energy range (blue line) and integrating over the narrow energy range shown by the highlighted region in panel a (orange line)

for the electron beam; therefore, the wave front of the electron waves is split into two paths, where interference between these paths will be formed on the detector screen. In contrast, the nanorod exploited in our simulations is a gold nanorod that is not charged. Moreover, the interference effects observed here do not take place in the absence of the laser illumination. The interspacing between the maxima in the transverse direction is, however, approximately $\delta k_y = 40\,k_0$, much beyond what is expected from the Kaptiza-Dirac effect as well (Fig. 8.18d). The value of δk_y is roughly equal to $2\pi\,d^{-1}$. These diffracted waves propagate in a broad angular range, both in the transverse and in the longitudinal directions. The reason such an effect is the scattering of the electron wave packet by the nanorod and the interference of the scattered waves by the unscattered waves, as is originally discussed by Dennis Gabor in his inline electron holography scheme [86].

Within only a short scattering angular range of ± 0.1 rad, the resonances observed along the longitudinal direction correspond to the one-dimensional theory exploited to explain PINEM effects [15]. In this one-dimensional model, the electron wave packet can be modelled by $\psi(z, t) = \psi_0(z - v_{\mathrm{el}}t, t)\exp(i(k_{\mathrm{el}}z - \omega_{\mathrm{el}}t))$, where $\hbar k_{\mathrm{el}} = m_0 v_{\mathrm{el}}$ and $\hbar \omega_{\mathrm{el}} = 0.5 m_0 v_{\mathrm{el}}^2$. Incorporating this wave packet into (8.1) and neglecting pondermotive interactions as well as the scalar potential, a simplified version of the Wolkow approximation is obtained, as

$$\psi_0(z', +\infty) = \psi_0(z', -\infty)$$
$$\times \exp\left\{ \exp\left(-\frac{(z' + v_{\mathrm{el}}\tau)^2}{4v_{\mathrm{el}}^2 \delta \tau^2} \right) \mathrm{Im}\left(\exp\left(i\frac{\omega_{\mathrm{ph}}}{v_{\mathrm{el}}}z' \right)\tilde{F}_z \right) \right\} \qquad (8.25)$$

where ι is the temporal delay between the electron and photon pulses, $z' = z - v_{\mathrm{el}}t$ is the coordination with respect to the centre of the moving electron wave packet, and $\tilde{F}_z$ defines the coupling strength between the electron and light [37] and is given by [15]

$$\tilde{F}_z = -i\frac{e}{\hbar \omega_{\mathrm{ph}}} \int_{-\infty}^{+\infty} dz\, \tilde{E}_z(z, t = 0)\exp\left(-i\frac{\omega_{\mathrm{ph}}}{v_{\mathrm{el}}}z \right) \qquad (8.26)$$

Furthermore, one might expand (8.25) by a Taylor series and use the Jacobi-Anger relation [as used in Sect. 8.2 to derive (8.13)] to obtain:

$$\psi(z, t) = \psi_0(z - v_{\mathrm{el}}t, -\infty)$$
$$\times \sum_{n=-\infty}^{+\infty} \xi_n(z - v_{\mathrm{el}}t)\exp\left\{ i\left(k_{\mathrm{el}} + n\frac{\omega_{\mathrm{ph}}}{v_{\mathrm{el}}} \right)z - i\left(\omega_{\mathrm{el}} + n\omega_{\mathrm{ph}} \right)t \right\} \qquad (8.27)$$

where $\xi_n(z - v_{el}t)$ is given by

$$\xi_n = \left(\frac{\tilde{F}_z}{|\tilde{F}_z|}\right)^n J_n\left\{|\tilde{F}_z| \exp\left(-\frac{(z' + v_{el}\tau)^2}{4v_{el}^2 \delta \tau^2}\right)\right\} \tag{8.28}$$

Thus, the argument of the Bessel function defines the coupling strength between the light and electron wave packet to provide electron-photon couplings of higher orders. After this near-field-mediated interaction, the electron wave packet evolves into a superposition of momentum states, where each state is defined by the momentum of $k_{el}^n = k_{el} + n\,\omega_{ph}v_{el}^{-1}$ and energy $\omega_{ph}^n = \omega_{el} + n\omega_{ph}$ [15].

The treatment stated above has been used to understand PINEM theory and to explain, for example, the generation of attosecond pulse trains and coherent phase modulations and the formation of Rabi oscillations in the electron wave packet in recent experiments as well [37, 77]. The application of such theory to the slow-electron regime, however, should be considered only with great care. This is due to the following facts:

1st—In deriving (8.25), the Coulomb gauge has been applied; however, as a general practice, the scalar potential has been neglected. This, by itself, introduces a discrepancy, as $\vec{\nabla} \cdot \vec{A} = 0$ implies that the vector potential is solenoidal, and thus it cannot explain the rotational terms in the near-field distribution. Adopting, however, a Coulomb gauge $(-i\hbar\vec{\nabla} + e\vec{A})^2$ term in the minimum coupling, the Hamiltonian is further expanded by $-\hbar^2\nabla^2 + e^2\vec{A} \cdot \vec{A} - i\hbar e\,\vec{A} \cdot \vec{\nabla} - i\hbar e\vec{\nabla} \cdot \vec{A}$, where the last term in this expansion is equal to zero. Neglecting the scalar potential, which is recast into $\vec{E}(\vec{r}, t) = -\dot{A} - \vec{\nabla}\varphi = -\dot{A}$, causes the electric field to also become solenoidal. However, the Gauss theory of Maxwell's equations is given by $\vec{\nabla} \cdot \vec{D} = \varepsilon_0 \vec{\nabla} \cdot \vec{E} + \vec{\nabla} \cdot \vec{P}$. As in most practical cases, there will be no impressed charges (this is true for the metallic nanorod introduced here; for carbon nanotubes and dielectric elements, a charging effect should be considered). Adopting the Coulomb gauge and neglecting the scalar potential at the same time imposes a severe restriction on the induced polarization as $\vec{\nabla} \cdot \vec{P} = 0$. Thus, the solenoidal aspects of the near-field distributions will be not modelled correctly.

2nd—In many experimental situations where transmission electron microscopes are exploited to study electron-light interaction effects, the collection angle for the transmitted electrons is restricted to a few milliradians. Additionally, the interaction of relativistic electrons by near-field distributions of the nanostructures and their scattering thereof is different from the slow electron wave packets studied above; as such, the one-dimensional model discussed above might be conceptually inadequate for those experiments involving relativistic beams.

3rd—In addition to the ponderomotive interaction (term related to $\vec{A} \cdot \vec{A}$) and the scalar potential, adopting the slowly varying approximation for deriving (8.25), two other terms are also neglected (terms related to $\vec{k}_{el} \cdot \vec{\nabla}\psi_0$ and $\vec{A} \cdot \vec{\nabla}\psi_0$). For high-energy electrons and moderate laser intensities, the term related to $\vec{A} \cdot \vec{k}_{el}$ has such high values that neglecting other terms seems rational. This is, however, not the case for slow electrons.

Therefore, the scattering of slow electron wave packets with near-field distributions demands a more complete theory to adequately model the effects of all terms in the Hamiltonian. Moreover, such a theory cannot be derived by only investigating the effect of each individual term on the electron wave packet, as mutual interactions and amplitude/phase modulations occurring by the influence of all terms may the interaction to follow specific paths, which is not addressable by adopting adiabatic theories. Nevertheless, to fully emphasize the importance of the scalar potential, three different systems were simulated: 1st—The interaction of an electron wave packet with a gold nanorod and laser excitation with the parameters specified in Fig. 8.17, considering the complete Hamiltonian, 2nd—considering only the scalar potential (the first and last terms in (8.2), leading to $-\hbar^2 (2m_0)^{-1} \nabla^2 \psi - e \varphi \psi = i\hbar\dot{\psi}$), and 3rd—assuming only interactions involving the vector potential as $-\hbar^2 (2m_0)^{-1} \nabla^2 \psi - i\hbar e m_0^{-1} \vec{A} \cdot \vec{\nabla} \psi = i\hbar\dot{\psi}$. By comparing these simulation results, as shown in Fig. 8.20, one can immediately see that considering only the vector potential as the interaction Hamiltonian does not adequately describe the physical effects observed considering the complete Hamiltonian.

The explanations above raise the question of under what conditions one should expect to observe PINEM effects of the sort explained above for electrons with kinetic energies as low as a few hundred electron volts or even, as shown in [48], $U_{el} = 40$ eV. To date, we noticed that instead of the vector potential, it is the scalar potential that is seemingly responsible for the electron-light interactions in the near-field region, as far as nanoscaled metallic particles are considered, where only electric dipole moments are excited (magnetic moments and toroidal moments are not excited).

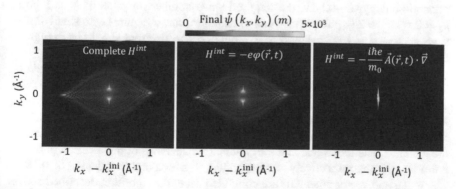

Fig. 8.20 Comparison between the distribution of the final electron wave packet in the momentum space. Various cases where the complete minimum-coupling Hamiltonian is considered (left), only the scalar potential is adopted (middle), and finally, only the interactions related to the vector potential are considered (left)

Thus, we can consider $\vec{E} = -\vec{\nabla}\varphi$ and write the solution to the envelope of the electron wave packet as

$$\psi(z', t) = \psi(z', t_0) \exp\left[\frac{-e}{i\hbar} \int_{t_0}^{t} \varphi(z' + v_{el}t', t')dt'\right]$$

$$= \psi(z', t_0) \exp\left[\frac{-e}{i\hbar} \int_{t_0}^{t} \left\{\frac{E_z(z' + v_{el}t', t')}{-ik_{ph}} - \frac{E_z^*(z' + v_{el}t', t')}{-ik_{ph}}\right\}dt'\right]$$

$$= \psi(z', t_0) \exp\left[\frac{ec}{i\hbar\omega_{ph}} \int_{t_0}^{t} \mathrm{Im}\{E_z(z' + v_{el}t', t')\}dt'\right] \tag{8.29}$$

Using the same approach as stated above, one might derive the coupling strength for the inelastic electron-light interaction involving the scalar potential as

$$\tilde{F}_z = -i\frac{e}{\hbar\omega_{ph}\beta_{el}} \int_{-\infty}^{+\infty} dz\, \tilde{E}_z(z, t = 0) \exp\left(-i\frac{\omega_{ph}}{v_{el}}z\right) \tag{8.30}$$

where $\beta_{el} = v_{el}/c$. Note that the momentum-matching condition remains the same $(k_z = \omega_{ph}v_{el}^{-1})$; however the coupling strength increases by a factor of $\beta_{el}^{-1} = c/v_{el}$, which shows the tremendous sensitivity of the slow electron wave packet to the electromagnetic interactions, at least when inelastic interactions are involved. However, the increase in the coupling strength is at the expense of a much harder criterion for satisfying the momentum-matching condition, as for the electrons with a kinetic energy of 400 eV, the required wave number for photons is as high as $k_z = 2 \times 10^8 \approx 25k_0$. To provide this momentum, nanostructures with small dimensions such as $d = \lambda_0/25$ should be exploited, or the electrons should be brought to the very proximity of the nanostructures.

In the following, we investigate the interaction of an electron wave packet at a kinetic energy of $U_{el} = 10$ keV with a gold nanorod with a diameter of $d = 200$nm, excited by a pulsed x-polarized laser excitation at a central carrier wavelength of $\lambda_{ph} = 800$ nm and a temporal broadening of 5.34 fs (two laser cycles) (Fig. 8.21a). We furthermore consider a focused aloof trajectory, only 1 nm away from the rim of the nanorod. The transverse and longitudinal broadenings of the electron pulse are 2 nm and 40 nm, respectively, corresponding to a temporal broadening of 0.68 fs. The scalar and vector potentials are computed using the algorithm described above. The scalar potential demonstrates an electric dipolar-like excitation of the sample (Fig. 8.21b), whereas the vector potential follows the symmetry of the excited electric field, along with a negligible value for its y-component (Fig. 8.21c).

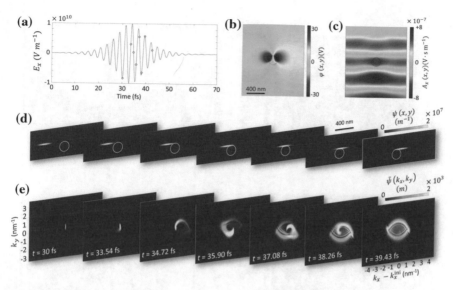

Fig. 8.21 Electron wave packet at a kinetic energy of 10 keV, interacting with a gold nanorod at a radius of 100 nm in an aloof trajectory. **a** Temporal broadening of the introduced laser pulse ($\lambda_0 = 800$ nm). Circular markers show the times at which the snapshots in panels d and e are represented. **b** Scalar potential and **c** x-component of the vector potential

Different from the diverging electron wave packet at a kinetic energy of 400 eV studied above, the focused high-energy electron wave packet experiences negligible amplitude modulations (Figs. 8.21d and 8.22a). In contrast, the phase of the high-energy electron wave packet and as a result its distribution in the momentum space is strongly modulated (Fig. 8.21e). The coupling strength $\tilde{F}_z$ for an electron wave packet at the energy of 10 keV is reduced by a factor of 5 with respect to the coupling strength for a 400 eV-electron wave packet; as a result, energy levels of up to the 50th order in the gain and up to 60th order in the loss spectra will be occupied, at similar laser intensities, compared to the 97th order for the 400 eV-electron wave packet (Fig. 8.22c).

The interaction time between the electron and the laser-induced near-field distribution of the nanostructure is approximately 6 fs. The temporal distribution of the electron wave packet is, however, less than that of a laser cycle, which enhances the streaking effects of the electron wave packet by the laser pulse. In this way, at each given time, the electron experiences either a repelling or attracting force towards the nanostructure, resulting in circular motions of the sort observed in the momentum representation (Fig. 8.21e). Nevertheless, in contrast to the free-space electron light interactions of the sort discussed in Sect. 8.2.2, these phase modulations happening for the electron wave packet do not average out.

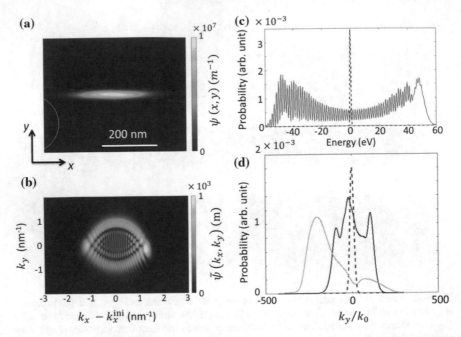

Fig. 8.22 Final distribution of an electron wavepacket at a kinetic energy of 10 keV interacting with the system described in the text. **a** Real-space and **b** reciprocal-space distributions, **c** energy-gain spectra and **d** experienced transverse recoil probability by integrating over the entire energy range (black line) and for $k_x = k_x^{\mathrm{ini}}$ (blue line). The red dashed line shows the probability distribution for a noninteracting electron wavepacket at the same kinetic energy

8.5   Summary

In this chapter we discussed possible cases where an adiabatic approximation for treating electron-light interactions breaks down. This happens particularly for cases where the interaction leads to non-negligible changes in the amplitude of the wave function as well as its phase. Moreover, we proposed and realized a numerical toolbox for simulating electron-light interactions from first principles by combining solvers for the time-dependent Schrödinger and Maxwell equations in a self-consistent way. Using this toolbox, the interference effects between quantum paths in the coherent Kapitza-Dirac effect were discussed and simulated. We also demonstrated that such interference paths can be used to manipulate the shape of the electron wave packet on demand. Additionally, we discussed the quantum effects in the interaction of slow-electron wave packets with dielectric laser accelerators. We demonstrated various effects, ranging from electron bunching to acceleration, that can be controlled by the broadening of the electron wave packet. In addition, laser-induced effects in the point-projection electron microscopy setup were discussed. We intensively discussed the similarities and differences between the PINEM effect involving slow and fast electrons and demonstrated that in contrast to what is routinely employed, it is the scalar potential that is responsible for the inelastic electron-light interactions when nanoparticles are employed as mediators. These findings will lead to further developments in the theory of coherent control and quantum random walks in a single-electron wave packet interacting with coherent electromagnetic radiation, initiated by controlling the laser parameters and designing appropriate nanoparticles.

References

1. M. Kozak, T. Eckstein, N. Schonenberger, P. Hommelhoff, Inelastic ponderomotive scattering of electrons at a high-intensity optical travelling wave in vacuum. Nat. Phys. Lett. advance online publication, 10/09/2017 [online], https://doi.org/10.1038/nphys4282, http://www.nature.com/nphys/journal/vaop/ncurrent/abs/nphys4282.html#supplementary-information
2. N. Talebi, Schrödinger electrons interacting with optical gratings: quantum mechanical study of the inverse Smith-Purcell effect. New J. Phys. **18**(12), 123006 (2016). https://doi.org/10.1088/1367-2630/18/12/123006
3. D.Y. Sergeeva, A.P. Potylitsyn, A.A. Tishchenko, M.N. Strikhanov, Smith-Purcell radiation from periodic beams. Opt. Express **25**(21), 26310–26328 (2017). https://doi.org/10.1364/Oe.25.026310. (in English)
4. S. Tsesses, G. Bartal, I. Kaminer, Light generation via quantum interaction of electrons with periodic nanostructures. Phys. Rev. A **95**(1), 013832 (2017). https://doi.org/10.1103/physreva.95.013832. (in English)
5. I. Kaminer et al., Quantum Cerenkov radiation: spectral cutoffs and the role of spin and orbital angular momentum. Phys. Rev. X **6**(1), 011006 (2016). https://doi.org/10.1103/physrevx.6.011006. (in English)
6. J. Peatross, C. Muller, K.Z. Hatsagortsyan, C.H. Keitel, Photoemission of a single-electron wave packet in a strong laser field. Phys. Rev. Lett. **100**(15), 153601 (2008). https://doi.org/10.1103/physrevlett.100.153601. (in English)
7. A.J. White, M. Sukharev, M. Galperin, Molecular nanoplasmonics: self-consistent electrodynamics in current-carrying junctions. Phys. Rev. B **86**(20), 205324 (2012). https://doi.org/10.1103/PhysRevB.86.205324
8. W. Kohn, L.J. Sham, Self-consistent equations including exchange and correlation effects. Phys. Rev. **140**(4A), 1133 (1965). https://doi.org/10.1103/PhysRev.140.A1133. (in English)
9. N. Talebi, Electron-light interactions beyond the adiabatic approximation: recoil engineering and spectral interferometry AU—Talebi. Nahid. Adv. Phys. X **3**(1), 1499438 (2018). https://doi.org/10.1080/23746149.2018.1499438
10. N. Talebi, C. Lienau, Interference between quantum paths in coherent Kapitza-Dirac effect. New J. Phys. (2019) [Online]. Available http://iopscience.iop.org/10.1088/1367-2630/ab3ce3
11. O. Smirnova, M. Spanner, M. Ivanov, Analytical solutions for strong field-driven atomic and molecular one- and two-electron continua and applications to strong-field problems. Phys. Rev. A **77**(3), 033407 (2008). https://doi.org/10.1103/physreva.77.033407. (in English)
12. D.M. Wolkow, On a mass of solutions of the Dirac equation. Z. Angew. Phys. **94**(3–4), 250–260 (1935). https://doi.org/10.1007/bf01331022. (in German)
13. E. Kasper, Generalization of Schrodingers wave mechanics for relativistic regions of validity. Z. Naturforsch. A, **A28**(2), 216–221 (1973) [Online]. Available: <Go to ISI>://WOS:A1973S611900009 (in German)
14. S.T. Park, Propagation of a relativistic electron wave packet in the Dirac equation. Phys. Rev. A **86**(6), 062105 (2012). https://doi.org/10.1103/physreva.86.062105. (in English)
15. S.T. Park, M.M. Lin, A.H. Zewail, Photon-induced near-field electron microscopy (PINEM): theoretical and experimental. New J. Phys. **12**, 123028 (2010). https://doi.org/10.1088/1367-2630/12/12/123028. (in English)

16. F.J.G. de Abajo, A. Asenjo-Garcia, M. Kociak, Multiphoton absorption and emission by inter-action of swift electrons with evanescent light fields. Nano Lett. **10**(5), 1859–1863 (2010). https://doi.org/10.1021/nl100613s. (in English)
17. D. Wolf et al., 3D magnetic induction maps of nanoscale materials revealed by electron holo-graphic tomography. Chem. Mater. **27**(19), 6771–6778 (2015). https://doi.org/10.1021/acs.chemmater.5b02723. (in English)
18. R.O. Jones, O. Gunnarsson, The density functional formalism, its applications and prospects. Rev. Mod. Phys. **61**(3), 689–746 (1989). https://doi.org/10.1103/RevModPhys.61.689. (in English)
19. EJ. Baerends, Perspective on self-consistent equations including exchange and correlation effects; W. Kohn, L.J. Sham, Phys. Rev. A **140**, 1133–1138 (in English), Theor. Chem. Acc. **103**(3–4), 265–269 (2000). Doi: https://doi.org/10.1007/s002140050031 (in English)
20. B. Walker, R. Gebauer, Ultrasoft pseudopotentials in time-dependent density-functional theory. J. Chem. Phys. **127**(16), 164106 (2007). https://doi.org/10.1063/1.2786999. (in English)
21. J. Harris, R.O. Jones, Pseudopotentials in density-functional theory. Phys. Rev. Lett. **41**(3), 191–194 (1978). https://doi.org/10.1103/PhysRevLett.41.191. (in English)
22. E. Runge, E.K.U. Gross, Density-functional theory for time-dependent systems. Phys. Rev. Lett. **52**(12), 997–1000 (1984). https://doi.org/10.1103/PhysRevLett.52.997. (in English)
23. X.S. Li, S.M. Smith, A.N. Markevitch, D.A. Romanov, R.J. Levis, H.B. Schlegel, A time-dependent Hartree-Fock approach for studying the electronic optical response of molecules in intense fields. Phys. Chem. Chem. Phys. **7**(2), 233–239 (2005). https://doi.org/10.1039/b415849k. (in English)
24. P.W. Hawkes, E. Kasper, *Principles of Electron Optics* (Academic Press, London, 1996)
25. N.W. Ashcroft, N.D. Mermin, *Solid State Physics* (Thomson Learning Inc., United States of America, 1976)
26. H. Tal-Ezer, R. Kosloff, An accurate and efficient scheme for propagating the time dependent Schrödinger equation. J. Chem. Phys. **81**(9), 3967–3971 (1984). https://doi.org/10.1063/1.448136
27. X.J. Shen, A. Lozano, W. Dong, H.F. Busnengo, X.H. Yan, towards bond selective chemistry from first principles: methane on metal surfaces. Phys. Rev. Lett. **112**(4), 046101 (2014). https://doi.org/10.1103/PhysRevLett.112.046101
28. L. Gaudreau et al., Coherent control of three-spin states in a triple quantum dot. Nat. Phys. **8**, 54. 11/27/2011 [online], https://doi.org/10.1038/nphys2149, https://www.nature.com/articles/nphys2149#supplementary-information
29. J. Hansom et al., Environment-assisted quantum control of a solid-state spin via coherent dark states. Nat. Phys. **10**, 725, 09/07/2014 [online], https://doi.org/10.1038/nphys3077, https://www.nature.com/articles/nphys3077#supplementary-information
30. I.S. Mark, Ultrafast nanoplasmonics under coherent control. New J. Phys. **10**(2), 025031 [online], http://stacks.iop.org/1367-2630/10/i=2/a=025031
31. R.P. Feynman, Space-time approach to non-relativistic quantum mechanics. Rev. Mod. Phys. **20**(2), 367–387 (1948). https://doi.org/10.1103/RevModPhys.20.367
32. M. Li et al., Classical-Quantum correspondence for above-threshold ionization. Phy. Rev. Lett. **112**(11), 113002 (2014). https://doi.org/10.1103/PhysRevLett.112.113002
33. D.B. Milošević, W. Becker, Improved strong-field approximation and quantum-orbit theory: application to ionization by a bicircular laser field. Phys. Rev. A **93**(6), 063418 (2016). https://doi.org/10.1103/PhysRevA.93.063418
34. A. Zaïr et al., Quantum path interferences in high-order harmonic generation. Phys. Rev. Lett. **100**(14), 143902 (2008). https://doi.org/10.1103/PhysRevLett.100.143902
35. P. Salieres et al., Feynman's path-integral approach for intense-laser-atom interactions. Science **292**(5518), 902–905 (2001). https://doi.org/10.1126/science.108836. (in English)
36. T.C. Weinacht, J. Ahn, P.H. Bucksbaum, Controlling the shape of a quantum wavefunction. Nature **397**, 233 (1999). https://doi.org/10.1038/16654
37. A. Feist, K.E. Echternkamp, J. Schauss, S.V. Yalunin, S. Schafer, C. Ropers, Quantum coherent optical phase modulation in an ultrafast transmission electron microscope. Nature **521**(7551), 200 (2015). https://doi.org/10.1038/nature14463. (in English)

38. K.E. Echternkamp, A. Feist, S. Schafer, C. Ropers, Ramsey-type phase control of free-electron beams. Nat. Phys. **12**(11), 1000 (2016). https://doi.org/10.1038/nphys3844. (in English)
39. H. Batelaan, Colloquium: Illuminating the Kapitza-Dirac effect with electron matter optics. Rev. Mod. Phys. **79**(3), 929–941 (2007). https://doi.org/10.1103/revmodphys.79.929. (in English)
40. P.L. Kapitza, P.A.M. Dirac, The reflection of electrons from standing light waves. Math. Proc. Cambridge Philos. Soc. **29**(2), 297–300 (2008). https://doi.org/10.1017/S0305004100011105
41. A. Howie, Photon interactions for electron microscopy applications. Eur. Phys. J. Appl. Phys. **54**(3), 33502 (2011). https://doi.org/10.1051/epjap/2010100353
42. H. Batelaan, The Kapitza-Dirac effect. Contemp. Phys. **41**(6), 369–381 (2000). https://doi.org/10.1080/00107510010001220. (in English)
43. F.J. García de Abajo, Optical excitations in electron microscopy. Rev. Mod. Phys. **82**(1), 209–275 (2010). https://doi.org/10.1103/revmodphys.82.209
44. R.F. Harrington, *Time-harmonic electromagnetic fields* (McGraw-Hill Book Company, New York, 1961)
45. A. Howie, Stimulated excitation electron microscopy and spectroscopy. Ultramicroscopy **151**, 116–121 (2015). https://doi.org/10.1016/j.ultramic.2014.09.006. (in English)
46. M. Kozak, T. Eckstein, N. Schonenberger, P. Hommelho, Inelastic ponderomotive scattering of electrons at a high-intensity optical travelling wave in vacuum. Nat. Phys. **14**(2), 121 (2018). https://doi.org/10.1038/nphys4282. (in English)
47. M. Kozak, N. Schonenberger, P. Hommelhoff, Ponderomotive generation and detection of attosecond free-electron pulse trains. Phys. Rev. Lett. **120**(10), 103203 (2018). https://doi.org/10.1103/physrevlett.120.103203. (in English)
48. J. Vogelsang et al., Plasmonic-nanofocusing-based electron holography. Acs Photonics **5**(9), 3584–3593 (2018). https://doi.org/10.1021/acsphotonics.8b00418
49. J. Kempe, Quantum random walks: an introductory overview. Contemp. Phys. **44**(4), 307–327 (2003). https://doi.org/10.1080/00107151031000110776. (in English)
50. S. Aaronson, A. Arkhipov, The computational complexity of linear optics. Acm S. Theory Comput. 333–342 (2011) [Online]. Available <GotoISI>://WOS:000297656800035. (in English)
51. N. Spagnolo et al., Experimental validation of photonic boson sampling. Nat. Photonics **8**(8), 615–620 (2014). https://doi.org/10.1038/Nphoton.2014.135. (in English)
52. L. Sansoni et al., Two-particle Bosonic-Fermionic Quantumwalk via integrated photonics. Phys. Rev. Lett. **108**(1), 010502 (2012). https://doi.org/10.1103/physrevlett.108.010502. (in English)
53. R. Garciamolina, A. Grasmarti, A. Howie, R.H. Ritchie, Retardation effects in the interaction of charged-particle beams with bounded condensed media. J. Phys. C. Solid State. **18**(27), 5335–5345 (1985). https://doi.org/10.1088/0022-3719/18/27/019. (in English)
54. F.J.G. de Abajo, A. Rivacoba, N. Zabala, N. Yamamoto, Boundary effects in cherenkov radiation. Phys. Rev. B **69**(15), 155420 (2004). https://doi.org/10.1103/physrevb.69.155420. (in English)
55. C. Luo, M. Ibanescu, S.G. Johnson, J.D. Joannopoulos, Cerenkov radiation in photonic crystals. Science **299**(5605), 368–371 (2003). https://doi.org/10.1126/science.1079549. (in English)
56. N. Yamamoto, F.J.G. de Abajo, V. Myroshnychenko, Interference of surface plasmons and Smith-Purcell emission probed by angle-resolved cathodoluminescence spectroscopy. Phys. Rev. B **91**(12), 125144 (2015). https://doi.org/10.1103/physrevb.91.125144. (in English)
57. K. Mizuno, J. Pae, T. Nozokido, K. Furuya, Experimental evidence of the inverse Smith-Purcell effect. Nature **328**(6125), 45–47 (1987). https://doi.org/10.1038/328045a0
58. A. Asenjo-Garcia, F.J.G. de Abajo, Plasmon electron energy-gain spectroscopy. New J. Phys. **15**, 103021 (2013). https://doi.org/10.1088/1367-2630/15/10/103021. (in English)
59. J.P. Verboncoeur, Particle simulation of plasmas: review and advances. Plasma Phys. Contr. F. **47**, A231–A260 (2005). https://doi.org/10.1088/0741-3335/47/5A/017. (in English)
60. A. Fallahi, F. Kartner, Field-based DGTD/PIC technique for general and stable simulation of interaction between light and electron bunches. J. Phys. B Mol. Opt. **47**(23), 234015 (2014). https://doi.org/10.1088/0953-4075/47/23/234015. (in English)

61. J.-L. Vay, Simulation of beams or plasmas crossing at relativistic velocity. Phys. Plasmas **15**(5), 056701 (2008). https://doi.org/10.1063/1.2837054

62. B. Naranjo, A. Valloni, S. Putterman, J.B. Rosenzweig, Stable charged-particle acceleration and focusing in a laser accelerator using spatial Harmonics. Phys. Rev. Lett. **109**(16), 164803 (2012). https://doi.org/10.1103/physrevlett.109.164803

63. J. Breuer, J. McNeur, P. Hommelhoff, Dielectric laser acceleration of electrons in the vicinity of single and double grating structures—theory and simulations. J. Phys. B: At. Mol. Opt. Phys. **47**(23), 234004 (2014). https://doi.org/10.1088/0953-4075/47/23/234004

64. M. Ferrario et al., IRIDE: Interdisciplinary research infrastructure based on dual electron linacs and lasers. Nucl. Instrum. Methods Phys. Res. Sect. A **740**, 138–146 (2014). https://doi.org/10.1016/j.nima.2013.11.040

65. E.A. Peralta et al., Demonstration of electron acceleration in a laser-driven dielectric microstructure. Nature **503**, 91, 11/06/2013 [online], https://doi.org/10.1038/nature12664, https://www.nature.com/articles/nature12664#supplementary-information

66. J. Breuer, P. Hommelhoff, Laser-based acceleration of nonrelativistic electrons at a dielectric structure. Phys. Rev. Lett. **111**(13), 134803 (2013). https://doi.org/10.1103/physrevlett.111.134803

67. P. Baum, On the physics of ultrashort single-electron pulses for time-resolved microscopy and diffraction. Chem. Phys. **423**, 55–61 (2013). https://doi.org/10.1016/j.chemphys.2013.06.012

68. L. Kasmi, D. Kreier, M. Bradler, E. Riedle, P. Baum, Femtosecond single-electron pulses generated by two-photon photoemission close to the work function. New J. Phys. **17**(3), 033008 (2015). https://doi.org/10.1088/1367-2630/17/3/033008

69. J. Hoffrogge et al., Tip-based source of femtosecond electron pulses at 30 keV. J. Appl. Phys. **115**(9), 094506 (2014). https://doi.org/10.1063/1.4867185

70. B. Piglosiewicz et al., Carrier-envelope phase effects on the strong-field photoemission of electrons from metallic nanostructures. Nat. Photonics **8**, 37, 11/10/2013 [online], https://doi.org/10.1038/nphoton.2013.288, https://www.nature.com/articles/nphoton.2013.288#supplementary-information

71. M. Aidelsburger, F.O. Kirchner, F. Krausz, P. Baum, Single-electron pulses for ultrafast diffraction. Proc. Natl. Acad. Sci. **107**(46), 19714–19719 (2010). https://doi.org/10.1073/pnas.1010165107

72. M. Krüger, M. Schenk, M. Förster, P. Hommelhoff, Attosecond physics in photoemission from a metal nanotip. J. Phys. B At. Mol. Opt. Phys. **45**(7), 074006 (2012). https://doi.org/10.1088/0953-4075/45/7/074006

73. G. Herink, D.R. Solli, M. Gulde, C. Ropers, Field-driven photoemission from nanostructures quenches the quiver motion. Nature **483**, 190, 03/07/2012 [online], https://doi.org/10.1038/nature10878, https://www.nature.com/articles/nature10878#supplementary-information

74. B. Barwick, C. Corder, J. Strohaber, N. Chandler-Smith, C. Uiterwaal, H. Batelaan, Laser-induced ultrafast electron emission from a field emission tip. New J. Phys. **9**(5), 142 (2007). https://doi.org/10.1088/1367-2630/9/5/142

75. B. Schröder, M. Sivis, R. Bormann, S. Schäfer, C. Ropers, An ultrafast nanotip electron gun triggered by grating-coupled surface plasmons. Appl. Phys. Lett. **107**(23), 231105 (2015). https://doi.org/10.1063/1.4937121

76. M. Müller, V. Kravtsov, A. Paarmann, M.B. Raschke, R. Ernstorfer, Nanofocused plasmon-driven sub-10 fs electron point Source. Acs Photonics **3**(4), 611–619 (2016). https://doi.org/10.1021/acsphotonics.5b00710

77. K.E. Echternkamp, G. Herink, S.V. Yalunin, K. Rademann, S. Schäfer, C. Ropers, Strong-field photoemission in nanotip near-fields: from quiver to sub-cycle electron dynamics. Appl. Phys. B J **122**(4), 80 (2016). https://doi.org/10.1007/s00340-016-6351-x

78. C. Kealhofer, W. Schneider, D. Ehberger, A. Ryabov, F. Krausz, P. Baum, All-optical control and metrology of electron pulses. Science **352**(6284), 429–433 (2016). https://doi.org/10.1126/science.aae0003

79. A. Gliserin, M. Walbran, P. Baum, A high-resolution time-of-flight energy analyzer for femtosecond electron pulses at 30 keV. Rev. Sci. Instrum. **87**(3), 033302 (2016). https://doi.org/10.1063/1.4942912

80. J. Vogelsang et al., Ultrafast electron emission from a sharp metal nanotaper driven by Adiabatic nanofocusing of surface plasmons. Nano Lett. **15**(7), 4685–4691 (2015). https://doi.org/10.1021/acs.nanolett.5b01513

81. A. Gliserin, A. Apolonski, F. Krausz, P. Baum, Compression of single-electron pulses with a microwave cavity. New J. Phys. **14**, 073055 (2012). https://doi.org/10.1088/1367-2630/14/7/073055. (in English)

82. P.G. Etchegoin, E.C. Le Ru, M. Meyer, An analytic model for the optical properties of gold. J. Chem. Phys. 125, **127**(18), 164705, 189901 (2006, 2007). Doi:https://doi.org/10.1063/1.2802403 (in English)

83. P.G. Etchegoin, E.C. Le Ru, M. Meyer, An analytic model for the optical properties of gold. J. Chem. Phys. **125**(16), 164705 (2006). Doi:https://doi.org/10.1063/1.2360270

84. R.M. Joseph, A. Taflove, FDTD Maxwell's equations models for nonlinear electrodynamics and optics. IEEE Trans. Antennas Propag. **45**(3), 364–374 (1997). https://doi.org/10.1109/8.558652

85. D. Ehberger et al., Highly coherent electron beam from a laser-triggered tungsten needle tip. Phys. Rev. Lett. **114**(22), 227601 (2015). https://doi.org/10.1103/physrevlett.114.227601. (in English)

86. D. Gabor, A new microscopic principle. Nature **161**(4098), 777–778 (1948). https://doi.org/10.1038/161777a0

Bibliography

1. N.W. Ashcroft, N.D. Mermin, in *Solid State Physics*. (Thomson Learning, Inc., United States of America, 1976)
2. J.M. Pitarke, V.M. Silkin, E.V. Chulkov, P.M. Echenique, Theory of surface plasmons and surface-plasmon polaritons (in English). Rep. Prog. Phys. **70**(1), 1–87 (2007). https://doi.org/10.1088/0034-4885/70/1/R01
3. R. Eisberg, R. Resnick, *Quantum Mechanics of Atoms, Molecules, Solids, Nuclei, and Particles (Lecture Notes in chemistry)* (John Wiley & Sons Inc, New York, United States of America, 1974)
4. N.J. Halas, S. Lal, W.S. Chang, S. Link, P. Nordlander, Plasmons in strongly coupled metallic nanostructures (in English). Chem. Rev. **111**(6), 3913–3961 (2011). https://doi.org/10.1021/cr200061k
5. J. Christensen, A. Manjavacas, S. Thongrattanasiri, F.H.L. Koppens, F.J.G. de Abajo, Graphene plasmon waveguiding and hybridization in individual and paired nanoribbons. Acs. Nano. **6**(1), 431–440 (2012). https://doi.org/10.1021/nn2037626
6. P. Nordlander, E. Prodan, Plasmon hybridization in nanoparticles near metallic surfaces. Nano Lett. **4**(11), 2209–2213 (2004). https://doi.org/10.1021/nl0486160
7. N. Liu, H. Guo, L. Fu, S. Kaiser, H. Schweizer, H. Giessen, Plasmon hybridization in stacked cut-wire metamaterials. Adv. Mater. **19**(21), 3628–3632 (2007). https://doi.org/10.1002/adma.200700123
8. X. Lu, M. Rycenga, S.E. Skrabalak, B. Wiley, Y. Xia, Chemical synthesis of novel plasmonic nanoparticles. Annu. Rev. Phys. Chem. **60**(1), 167–192 (2009). https://doi.org/10.1146/annurev.physchem.040808.090434
9. M. Bosman et al., Surface plasmon damping quantified with an electron nanoprobe. Sci. Rep. **3**, 1312 (2013). https://doi.org/10.1038/srep01312, https://www.nature.com/articles/srep01312#supplementary-information
10. B.J. Wiley, S.H. Im, Z.Y. Li, J. McLellan, A. Siekkinen, Y.N. Xia, Maneuvering the surface plasmon resonance of silver nanostructures through shape-controlled synthesis (in English). J. Phys. Chem. B **110**(32), 15666–15675 (2006). https://doi.org/10.1021/jp0608628
11. V. Giannini, A.I. Fernandez-Dominguez, S.C. Heck, S.A. Maier, Plasmonic nanoantennas: fundamentals and their use in controlling the radiative properties of nanoemitters (in English). Chem. Rev. **111**(6), 3888–3912 (2011). https://doi.org/10.1021/cr1002672
12. V. Dziom et al., Observation of the universal magnetoelectric effect in a 3D topological insulator. Nat. Commun. **8**, 15197 (2017). https://doi.org/10.1038/ncomms15197, https://www.nature.com/articles/ncomms15197#supplementary-information
13. T. Morimoto, A. Furusaki, N. Nagaosa, Topological magnetoelectric effects in thin films of topological insulators (in English). Phys. Rev. B **92**(8), 085113 (2015). https://doi.org/10.1103/physrevb.92.085113

© Springer Nature Switzerland AG 2019
N. Talebi, *Near-Field-Mediated Photon–Electron Interactions*,
Springer Series in Optical Sciences 228,
https://doi.org/10.1007/978-3-030-33816-9

14. T. Low et al., Polaritons in layered two-dimensional materials (in English). Nat. Mater. **16**(2), 182–194 (2017). https://doi.org/10.1038/NMAT4792

15. D.N. Basov, M.M. Fogler, F.J.G. de Abajo, Polaritons in van der Waals materials (in English). Science **354**(6309), aag1992. https://doi.org/10.1126/science.aag1992

16. J.S. Wu, D.N. Basov, M.M. Fogler, Topological insulators are tunable waveguides for hyperbolic polaritons (in English). Phys. Rev. B **92**(20), 205430 (2015). https://doi.org/10.1103/physrevb.92.205430

17. T. Stauber, Plasmonics in Dirac systems: from graphene to topological insulators (in English). J. Phys. Condens. Mat. **26**(12), 123201 (2014). https://doi.org/10.1088/0953-8984/26/12/123201

18. A. Manjavacas, F.J.G. de Abajo, Tunable plasmons in atomically thin gold nanodisks (in English). Nat. Commun. **5**, 3548 (2014). https://doi.org/10.1038/ncomms4548

19. F.J. García de Abajo, A. Manjavacas, Plasmonics in atomically thin materials. *Faraday Discuss.* **178**(0), 87–107 (2015). https://doi.org/10.1039/c4fd00216d, https://doi.org/10.1039/c4fd00216d

20. A. Woessner et al., Highly confined low-loss plasmons in graphene-boron nitride heterostructures (in English). Nat. Mater. **14**(4), 421–425 (2015). https://doi.org/10.1038/NMAT4169

21. F.J.G. de Abajo, Graphene plasmonics: challenges and opportunities (in English). Acs Photonics **1**(3), 135–152 (2014). https://doi.org/10.1021/ph400147y

22. Y. Ding et al., Effective Electro-optical modulation with high extinction ratio by a Graphene-Silicon microring resonator. Nano Lett. **15**(7), 4393–4400 (2015). https://doi.org/10.1021/acs.nanolett.5b00630

23. E.W. Hill, A. Vijayaragahvan, K. Novoselov, Graphene sensors (in English). IEEE Sens. J. **11**(12), 3161–3170 (2011). https://doi.org/10.1109/Jsen.2011.2167608

24. A. Vakil, N. Engheta, Transformation optics using graphene. Science **332**(6035), 1291–1294 (2011). https://doi.org/10.1126/science.1202691

25. A. Poddubny, I. Iorsh, P. Belov, Y. Kivshar, Hyperbolic metamaterials (in English). Nat. Photonics **7**(12), 948–957 (2013). https://doi.org/10.1038/Nphoton.2013.243

26. R.S. Kshetrimayum, A brief intro to metamaterials. IEEE Potentials **23**(5), 44–46 (2005). https://doi.org/10.1109/MP.2005.1368916

27. Z. Jacob, I.I. Smolyaninov, E.E. Narimanov, Broadband Purcell effect: radiative decay engineering with metamaterials (in English). Appl. Phys. Lett. **100**(18), 181105 (2012). https://doi.org/10.1063/1.4710548

28. Y.R. He, S.L. He, X.D. Yang, Optical field enhancement in nanoscale slot waveguides of hyperbolic metamaterials (in English). Opt. Lett. **37**(14), 2907–2909 (2012). [Online]. Available: <GotoISI>://WOS:000306709900046

29. E.E. Narimanov, A.V. Kildishev, METAMATERIALS naturally hyperbolic (in English). Nat. Photonics **9**(4), 214–216 (2015). https://doi.org/10.1038/nphoton.2015.56

30. F. Wilczek, Two applications of axion electrodynamics. Phys. Rev. Lett. **58**(18), 1799–1802 (1987). https://doi.org/10.1103/physrevlett.58.1799

31. L. Novotny, The history of near-field optics (in English). Prog. Opt. **50**, 137–184 (2007). https://doi.org/10.1016/S0079-6638(07)50005-3

32. B. Knoll, F. Keilmann, Near-field probing of vibrational absorption for chemical microscopy. Nature **399**(6732), 134–137 (1999). https://doi.org/10.1038/20154

33. B. Knoll, F. Keilmann, Enhanced dielectric contrast in scattering-type scanning near-field optical microscopy. Optics Commun. **182**(4–6), 321–328 (2000). https://doi.org/10.1016/s0030-4018(00)00826-9

34. J. Dorfmuller et al., Near-field dynamics of optical Yagi-Uda nanoantennas (in English). Nano Lett. **11**(7), 2819–2824 (2011). https://doi.org/10.1021/nl201184n

35. R.C. Dunn, Near-field scanning optical microscopy. Chem. Rev. **99**(10), 2891–2928 (1999). https://doi.org/10.1021/cr980130e

36. M.I. Stockman, Nanoplasmonics: past, present, and glimpse into future (in English). Opt. Express 19(22), 22029–22106 (2011). https://doi.org/10.1364/Oe.19.022029
37. M.I. Stockman, Nanofocusing of optical energy in tapered plasmonic waveguides. Phys. Rev. Lett. 93, 137404 (2004) (in English); 106(1), 019901 (2011). https://doi.org/10.1103/physrevlett.106.019901
38. M.I. Stockman, Nanofocusing of optical energy in tapered plasmonic waveguides (in English). Phys. Rev. Lett. 93(13), 137404 (2004). https://doi.org/10.1103/physrevlett.93.137404
39. L. Douillard et al., Short range plasmon resonators probed by photoemission electron microscopy (in English). Nano Lett. 8(3), 935–940 (2008). https://doi.org/10.1021/nl080053v
40. D. Podbiel et al., Imaging the nonlinear plasmoemission dynamics of electrons from strong plasmonic fields (in English). Nano Lett. 17(11), 6569–6574 (2017). https://doi.org/10.1021/acs.nanolett.7b02235
41. W. Quan et al., Quantum interference in laser-induced nonsequential double ionization (in English). Phys. Rev. A 96(3), 032511 (2017). https://doi.org/10.1103/physreva.96.032511
42. C.F.D. Faria, T. Shaaran, X. Liu, W. Yang, Quantum interference in laser-induced nonsequential double ionization in diatomic molecules: role of alignment and orbital symmetry (in English). Phys. Rev. A 78(4), 043407 (2008). https://doi.org/10.1103/physreva.78.043407
43. D. Pengel, S. Kerbstadt, L. Englert, T. Bayer, M. Wollenhaupt, Control of three-dimensional electron vortices from femtosecond multiphoton ionization. Phys. Rev. A 96(4), 043426 (2017). https://doi.org/10.1103/physreva.96.043426
44. L. Seiffert, T. Paschen, P. Hommelhoff, T. Fennel, "High-order above-threshold photoemission from nanotips controlled with two-color laser fields," (in English). J. Phys. B-at Mol. Opt. 51(13), 134001 (2018). https://doi.org/10.1088/1361-6455/aac34f
45. T. Coenen, N.M. Haegel, Cathodoluminescence for the 21st century: learning more from light (in English). Appl. Phys. Rev. 4(3), 031103 (2017). https://doi.org/10.1063/1.4985767
46. M. Kociak, L.F. Zagonel, Cathodoluminescence in the scanning transmission electron microscope (in English). Ultramicroscopy 176, 112–131 (2017). https://doi.org/10.1016/j.ultramic.2017.03.014
47. R. Gómez-Medina, N. Yamamoto, M. Nakano, F.J. García de Abajo, Mapping plasmons in nanoantennas via cathodoluminescence. New J. Phys. 10(10), 105009 (2008). https://doi.org/10.1088/1367-2630/10/10/105009
48. V.L. Ginzburg, I.M. Frank, Radiation of a uniformly moving electron due to its transition from one medium into another. J. Phys. (USSR) 9, 353–362 (1945)
49. T. Coenen, B.J.M. Brenny, E.J. Vesseur, A. Polman, Cathodoluminescence microscopy: optical imaging and spectroscopy with deep-subwavelength resolution. MRS Bull. 40(4), 359–365 (2015). https://doi.org/10.1557/mrs.2015.64
50. A. Petersson, A. Gustafsson, L. Samuelson, S. Tanaka, Y. Aoyagi, Cathodoluminescence spectroscopy and imaging of individual GaN dots (in English). Appl. Phys. Lett. 74(23), 3513–3515 (1999). https://doi.org/10.1063/1.124147
51. L. Kiewidt, M. Karamehmedović, C. Matyssek, W. Hergert, L. Mädler, T. Wriedt, Numerical simulation of electron energy loss spectroscopy using a generalized multipole technique. Ultramicroscopy 133, 101–108 (2013). https://doi.org/10.1016/j.ultramic.2013.07.001
52. T. Wriedt, Y. Eremin, The Generalized Multipole Technique for Light Scattering: Recent Developments, ed. (Springer International Publishing, Switzerland, 2018)
53. Q. Liang et al., Investigating hybridization schemes of coupled split-ring resonators by electron impacts. Opt. Express 23(16), 20721–20731 (2015). https://doi.org/10.1364/oe.23.020721
54. C. Matyssek, J. Niegemann, W. Hergert, K. Busch, Computing electron energy loss spectra with the discontinuous Galerkin time-domain method. Photonics Nanostruct. Fundam. Appli. 9(4), 367–373 (2011). https://doi.org/10.1016/j.photonics.2011.04.003
55. Y. Cao, A. Manjavacas, N. Large, P. Nordlander, electron energy-loss spectroscopy calculation in finite-difference time-domain package. Acs Photonics 2(3), 369–375 (2015). https://doi.org/10.1021/ph500408e

56. S. Guo, N. Talebi, A. Campos, M. Kociak, P.A. van Aken, Radiation of dynamic toroidal moments. Acs Photonics (2019), https://doi.org/10.1021/acsphotonics.8b01422

57. N. Talebi, S. Guo, A. van Aken Peter, Theory and applications of toroidal moments in electrodynamics: their emergence, characteristics, and technological relevance. Nanophotonics **7**, 93 (2018)

58. S. Guo, N. Talebi, P.A. van Aken, Long-range coupling of toroidal moments for the visible. Acs Photonics **5**(4), 1326–1333 (2018). https://doi.org/10.1021/acsphotonics.7b01313

59. O.L. Krivanek et al., Vibrational spectroscopy in the electron microscope. Nature **514**, 209 (2014). https://doi.org/10.1038/nature13870

60. N. Talebi Sarvari, Method and devices for time-resolved pump-probe electron microscopy. USA Patent Appl. US20170271123A1 (2018)

61. K. Mizuno, J. Pae, T. Nozokido, K. Furuya, Experimental evidence of the inverse Smith-Purcell effect. Nature **328**(6125), 45–47 (1987). https://doi.org/10.1038/328045a0

62. J. Vogelsang et al., Plasmonic-nanofocusing-based electron holography. Acs Photonics **5**(9), 3584–3593 (2018). https://doi.org/10.1021/acsphotonics.8b00418

63. S.N. Lyle, *Self Force and Inertia, Old Light on New Ideas* (Springer Verlag, Heidelberg, 2010)

64. J.D. Jackson, *Classical Electrodynamics* (John Wiley and Sons Inc, USA, 1998)

65. I.B. Zeldovich, Electromagnetic interaction with parity violation (in English). Sov. Phys. Jetp-Ussr **6**(6), 1184–1186 (1958). [Online]. Available: <GotoISI>://WOS:A1958WT97900044

66. V.M. Dubovik, V.V. Tugushev, Toroid moments in electrodynamics and solid-state physics (in English). Phys. Rep. **187**(4), 145–202 (1990). https://doi.org/10.1016/0370-1573(90)90042-Z

67. N. Papasimakis, V.A. Fedotov, V. Savinov, T.A. Raybould, N.I. Zheludev, Electromagnetic toroidal excitations in matter and free space (in English). Nat. Mater. **15**(3), 263–271 (2016). https://doi.org/10.1038/NMAT4563

68. V.M. Dubovik, L.A. Tosunian, V.V. Tugushev, Axial toroidal moments in electrodynamics and solid-state physics (in Russian). Zh Eksp Teor Fiz+, **90**(2), 590–605 (1986). [Online]. Available: <GotoISI>://WOS:A1986A438000018

69. K. Marinov, A.D. Boardman, V.A. Fedotov, N. Zheludev, Toroidal metamaterial (in English). New J. Phys. **9**, 324 (2007). https://doi.org/10.1088/1367-2630/9/9/324

70. N.A. Spaldin, M. Fiebig, M. Mostovoy, The toroidal moment in condensed-matter physics and its relation to the magnetoelectric effect (in English). J. Phys. Condens. Mat. **20**(43), 434203 (2008). https://doi.org/10.1088/0953-8984/20/43/434203

71. A.S. Zimmermann, D. Meier, M. Fiebig, Ferroic nature of magnetic toroidal order (in English). Nat. Commun. **5**, 4796 (2014). https://doi.org/10.1038/ncomms5796

72. V.A. Fedotov, A.V. Rogacheva, V. Savinov, D.P. Tsai, N.I. Zheludev, Resonant transparency and non-trivial non-radiating excitations in toroidal metamaterials (in English). Sci. Rep. **3**, 2967 (2013). https://doi.org/10.1038/srep02967

73. V. Savinov, V.A. Fedotov, N.I. Zheludev, Toroidal dipolar excitation and macroscopic electromagnetic properties of metamaterials (in English). Phys. Rev. B **89**(20), 205112 (2014). https://doi.org/10.1103/physrevb.89.205112

74. T. Kaelberer, V.A. Fedotov, N. Papasimakis, D.P. Tsai, N.I. Zheludev, Toroidal dipolar response in a metamaterial (in English). Science **330**(6010), 1510–1512 (2010). https://doi.org/10.1126/science.1197172

75. Y.W. Huang et al., Design of plasmonic toroidal metamaterials at optical frequencies (in English). Opt. Express **20**(2), 1760–1768 (2012). https://doi.org/10.1364/Oe.20.001760

76. G. Thorner, J.M. Kiat, C. Bogicevic, I. Kornev, Axial hypertoroidal moment in a ferroelectric nanotorus: a way to switch local polarization (in English). Phys. Rev. B **89**(22), 220103 (2014). https://doi.org/10.1103/physrevb.89.220103

77. J.D. Jackson, From Lorenz to Coulomb and other explicit gauge transformations (in English). Am. J. Phys. **70**(9), 917–928 (2002). https://doi.org/10.1119/1.1491265

78. C. Vrejoiu, Electromagnetic multipoles in Cartesian coordinates (in English). J. Phys. a-Math. Gen. **35**(46), 9911–9922 (2002); Pii S0305–4470(02), 39273–4. https://doi.org/10.1088/0305-4470/35/46/313

79. A. Gongora T., E. Ley-Koo, Complete electromagnetic multipole expansion including toroidal moments. *Revista Mexicana De Fisica E* **52**, 188–197 (2006)

80. C.G. Gray, Multipole expansions of electromagnetic-fields using debye potentials (in English). Am. J. Phys. **46**(2), 169–179 (1978). https://doi.org/10.1119/1.11364

81. N.A. Spaldin, M. Fechner, E. Bousquet, A. Balatsky, L. Nordstrom, Monopole-based formalism for the diagonal magnetoelectric response (in English). Phys. Rev. B **88**(9), 094429 (2013). https://doi.org/10.1103/physrevb.88.094429

82. J. Preskill, Magnetic monopoles (in English). Annu. Rev. Nucl. Part S **34**, 461–530 (1984). [Online]. Available: <GotoISI>://WOS:A1984TU83600012

83. Y.A. Artamonov, A.A. Gorbatsevich, Symmetry and dynamics of systems with toroidal moments (in Russian). Zh Eksp Teor Fiz+ **89**(3), 1078–1093 (1985). [Online]. Available: <GotoISI>://WOS:A1985ASL8100033

84. E.E. Radescu, G. Vaman, Exact calculation of the angular momentum loss, recoil force, and radiation intensity for an arbitrary source in terms of electric, magnetic, and toroid multipoles. Phys. Rev. E Stat. Nonlinear Soft Mat. Phys. **65**(4), 046609 (2002). https://doi.org/10.1103/physreve.65.046609

85. A.S. Schwanecke, V.A. Fedotov, V.V. Khardikov, S.L. Prosvirnin, Y. Chen, N.I. Zheludev, Nanostructured metal film with asymmetric optical transmission (in English). Nano Lett. **8**(9), 2940–2943 (2008). https://doi.org/10.1021/nl801794d

86. S. Nanz, *Toroidal Multipole Moments in Classical Electrodynamics* (Springer Spektrum, Wiesbaden, 2016)

87. N. Papasimakis, V.A. Fedotov, V. Savinov, T.A. Raybould, N.I. Zheludev, Electromagnetic toroidal excitations in matter and free space. Nat. Mater. **15**(3), 263–271 (2016). https://doi.org/10.1038/nmat4563

88. N. Papasimakis, V.A. Fedotov, K. Marinov, N.I. Zheludev, Gyrotropy of a metamolecule: wire on a torus. Phys. Rev. Lett. **103**(9), 093901 (2009). https://doi.org/10.1103/physrevlett.103.093901

89. K. Marinov, A.D. Boardman, V.A. Fedotov, N. Zheludev, Toroidal metamaterial. New J. Phys. **9**(9), 324 (2007). http://stacks.iop.org/1367-2630/9/i=9/a=324

90. Y.-W. Huang et al., Design of plasmonic toroidal metamaterials at optical frequencies. Opt. Express **20**(2), 1760–1768 (2012). https://doi.org/10.1364/oe.20.001760

91. T.A. Raybould et al., Toroidal circular dichroism. Phys. Rev. B **94**(3), 035119 (2016). http://link.aps.org/doi/10.1103/PhysRevB.94.035119

92. S. Han, L. Cong, F. Gao, R. Singh, H. Yang, Observation of Fano resonance and classical analog of electromagnetically induced transparency in toroidal metamaterials. Annalen der Physik **528**(5), 352–357 (2016). https://doi.org/10.1002/andp.201600016

93. Y. Fan, Z. Wei, H. Li, H. Chen, C.M. Soukoulis, Low-loss and high-Qplanar metamaterial with toroidal moment. Phys. Rev. B **87**(11) (2013). https://doi.org/10.1103/physrevb.87.115417

94. A.A. Basharin, V. Chuguevsky, N. Volsky, M. Kafesaki, E.N. Economou, Extremely high Q-factor metamaterials due to anapole excitation. Phys. Rev. B **95**(3), 035104 (2017). http://link.aps.org/doi/10.1103/PhysRevB.95.035104

95. Y. Bao, X. Zhu, Z. Fang, Plasmonic toroidal dipolar response under radially polarized excitation. Sci. Rep. **5**, 11793 (2015). https://doi.org/10.1038/srep11793 http://www.nature.com/articles/srep11793#supplementary-information

96. C. Tang et al., Toroidal dipolar response in metamaterials composed of metal–dielectric–metal sandwich magnetic resonators. IEEE Photonics J. **8**(3), 1–9 (2016). https://doi.org/10.1109/JPHOT.2016.2574865

97. Z.-G. Dong et al., Optical toroidal dipolar response by an asymmetric double-bar metamaterial. Appl. Phys. Lett. **101**(14), 144105 (2012). https://doi.org/10.1063/1.4757613

98. J. Li et al., Optical responses of magnetic-vortex resonance in double-disk metamaterial variations. Phys. Lett. A **378**(26–27), 1871–1875 (2014). https://doi.org/10.1016/j.physleta.2014.04.049

99. V.A. Fedotov, A.V. Rogacheva, V. Savinov, D.P. Tsai, N.I. Zheludev, Resonant transparency and non-trivial non-radiating excitations in toroidal metamaterials. Sci. Rep. **3**, 2967 (2013). https://doi.org/10.1038/srep02967

100. L.-Y. Guo, M.-H. Li, X.-J. Huang, H.-L. Yang, Electric toroidal metamaterial for resonant transparency and circular cross-polarization conversion. Appl. Phys. Lett. **105**(3), 033507 (2014). https://doi.org/10.1063/1.4891643

101. P.C. Wu et al., Vertical split-ring resonators for plasmon coupling, sensing and metasurface **9544**, 954423–954423–4. [Online]. (2015). http://dx.doi.org/10.1117/12.2189249, http://dx.doi.org/10.1117/12.2189249

102. A.E. Miroshnichenko et al., Nonradiating anapole modes in dielectric nanoparticles. Nat. Commun. 6, 8069 (2015). https://doi.org/10.1038/ncomms9069, https://www.nature.com/articles/ncomms9069#supplementary-information

103. A.A. Basharin et al., Dielectric metamaterials with toroidal dipolar response. Phys. Rev. X **5**(1), 011036 (2015). http://link.aps.org/doi/10.1103/PhysRevX.5.011036

104. J. Li, J. Shao, Y.-H. Wang, M.-J. Zhu, J.-Q. Li, Z.-G. Dong, Toroidal dipolar response by a dielectric microtube metamaterial in the terahertz regime. Opt. Express **23**(22), 29138–29144 (2015). https://doi.org/10.1364/oe.23.029138

105. W. Liu, J. Zhang, A.E. Miroshnichenko, Toroidal dipole-induced transparency in core–shell nanoparticles. Laser Photonics Rev. **9**(5), 564–570 (2015). https://doi.org/10.1002/lpor.201500102

106. Q. Zhang, J.J. Xiao, X.M. Zhang, D. Han, L. Gao, Core–shell-structured dielectric-metal circular nanodisk antenna: gap plasmon assisted magnetic toroid-like cavity modes. Acs Photonics **2**(1), 60–65 (2015). https://doi.org/10.1021/ph500229p

107. J. Li, Y. Zhang, R. Jin, Q. Wang, Q. Chen, Z. Dong, Excitation of plasmon toroidal mode at optical frequencies by angle-resolved reflection. Opt. Lett. **39**(23), 6683–6686 (2014). https://doi.org/10.1364/ol.39.006683

108. P.C. Wu et al., Plasmon coupling in vertical split-ring resonator metamolecules. Sci. Rep. **5**, 9726 (2015). https://doi.org/10.1038/srep09726

109. Z.-G. Dong, J. Zhu, X. Yin, J. Li, C. Lu, X. Zhang, All-optical hall effect by the dynamic toroidal moment in a cavity-based metamaterial. Phys. Rev. B **87**(24), 245429 (2013). http://link.aps.org/doi/10.1103/PhysRevB.87.245429

110. P.R. Wu et al., Horizontal toroidal response in three-dimensional plasmonic (conference presentation) **9921**, 992120–992121. http://dx.doi.org/10.1117/12.2236879, http://dx.doi.org/10.1117/12.2236879

111. C.Y. Liao et al., Optical toroidal response in three-dimensional plasmonic metamaterial **9547**, 954724–954724–4 (2015). http://dx.doi.org/10.1117/12.2189052, http://dx.doi.org/10.1117/12.2189052

112. V.C. Alexey, A. Basharin, N. Volsky, M. Kafesaki, E.N. Economou, Extremely High Q-factor metamaterials due to anapole excitation. [Online]. Available: arXiv:1608.03233[physics.class-ph]

113. F. Roder, H. Lichte, Inelastic electron holography—first results with surface plasmons (in English). Eur. Phys. J. Appl. Phys. **54**(3), 33504 (2011). https://doi.org/10.1051/epjap/2010100378

114. X. Zhang et al., Asymmetric excitation of surface plasmons by dark mode coupling. Sci. Adv. **2**(2) (2016). https://doi.org/10.1126/sciadv.1501142

115. S.J. Barrow, D. Rossouw, A.M. Funston, G.A. Botton, P. Mulvaney, Mapping bright and dark modes in gold nanoparticle chains using electron energy loss spectroscopy. Nano Lett. **14**(7), 3799–3808 (2014). https://doi.org/10.1021/nl5009053

116. A.L. Koh et al., Electron energy-loss spectroscopy (EELS) of surface plasmons in single silver nanoparticles and dimers: influence of beam damage and mapping of dark modes. ACS Nano **3**(10), 3015–3022 (2009). https://doi.org/10.1021/nn900922z

117. G. Fletcher, M.D. Arnold, T. Pedersen, V.J. Keast, M.B. Cortie, Multipolar and dark-mode plasmon resonances on drilled silver nano-triangles. Opt. Express **23**(14), 18002–18013 (2015). https://doi.org/10.1364/oe.23.018002

118. S.A. Maier, Plasmonics: the benefits of darkness. Nat. Mater. **8**(9), 699–700. [Online]. (2009). https://doi.org/10.1038/nmat2522, http://dx.doi.org/10.1038/nmat2522

119. W.H. Yang, C. Zhang, S. Sun, J. Jing, Q. Song, S. Xiao, Dark plasmonic mode based perfect absorption and refractive index sensing. Nanoscale **9**(26), 8907–8912 (2017). https://doi.org/10.1039/c7nr02768k

120. W. Zhou, T.W. Odom, Tunable subradiant lattice plasmons by out-of-plane dipolar interactions. Nat. Nanotechnol. **6**, 423 (2011). https://doi.org/10.1038/nnano.2011.72, https://www.nature.com/articles/nnano.2011.72#supplementary-information

121. M.-W. Chu, V. Myroshnychenko, C.H. Chen, J.-P. Deng, C.-Y. Mou, F.J. García de Abajo, Probing bright and dark surface-plasmon modes in individual and coupled noble metal nanoparticles using an electron beam. Nano Lett. **9**(1), 399–404 (2009). https://doi.org/10.1021/nl803270x

122. F.-P. Schmidt, A. Losquin, F. Hofer, A. Hohenau, J.R. Krenn, M. Kociak, How dark are radial breathing modes in plasmonic nanodisks? Acs Photonics **5**(3), 861–866 (2018). https://doi.org/10.1021/acsphotonics.7b01060

123. M. Gupta et al., Sharp toroidal resonances in planar terahertz metasurfaces. Adv. Mater. **28**(37), 8206–8211 (2016). https://doi.org/10.1002/adma.201601611

124. B. Han, X. Li, C. Sui, J. Diao, X. Jing, Z. Hong, Analog of electromagnetically induced transparency in an E-shaped all-dielectric metasurface based on toroidal dipolar response. Opt. Mater. Express **8**(8), 2197–2207 (2018). https://doi.org/10.1364/ome.8.002197

125. D.W. Watson, S.D. Jenkins, J. Ruostekoski, V.A. Fedotov, N.I. Zheludev, Toroidal dipole excitations in metamolecules formed by interacting plasmonic nanorods. Phys. Rev. B **93**(12) (2016). https://doi.org/10.1103/physrevb.93.125420

126. L. Zhu et al., A low-loss electromagnetically induced transparency (EIT) metamaterial based on coupling between electric and toroidal dipoles. RSC Adv. **7**(88), 55897–55904 (2017). https://doi.org/10.1039/c7ra11175d

127. L. Wei, Z. Xi, N. Bhattacharya, H.P. Urbach, Excitation of the radiationless anapole mode. Optica **3**(8), 799–802 (2016). https://doi.org/10.1364/optica.3.000799

128. J.S. Totero Góngora, A.E. Miroshnichenko, Y.S. Kivshar, A. Fratalocchi, Anapole nanolasers for mode-locking and ultrafast pulse generation. Nat. Commun. **8**, 15535 (2017). https://doi.org/10.1038/ncomms15535, http://dharmasastra.live.cf.private.springer.com/articles/ncomms15535#supplementary-information

129. S.-J. Kim, S.-E. Mun, Y. Lee, H. Park, J. Hong, B. Lee, Nanofocusing of toroidal dipole for simultaneously enhanced electric and magnetic fields using plasmonic waveguide. J. Lightwave Technol. **36**(10), 1882–1889 (2018). http://jlt.osa.org/abstract.cfm?URI=jlt-36-10-1882

130. A. Ahmadivand et al., Rapid detection of infectious envelope proteins by magnetoplasmonic toroidal metasensors. ACS Sens. **2**(9), 1359–1368 (2017). https://doi.org/10.1021/acssensors.7b00478

131. P.C. Wu et al., Optical anapole metamaterial. ACS Nano **12**(2), 1920–1927 (2018). https://doi.org/10.1021/acsnano.7b08828

132. Y. Wu, G. Li, J.P. Camden, Probing nanoparticle plasmons with electron energy loss spectroscopy. Chem. Rev. (2017). https://doi.org/10.1021/acs.chemrev.7b00354

133. M. Kociak, L.F. Zagonel, Cathodoluminescence in the scanning transmission electron microscope. Ultramicroscopy **174**, 50–69 (2017). https://doi.org/10.1016/j.ultramic.2016.11.018

134. A. Losquin et al., Unveiling nanometer scale extinction and scattering phenomena through combined electron energy loss spectroscopy and cathodoluminescence measurements. Nano Lett. **15**(2), 1229–1237 (2015). https://doi.org/10.1021/nl5043775

135. N. Liu, H. Giessen, Coupling effects in optical metamaterials (in English). Angew. Chem. Int. Edit **49**(51), 9838–9852 (2010). https://doi.org/10.1002/anie.200906211

136. J.A. Heras, Electric and magnetic fields of a toroidal dipole in arbitrary motion (in English). Phys. Lett. A **249**(1–2), 1–9 (1998). https://doi.org/10.1016/S0375-9601(98)00712-9

137. C. Ropers, C.C. Neacsu, T. Elsaesser, M. Albrecht, M.B. Raschke, C. Lienau, Grating-coupling of surface plasmons onto metallic tips: a nanoconfined light source. Nano Lett. **7**(9), 2784–2788 (2007). https://doi.org/10.1021/nl071340m

138. S. Grésillon et al., Experimental observation of localized optical excitations in random metal-dielectric films. Phys. Rev. Lett. **82**(22), 4520–4523 (1999). https://doi.org/10.1103/physrevlett.82.4520

139. P.G. Etchegoin, E.C.L. Ru, M. Meyer, An analytic model for the optical properties of gold. J. Chem. Phys. **125**(16), 164705 (2006). https://doi.org/10.1063/1.2360270

140. F. Huth et al., Resonant antenna probes for tip-enhanced infrared near-field microscopy. Nano Lett. **13**(3), 1065–1072 (2013). https://doi.org/10.1021/nl304289g

141. L.B. Felsen, N. Marcuvitz, *Radiation and Scattering of Waves*. (Prentice Hall, Englewood Cliff, New Jersey, 1973)

142. J.J. Bowman, T.B.A. Senior, P.L.E. Uslenghi, *Electromagnetic and Acoustic Scattering by Simple Shapes*. (Hemisphere, New York, 1987)

143. M.A. Lyalinov, Electromagnetic scattering by a circular impedance cone: diffraction coefficients and surface waves. IMA J. Appl. Math. **79**(3), 393–430 (2014). https://doi.org/10.1093/imamat/hxs072

144. K. Kurihara, A. Otomo, A. Syouji, J. Takahara, K. Suzuki, S. Yokoyama, Superfocusing modes of surface plasmon polaritons in conical geometry based on the quasi-separation of variables approach. J. Phys. A Math. Theor. **40**(41), 12479–12503 (2007). https://doi.org/10.1088/1751-8113/40/41/015

145. J.C. Ashley, L.C. Emerson, Dispersion relations for non-radiative surface plasmons on cylinders. Surf. Sci. **41**(2), 615–618 (1974). https://doi.org/10.1016/0039-6028(74)90080-6

146. C.A. Pfeiffer, E.N. Economou, K.L. Ngai, Surface polaritons in a circularly cylindrical interface: surface plasmons. Phys. Rev. B **10**(8), 3038–3051. https://doi.org/10.1103/physrevb.10.3038

147. L. Novotny, C. Hafner, Light propagation in a cylindrical waveguide with a complex, metallic, dielectric function. Phys. Rev. E **50**(5), 4094–4106 (1994). https://doi.org/10.1103/physreve.50.4094

148. N. Issa, R. Guckenberger, Optical nanofocusing on tapered metallic waveguides. Plasmonics **2**, 31–37 (2007)

149. M. Esmann et al., K-space imaging of the eigenmodes on a sharp gold taper for near-field scanning optical microscopy. Beilstein J. Nanotechnol. **4**, 603–610 (2013)

150. J.A. Stratton, *Electromagnetic Theory*. (Wiley, 2007)

151. R.F. Harrington, *Time-Harmonic Electromagnetic Fields* (McGraw-Hill Book Company, New York, 1961)

152. J.C. Ashley, L.C. Emerson, Dispersion-relations for non-radiative surface plasmons on cylinders. Surf. Sci. **41**(2), 615–618 (1974). https://doi.org/10.1016/0039-6028(74)90080-6

153. C.A. Pfeiffer, E.N. Economou, K.L. Ngai, Surface polaritons in a circularly cylindrical interface—surface plasmons. Phys. Rev. B **10**(8), 3038–3051 (1974). https://doi.org/10.1103/physrevb.10.3038

154. F.J.G. de Abajo, M. Kociak, Probing the photonic local density of states with electron energy loss spectroscopy (in English). Phys. Rev. Lett. **100**(10), 4, 106804 (2008). https://doi.org/10.1103/physrevlett.100.106804

155. C.C. Neacsu, S. Berweger, R.L. Olmon, L.V. Saraf, C. Ropers, M.B. Raschke, Near-field localization in plasmonic superfocusing: a nanoemitter on a tip. Nano Lett. **10**(2), 592–596 (2010). https://doi.org/10.1021/nl903574a

156. M. Esmann et al., k-space imaging of the eigenmodes of sharp gold tapers for scanning near-field optical microscopy (in English). Beilstein J. Nanotech. **4**, 603–610 (2013). https://doi.org/10.3762/Bjnano.4.67

157. S. Thomas, G. Wachter, C. Lemell, J. Burgdörfer, P. Hommelhoff, Large optical field enhancement for nanotips with large opening angles. New J. Phys. **17**(6), 063010 (2015). http://stacks.iop.org/1367-2630/17/i=6/a=063010

158. B. Schröder et al., Real-space imaging of nanotip plasmons using electron energy loss spectroscopy. Phys. Rev. B **92**(8), 085411 (2015). https://doi.org/10.1103/PhysRevB.92.085411

159. S.V. Yalunin, B. Schröder, C. Ropers, Theory of electron energy loss near plasmonic wires, nanorods, and cones. Phys. Rev. B **93**(11), 115408 (2016). http://link.aps.org/doi/10.1103/PhysRevB.93.115408

160. A.J. Babadjanyan, N.L. Margaryan, K.V. Nerkararyan, Superfocusing of surface polaritons in the conical structure. J. Appl. Phys. **87**(8), 3785–3788 (2000). https://doi.org/10.1063/1.372414

161. M. Stockman, Nanofocusing of optical energy in tapered plasmonic waveguides. Phys. Rev. Lett. **93**(13) (2004). https://doi.org/10.1103/physrevlett.93.137404

162. M.S. Jang, H. Atwater, Plasmonic rainbow trapping structures for light localization and spectrum splitting. Phys. Rev. Lett. **107**(20), 207401 (2011). http://link.aps.org/doi/10.1103/PhysRevLett.107.207401

163. P.B. Johnson, R.W. Christy, Optical constants of the noble metals. Phys. Rev. B **6**(12), 4370–4379 (1972). http://link.aps.org/doi/10.1103/PhysRevB.6.4370

164. V. Kravtsov, J.M. Atkin, M.B. Raschke, Group delay and dispersion in adiabatic plasmonic nanofocusing. Opt. Lett. **38**(8), 1322–1324 (2013). https://doi.org/10.1364/ol.38.001322

165. G. Richter, K. Hillerich, D.S. Gianola, R. Mönig, O. Kraft, C.A. Volkert, Ultrahigh strength single crystalline nanowhiskers grown by physical vapor deposition. Nano Lett. **9**(8), 3048–3052 (2009). https://doi.org/10.1021/nl9015107

166. X. Zhou, A. Hörl, A. Trügler, U. Hohenester, T.B. Norris, A.A. Herzing, Effect of multipole excitations in electron energy-loss spectroscopy of surface plasmon modes in silver nanowires. J. Appl. Phys. **116**(22), 223101 (2014). https://doi.org/10.1063/1.4903535

167. R.F. Harrington, *Time-Harmonic Electromagnetic Fields* (Wiley-IEEE Press, New York, 2001)

168. M. Fiebig, Revival of the magnetoelectric effect. J. Phys. D Appl. Phys. **38**(8), R123–R152 (2005). https://doi.org/10.1088/0022-3727/38/8/r01

169. S. Bassiri, C.H. Papas, N. Engheta, Electromagnetic-wave propagation through a dielectric-chiral interface and through a chiral slab (in English). J. Opt. Soc. Am. A **5**(9), 1450–1459 (1988). https://doi.org/10.1364/Josaa.5.001450

170. K. Robbie, M.J. Brett, A. Lakhtakia, Chiral sculptured thin films. Nature **384**(6610), 616 (1996). https://doi.org/10.1038/384616a0

171. A. Karch, Surface plasmons and topological insulators. Phys. Rev. B **83**(24), 245432 (2011). https://doi.org/10.1103/physrevb.83.245432

172. B.D.H. Tellegen, The Gyrator, a new electric network element (in English). Philips Res. Rep. **3**(2), 81–101. [Online]. (1948). Available: <GotoISI>://WOS:A1948UY40600001

173. B.D.H. Tellegen, The synthesis of passive, resistanceless 4-Poles that may violate the reciprocity relation (in English). Philips Res. Rep. **3**(5), 321–337. [Online]. (1948). Available: <GotoISI>://WOS:A1948UY40900001

174. A.M. Essin, J.E. Moore, D. Vanderbilt, Magnetoelectric polarizability and axion electrodynamics in crystalline insulators. Phys. Rev. Lett. **102**(14), 146805 (2009). https://doi.org/10.1103/physrevlett.102.146805

175. F.W. Hehl, Y.N. Obukhov, J.-P. Rivera, H. Schmid, Relativistic analysis of magnetoelectric crystals: extracting a new 4-dimensional P odd and T odd pseudoscalar from Cr2O3 data. Phys. Lett. A **372**(8), 1141–1146 (2008). https://doi.org/10.1016/j.physleta.2007.08.069

176. J. Lekner, Optical properties of isotropic chiral media. Pure Appl. Opt. J. Eur. Opt. Soc. Part A **5**(4), 417–443 (1996). https://doi.org/10.1088/0963-9659/5/4/008

177. C.M. Soukoulis, M. Wegener, Past achievements and future challenges in the development of three-dimensional photonic metamaterials. Nat. Photonics Rev. **5**, 523 (2011). https://doi.org/10.1038/nphoton.2011.154

178. M.-C. Chang, M.-F. Yang, Optical signature of topological insulators. Phys. Rev. B **80**(11), 113304 (2009). https://doi.org/10.1103/physrevb.80.113304

179. M. Kozak, T. Eckstein, N. Schonenberger, P. Hommelhoff, Inelastic ponderomotive scattering of electrons at a high-intensity optical travelling wave in vacuum. Nat. Phys. Lett. Advance online publication 10/09/2017. https://doi.org/10.1038/nphys4282, http://www.nature.com/nphys/journal/vaop/ncurrent/abs/nphys4282.html#supplementary-information

180. D.Y. Sergeeva, A.P. Potylitsyn, A.A. Tishchenko, M.N. Strikhanov, Smith-Purcell radiation from periodic beams (in English). Opt. Express **25**(21), 26310–26328 (2017). https://doi.org/10.1364/Oe.25.026310

181. S. Tsesses, G. Bartal, I. Kaminer, Light generation via quantum interaction of electrons with periodic nanostructures (in English). Phys. Rev. A **95**(1), 013832 (2017). https://doi.org/10.1103/physreva.95.013832

182. I. Kaminer et al., Quantum cerenkov radiation: spectral cutoffs and the role of spin and orbital angular momentum (in English). Phys. Rev. X **6**(1), 011006 (2016). https://doi.org/10.1103/physrevx.6.011006

183. J. Peatross, C. Muller, K.Z. Hatsagortsyan, C.H. Keitel, Photoemission of a single-electron wave packet in a strong laser field (in English). Phys. Rev. Lett. **100**(15), 153601 (2018). https://doi.org/10.1103/physrevlett.100.153601

184. A.J. White, M. Sukharev, M. Galperin, Molecular nanoplasmonics: self-consistent electrodynamics in current-carrying junctions. Phys. Rev. B **86**(20), 205324 (2012). https://link.aps.org/doi/10.1103/PhysRevB.86.205324

185. W. Kohn, L.J. Sham, Self-consistent equations including exchange and correlation effects (in English). Phys. Rev. **140**(4A), 1133 (1965). https://doi.org/10.1103/physrev.140.a1133

186. O. Smirnova, M. Spanner, M. Ivanov, Analytical solutions for strong field-driven atomic and molecular one- and two-electron continua and applications to strong-field problems (in English). Phys. Rev. A **77**(3), 033407 (2008). https://doi.org/10.1103/physreva.77.033407

187. D.M. Wolkow, On a mass of solutions of the Dirac equation (in German). Zeitschrift Fur Physik **94**(3–4), 250–260 (1935). https://doi.org/10.1007/bf01331022

188. E. Kasper, Generalization of Schrodingers wave mechanics for relativistic regions of validity (in German). Z Naturforsch A **A28**(2), 216–221. [Online]. (1973). Available: <GotoISI>://WOS:A1973S611900009

189. S.T. Park, "Propagation of a relativistic electron wave packet in the Dirac equation (in English). Phys. Rev. A **86**(6), 062105 (2012). https://doi.org/10.1103/physreva.86.062105

190. F.J.G. de Abajo, A. Asenjo-Garcia, M. Kociak, Multiphoton absorption and emission by interaction of swift electrons with evanescent light fields (in English). Nano Lett. **10**(5), 1859–1863 (2010). https://doi.org/10.1021/nl100613s

191. D. Wolf et al., 3D magnetic induction maps of nanoscale materials revealed by electron holographic tomography (in English). Chem. Mater. **27**(19), 6771–6778 (2015). https://doi.org/10.1021/acs.chemmater.5b02723

192. R.O. Jones, O. Gunnarsson, The density functional formalism, its applications and prospects (in English). Rev. Mod. Phys. **61**(3), 689–746 (1989). https://doi.org/10.1103/RevModPhys.61.689

193. E.J. Baerends, Perspective on self-consistent equations including exchange and correlation effects (in English). Theor. Chem. Acc **103**(3–4), 265–269 (2000); W. Kohn, L.J. Sham, Phys. Rev. A **140**, 133–1138 (1965). https://doi.org/10.1007/s002140050031

194. B. Walker, R. Gebauer, Ultrasoft pseudopotentials in time-dependent density-functional theory (in English). J. Chem. Phys. **127**(16), 164106 (2007). https://doi.org/10.1063/1.2786999

195. J. Harris, R.O. Jones, Pseudopotentials in density-functional theory (in English). Phys. Rev. Lett. **41**(3), 191–194 (1978). https://doi.org/10.1103/PhysRevLett.41.191

196. E. Runge, E.K.U. Gross, Density-functional theory for time-dependent systems (in English). Phys. Rev. Lett. **52**(12), 997–1000 (1984). https://doi.org/10.1103/PhysRevLett.52.997

197. X.S. Li, S.M. Smith, A.N. Markevitch, D.A. Romanov, R.J. Levis, H.B. Schlegel, A time-dependent Hartree-Fock approach for studying the electronic optical response of molecules in intense fields (in English). Phys. Chem. Chem. Phys. **7**(2), 233–239 (2005). https://doi.org/10.1039/b415849k

198. P.W. Hawkes, E. Kasper, *Principles of Electron Optics*. (Academic Press, London, 1996)

199. H. Tal-Ezer, R. Kosloff, An accurate and efficient scheme for propagating the time dependent Schrödinger equation. J. Chem. Phys. **81**(9), 3967–3971 (1984). https://doi.org/10.1063/1.448136

200. N. Talebi, C. Lienau, Interference between quantum paths in coherent Kapitza-Dirac effect. New J. Phys. (2019). http://iopscience.iop.org/10.1088/1367-2630/ab3ce3

201. X.J. Shen, A. Lozano, W. Dong, H.F. Busnengo, X.H. Yan, Towards bond selective chemistry from first principles: methane on metal surfaces. Phys. Rev. Lett. **112**(4), 046101 (2014). https://link.aps.org/doi/10.1103/PhysRevLett.112.046101

202. L. Gaudreau et al., Coherent control of three-spin states in a triple quantum dot. Nat. Phys. **8**, 54 (2011). https://doi.org/10.1038/nphys2149, https://www.nature.com/articles/nphys2149#supplementary-information

203. J. Hansom et al., Environment-assisted quantum control of a solid-state spin via coherent dark states. Nat. Phys. **10**, 725 (2014). https://doi.org/10.1038/nphys3077, https://www.nature.com/articles/nphys3077#supplementary-information

204. I.S. Mark, Ultrafast nanoplasmonics under coherent control. New J. Phys. **10**(2), 025031 (2008). http://stacks.iop.org/1367-2630/10/i=2/a=025031

205. R.P. Feynman, Space-time approach to non-relativistic quantum mechanics. Rev. Mod. Phys. **20**(2), 367–387 (1948). https://link.aps.org/doi/10.1103/RevModPhys.20.367

206. M. Li et al., Classical-quantum correspondence for above-threshold ionization. Phys. Rev. Lett. **112**(11), 113002 (2014). https://link.aps.org/doi/10.1103/PhysRevLett.112.113002

207. D.B. Milošević, W. Becker, Improved strong-field approximation and quantum-orbit theory: application to ionization by a bicircular laser field. Phys. Rev. A **93**(6), 063418 (2016). https://link.aps.org/doi/10.1103/PhysRevA.93.063418

208. A. Zaïr et al., Quantum path interferences in high-order harmonic generation. Phys. Rev. Lett. **100**(14), 143902 (2008). https://link.aps.org/doi/10.1103/PhysRevLett.100.143902

209. P. Salieres et al., Feynman's path-integral approach for intense-laser-atom interactions (in English). Science **292**(5518), 902–905 (2001). https://doi.org/10.1126/science.108836

210. T.C. Weinacht, J. Ahn, P.H. Bucksbaum, Controlling the shape of a quantum wavefunction. Nature **397**, 233 (1999). https://doi.org/10.1038/16654

211. P.L. Kapitza, P.A.M. Dirac, The reflection of electrons from standing light waves. Math. Proc. Cambridge Philos. Soc. **29**(2), 297–300 (2008). https://doi.org/10.1017/S0305004100011105

212. A. Howie, Photon interactions for electron microscopy applications. Eur. Phys. J. Appl. Phys. **54**(3), 33502 (2011). https://doi.org/10.1051/epjap/2010100353

213. H. Batelaan, The Kapitza-Dirac effect (in English). Contemp. Phys. **41**(6), 369–381 (2000). https://doi.org/10.1080/00107510010001220

214. M. Kozak, T. Eckstein, N. Schonenberger, P. Hommelho, Inelastic ponderomotive scattering of electrons at a high-intensity optical travelling wave in vacuum (in English). Nat. Phys. **14**(2), 121 (2018). https://doi.org/10.1038/nphys4282

215. M. Kozak, N. Schonenberger, P. Hommelhoff, Ponderomotive generation and detection of attosecond free-electron pulse trains (in English). Phys. Rev. Lett. **120**(10), 103203 (2018). https://doi.org/10.1103/physrevlett.120.103203

216. J. Kempe, Quantum random walks: an introductory overview (in English). Contemp. Phys. **44**(4), 307–327 (2003). https://doi.org/10.1080/00107151031000110776

217. N. Spagnolo et al., Experimental validation of photonic boson sampling (in English). Nat. Photonics **8**(8), 615–620 (2014). https://doi.org/10.1038/Nphoton.2014.135

218. L. Sansoni et al., Two-particle bosonic-fermionic quantumwalk via integrated photonics (in English). Phys. Rev. Lett. **108**(1), 010502 (2012). https://doi.org/10.1103/physrevlett.108.010502

219. C. Luo, M. Ibanescu, S.G. Johnson, J.D. Joannopoulos, Cerenkov radiation in photonic crystals (in English). Science **299**(5605), 368–371 (2003). https://doi.org/10.1126/science.1079549

220. A. Howie, Photon interactions for electron microscopy applications (in English). Eur. Phys. J. Appl. Phys. **54**(3), 33502 (2011). https://doi.org/10.1051/epjap/2010100353

221. A. Asenjo-Garcia, F.J.G. de Abajo, Plasmon electron energy-gain spectroscopy (in English). New J. Phys. **15**, 103021 (2013). https://doi.org/10.1088/1367-2630/15/10/103021

222. A. Fallahi, F. Kartner, Field-based DGTD/PIC technique for general and stable simulation of interaction between light and electron bunches (in English). J. Phys. B Mol. Opt. **47**(23), 234015 (2014). https://doi.org/10.1088/0953-4075/47/23/234015

223. J.-L. Vay, Simulation of beams or plasmas crossing at relativistic velocity. Phys. Plasmas **15**(5), 056701 (2008). https://doi.org/10.1063/1.2837054

224. B. Naranjo, A. Valloni, S. Putterman, J.B. Rosenzweig, Stable charged-particle acceleration and focusing in a laser accelerator using spatial harmonics. Phys. Rev. Lett. **109**(16), 164803 (2012). https://doi.org/10.1103/physrevlett.109.164803

225. J. Breuer, J. McNeur, P. Hommelhoff, Dielectric laser acceleration of electrons in the vicinity of single and double grating structures—theory and simulations. J. Phys. B Atom. Mol. Opt. Phys. **47**(23), 234004 (2014). https://doi.org/10.1088/0953-4075/47/23/234004

226. M. Ferrario et al., IRIDE: interdisciplinary research infrastructure based on dual electron linacs and lasers. Nucl. Instrum. Methods Phys. Res. Sect. A Accelerators Spectrometers Detectors Assoc. Equip. **740**, 138–146 (2014). https://doi.org/10.1016/j.nima.2013.11.040

227. E.A. Peralta et al., Demonstration of electron acceleration in a laser-driven dielectric microstructure. Nature **503**, 91 (2013). https://doi.org/10.1038/nature12664, https://www.nature.com/articles/nature12664#supplementary-information

228. J. Breuer, P. Hommelhoff, Laser-based acceleration of nonrelativistic electrons at a dielectric structure. Phys. Rev. Lett. **111**(13), 134803 (2013). https://doi.org/10.1103/physrevlett.111.134803

229. L. Kasmi, D. Kreier, M. Bradler, E. Riedle, P. Baum, Femtosecond single-electron pulses generated by two-photon photoemission close to the work function. New J. Phys. **17**(3), 033008 (2015). https://doi.org/10.1088/1367-2630/17/3/033008

230. J. Hoffrogge et al., Tip-based source of femtosecond electron pulses at 30 keV. J. Appl. Phys. **115**(9), 094506 (2014). https://doi.org/10.1063/1.4867185

231. B. Piglosiewicz et al., Carrier-envelope phase effects on the strong-field photoemission of electrons from metallic nanostructures. Nat. Photonics **8**, 37 (2013). https://doi.org/10.1038/nphoton.2013.288, https://www.nature.com/articles/nphoton.2013.288#supplementary-information

232. M. Aidelsburger, F.O. Kirchner, F. Krausz, P. Baum, Single-electron pulses for ultrafast diffraction. Proc. Nat. Acad. Sci. **107**(46), 19714–19719 (2010). https://doi.org/10.1073/pnas.1010165107

233. M. Krüger, M. Schenk, M. Förster, P. Hommelhoff, Attosecond physics in photoemission from a metal nanotip. J. Phys. B Atom. Mol. Opt. Phys. **45**(7), 074006 (2012). https://doi.org/10.1088/0953-4075/45/7/074006

234. G. Herink, D.R. Solli, M. Gulde, C. Ropers, Field-driven photoemission from nanostructures quenches the quiver motion. Nature **483**, 190 (2012). https://doi.org/10.1038/nature10878, https://www.nature.com/articles/nature10878#supplementary-information

235. B. Barwick, C. Corder, J. Strohaber, N. Chandler-Smith, C. Uiterwaal, H. Batelaan, Laser-induced ultrafast electron emission from a field emission tip. New J. Phys. **9**(5), 142 (2007). https://doi.org/10.1088/1367-2630/9/5/142

236. B. Schröder, M. Sivis, R. Bormann, S. Schäfer, C. Ropers, An ultrafast nanotip electron gun triggered by grating-coupled surface plasmons. Appl. Phys. Lett. **107**(23), 231105 (2015). https://doi.org/10.1063/1.4937121

237. M. Müller, V. Kravtsov, A. Paarmann, M.B. Raschke, R. Ernstorfer, Nanofocused plasmon-driven sub-10 fs electron point source. Acs Photonics **3**(4), 611–619 (2016). https://doi.org/10.1021/acsphotonics.5b00710

238. K.E. Echternkamp, G. Herink, S.V. Yalunin, K. Rademann, S. Schäfer, C. Ropers, Strong-field photoemission in nanotip near-fields: from quiver to sub-cycle electron dynamics. Appl. Phys. B J. **122**(4), 80 (2016). https://doi.org/10.1007/s00340-016-6351-x

239. A. Gliserin, M. Walbran, P. Baum, A high-resolution time-of-flight energy analyzer for femtosecond electron pulses at 30 keV. Rev. Sci. Instrum. **87**(3), 033302 (2016). https://doi.org/10.1063/1.4942912

240. J. Vogelsang et al., Ultrafast electron emission from a sharp metal nanotaper driven by adiabatic nanofocusing of surface plasmons. Nano Lett. **15**(7), 4685–4691 (2015). https://doi.org/10.1021/acs.nanolett.5b01513

241. P.G. Etchegoin, E.C. Le Ru, M. Meyer, An analytic model for the optical properties of gold (in English). J. Chem. Phys. **125**, **127**(18), 164705, 189901 (2006, 2007). https://doi.org/10. 1063/1.2802403

242. P.G. Etchegoin, E.C. Le Ru, M. Meyer, An analytic model for the optical properties of gold (in English), J. Chem. Phys. **125**(16), 164705. https://doi.org/10.1063/1.2360270

243. R.M. Joseph, A. Taflove, FDTD Maxwell's equations models for nonlinear electrodynamics and optics. IEEE Trans. Antennas Propag. **45**(3), 364–374 (1997). https://doi.org/10.1109/8. 558652

Index

© Springer Nature Switzerland AG 2019
N. Talebi, *Near-Field-Mediated Photon–Electron Interactions*,
Springer Series in Optical Sciences 228,
https://doi.org/10.1007/978-3-030-33816-9

Printed in the United States
By Bookmasters